New Techniques for
MODELLING
the
MANAGEMENT
of
STORMWATER
QUALITY IMPACTS

Edited by

William James

LEWIS PUBLISHERS
Boca Raton   Ann Arbor   London   Tokyo

Library of Congress Cataloging-in-Publication Data

Catalog record is available from the Library of Congress.

International Standard Book Number   0-87371-898-4

Reference Data:

New Techniques for Modelling the Management of Stormwater Quality Impacts, Proceedings of the Stormwater and Water Quality Management Modelling Conference, Toronto, Ontario, February 26-27, 1992.

Compiled by the School of Engineering, University of Guelph and Computational Hydraulics International, Guelph, Ontario, Canada.

PRINTED IN THE UNITED STATES OF AMERICA
1 2 3 4 5 6 7 8 9 0

Printed on acid-free paper

Preface

The development of models for evaluation of stormwater management practices dates from the late 1960's, the heyday of the so-called "big-batch" mainframes. This was soon after Norm Crawford and the late Ray Linsley published the Stanford Watershed Model, which first showed that continuous runoff and pollutant flux intensities could be reasonably accurately computed using an approach based on a time-step-wise materials budget, with component process models. Execution of the code took an overnight run on a computer occupying a large room and weighing several tonnes; stormwater management modelling was an expensive, academic exercise.

During the ensuing two decades, water resources engineers quickly adopted emerging computer technology. Microprocessor techniques are now widely used in field data acquisition, laboratory analysis, and the design office. Today, an average-sized continuous SWMM problem can be run on a desktop computer in less than (say) ten minutes; indeed wherever municipal and environmental engineering work is being done, microcomputer-based data processing is a sine-qua-non.

Of course, innovation is proceeding unabated in the information and computer sciences, and new tools beckon, including geographical information systems, integrated distributed-information systems, expert systems, computer animation, and multi-media systems. True or not, it is a matter of widespread belief that engineering software of dubious quality is being unjustifiably integrated with new, powerful systemsware, and aggressively marketed. So there is a risk that the poor choices (such as an unfortunate mix of technologies), which are naturally being made everywhere, are widely promoted, developed and adopted.

That is why a conference on **Emerging Technologies and Stormwater and Water Quality Management Modelling** was sponsored by Computational Hydraulics International in Toronto in February, 1992. The intention was to provide an open forum for discussion of new developments in modelling, BMPs, information management, user interfacing, and instrumentation for reducing the impacts of urbanisation on aquatic ecosystems. The conference was part of a long series of similar meetings, but was, if anything, better attended than any before.

This book is the collection of most of the papers, plus two other papers not presented; it introduces current work on modelling, data management, geographical information systems, and process control for the evaluation of best management practices for reducing the impacts on aquatic ecosystems of urban water utilization, especially stormwater and combined sewer overflows.

There are three groups of papers:
1. ecosystems and environmental modelling contexts,
2. best management practices, including real-time control, and
3. applications of geographical information systems.

1. The ecosystems and environmental modelling context.

In Chapter 1, **James** provides a history and literature review of the evolution of the USEPA Stormwater Management Model. The need to move from a regulatory basis for system management to a process-based management system is addressed by **Imhoff** and **Annable** in Chapter 2; the new philosophy should be one of systems integration based on logical, interactive ecosystem units. An application of modelling to quantify point and non-point pollution discharges, to assess the relative impacts on the Little River's ecosystem, to develop a strategy for mitigating these effects, and to improve the health of the stream, is presented in Chapter 3 by **Mahood** and **Zukovs**. In Chapter 4, **Snodgrass,**

Maunder, Schiefer and **Whistler** present an integrated approach for using modelling tools, monitoring data, and planning studies for evaluating options for managing the impacts of urbanization on the health of a watershed. **Driscoll** describes in Chapter 5 a decision support system for assessing the impacts of highway runoff on receiving waters; the technique is appropriate for assessing stormwater impacts from urban sources in general. In Chapter 6, **Weatherbe, Schroeter, Draper** and **Whiteley** describe a linked cascade of modelling procedures to assess changes created by urbanisation in the quantity and quality of streamflow. A new analysis methodology for the preliminary planning of urban runoff quantity and quality control systems is introduced by **Li** and **Adams** in Chapter 7. **Kouwen** and **Soulis** describe in Chapter 8 how remote sensing may be used to disaggregate a catchment into six different hydrologic response classes to compute floods. In Chapter 9, **Hodge** and **Armstrong** summarize a multiple regression model to estimate stormwater pollutant loading based on water quality and land-use data collected in 1989-90 in Alameda County, California. Finally, **Irvine, Loganathan, Pratt** and **Sikka** describe in Chapter 10 a calibration procedure for PCSWMM for metals, PCBs and HCB in combined sewer overflows in Buffalo, New York.

2. Best Management Practices including real-time-control.

Three papers were presented on instrumentation, data acquisition and real-time control, and four papers on various aspects of stormwater detention ponds.

In Chapter 11, **Robinson** and **Stirrup** describe the design, software and installation of a real-time rain and flow monitoring network, a microcomputer controller in a combined sewer regulator, and estimation of pollution loads in Hamilton using continuous SWMM modelling. **Kwan, Yue** and **Hodgson** discuss the use of SWMM to utilize real-time control for maximizing storage in the existing trunk sewer system in Edmonton in Chapter 12. In Chapter 13 **Carvalho** and **James** describe weather instrumentation and software for real-time data acquisition and an expert system shell for modelling crop management for agricultural non-point source water concerns, such as nitrate washoff.

Licsko, Whiteley and **Corsi** found in Chapter 14 that wet and dry ponds in Guelph remove 95% suspended solids, 70% total phosphorous and 30% BOD, and were able to make several other generalisations about the quality of stormwater in urban areas. In Chapter 15 **Droste, Rowney** and **MacRae** investigate the sensitivity of the performance of batch and continuous plug flow operation modes of pond configurations on the removal of fecal coliforms and suspended solids. An overview of a simulation technique for assessing the performance of infiltration basins is presented in Chapter 16 by **Thomson**. Finally in Chapter 17 **Woodbury** and **Padmanabhan** introduce modelling techniques for predicting water quality conditions in proposed reservoirs.

3. Applications of Geographical Information Systems.

It is possible that all engineering involves the creation of devices or processes that are solutions to real problems; in other words, problems that relate to information connected to spatial co-ordinates. GIS is software that manages such information; it differs from time series data management. There is no doubt that GIS make environmental engineering much easier, and this book concludes with six papers demonstrating this point.

In Chapter 18 **Ribeiro** provides an overview of interactive computer-aided infrastructure design using GIS. **Shamsi** and **Schneider** discuss the integration of SWMM with both a watershed GIS and USGS digital elevation models for the design of an interceptor in Allegheny County in Chapter 19. An application of GIS to modelling of urban hydrology is presented in Chapter 20 by **Muzik**. In Chapter 21, **Rudra, Dickinson** and **Sharma** present a method to integrate the GAMES distributed watershed model and ArcInfo for management of soil erosion and fluvial sedimentation. Similarly, integrating a GIS with HEC-2 for flood plain management is described in Chapter 22 by **DePodesta, Nimmrichter** and **Moore**. Finally **Zhen, James** and **Wang** describe an effective method for handling gradient data in ArcInfo for resource management modelling.

This book should interest practising engineers in municipal and environmental engineering, whether they are consultants, or with local, state or federal organizations. Innovative and emerging technology is described in each of the 23 chapters, and the ideas are of especial interest to researchers in civil engineering. Graduate students will also find the book valuable, as the material covers most areas that excite urban developers today. Detailed indexes, lists of acronyms and programs and models, and a full glossary for readers not yet familiar with all the terminology, are provided at the end of the book.

And yes, this is an excellent book (perhaps because many of the authors are former and present students and graduate students - **Bill Annable, Dave Maunder, Kim Irvine, Mark Robinson, Mark Stirrup, Luis Carvalho, John Licsko, Gus Ribeiro, Ken De Podesta** and **Pete Nimmrichter**). Thanks to them and to all co-workers and colleagues who contributed!

William James
Guelph.

About the Editor

William James received the B.Sc. degree in Civil Engineering from the University of Natal in 1958, the Diploma of Hydraulic Engineering from Delft Technological University, Holland in 1962, the Ph.D degree from Aberdeen University, Scotland in 1965, and the D.Sc. degree from the University of Natal in 1986 for contributions to hydraulic engineering. He started his professional career as a Provincial Water Engineer in Natal in 1959. With time out for graduate studies, he has also worked with city engineers, as a consulting engineer and professor. From 1965 to 1970 he was lecturer and senior lecturer in charge of Hydraulics in the Civil Engineering Department at the University of Natal. In 1971 he joined the Civil Engineering Department at McMaster University in Hamilton, Ontario and was professor of Civil Engineering until 1986. He was appointed Cudworth Professor of Computational Hydrology in the Civil Engineering Department at the University of Alabama in Tuscaloosa, Alabama, and then Chair of Civil Engineering at Wayne State University, Michigan. He has been visiting professor at the Universities of Lund and Lulea in Sweden, Queen's in Canada, and the University of the Witwatersrand in South Africa.

He is currently Professor and Director of the School of Engineering at the University of Guelph, and, for the past five years, has been involved in studies of complex metropolitan drainage and polluted surface water systems. These systems have included thunderstorm dynamics and steep rainfall gradients; surface pollutant build-up due to atmospheric fallout and anthropogenic activities; pollutant and particulate washoff and transport; flow and pollutant routing through partially-surcharging drainage networks of great complexity; pollutant removal in storage-treatment systems; dispersal in receiving waters; and transients in coastal and tidal waters.

He has published 157 scientific papers and 175 technical reports and books. Associated with this work, Dr. James has organized and presented professional seminars in Canada, the U.S. and overseas in Australia, Europe and South Africa, and is active on research committees of the American Society of Civil Engineering relating to stormwater management modelling. He has extensive consulting experience through Computational Hydraulics International in Canada and the U.S.A.

Dr. James has organized or helped organize eight international meetings, edited nine sets of conference proceedings and is a Fellow of both the Canadian and American Society of Civil Engineers. He has served on numerous committees in his area. He has led lecture tours in Australia, Sweden and Finland, South Africa and the east (Pakistan and Bangladesh), and has lectured widely in Canada, Australia, South Africa, Scandinavia and the U.S.A.

Acknowledgements

This book and the conference were self-supporting and no financial support was sought or obtained. Thus no acknowledgements for financial support are necessary, for a change.

Sincere thanks are, however, due to all the authors and participants in the *Stormwater and Water Quality Management Modelling Conference* held in Toronto February 26-27, 1992; they have made the publication of this book possible.

Special thanks once again to Dr. Lyn James for organizing the conference and collecting the papers. Thanks also to the ASCE Urban Water Resources Council, the US Environmental Protection Agency and the Ontario Ministry of the Environment for sanctioning this conference.

Tony Kuch, graduate student at the School of Engineering at the University of Guelph, did yeoman work on the word processor. His cheerful, imaginative help was amazing. Three months from conference to publisher is not perfect, but passing excellent.

Contents

List of Figures

Chapter 4

Chapter 5

Chapter 16

Chapter 1

Introduction to the SWMM Environment

William James
Director, School of Engineering,
University of Guelph,
Guelph, ON Canada N1G 2W1

This necessarily brief review features a few highlights from the perspective of an academic marginally involved in the development of SWMM. As a historical sketch it is admittedly poor, covering only what the writer considers to be a few central activities. It was written to help new users understand the peculiarities of the code and manuals, and to plot a path through part of the confusing serial literature. Textbooks, manuals and reports are not covered.

Most urban stormwater code has been written and distributed for commercial reasons. The development of SWMM, its ancestry and its continuing support, on the other hand, is probably unique. Apart from intermittent support of the USEPA, research groups at several different universities, including those of the author, the University of Florida, and Oregon State University, and engineers at agencies and in consulting offices, have spasmodically contributed ideas or more materially to the evolution of SWMM.

0-87371-898-4/93/$0.00 + $.50
© 1993 by Lewis Publishers

1

The SWMM environment is a natural consequence of active participation in scientific, technical and engineering conferences, symposia, seminars, workshops and other meetings.   Besides workshops and short courses given by (i) the University of Florida, (ii) the USEPA, and (iii) the writer, the more common meetings include:

1. An approximately biennial series of Specialty and Engineering Foundation conferences (both involving the Urban Water Resources Research Council of the American Society of Civil Engineers or ASCE);

2. A series of regular international conferences every three years and;

3. Irregular six-monthly user group conferences in the US and Canada, whose papers are listed at the end of the chapter.

A study of the lists of committee members, invited speakers, authors of papers, and the content of some papers, will inevitably detect some repetition and find a remarkable amount of consistency over the years.   Fortunately other consulting engineers, planners, geographers, aquatic biologists and related professionals concerned with urban development and its impact on aquatic environments have contributed equally to the rich literature.

Since the proceedings of the water quality management modellers user conferences are generally more difficult to locate in libraries or elsewhere, this review includes a list of most of the papers presented in that series. The papers cover many topics of intense interest to readers of these proceedings, and nicely encapsulate the changing emphasis over the past two decades. Topics range from concerns with water flows in a remote-batch-mainframe environment, to interdisciplinary ecosystems concerns in the evolving networked-workstation design environment. A crude guide to the papers is provided.

1.1 Introduction

The public domain program known as SWMM was originally the result of generous funding provided by the USEPA. The early contractors were: 1. Metcalf and Eddy Inc. of Palo Alto, 2. University of Florida, and 3. Water Resources Engineers Inc. of Walnut Creek, California. It's important to note that the code has been under more-or-less continuous development for about three decades, almost exclusively at the University of Florida (until 1991) under Wayne Huber's leadership. The manuals were also written there, and produced occasionally in response to EPA contract agreements. But, under Reaganomics, USEPA support fell away over the years, and in the late 80s, it virtually dried up. Nevertheless a few die-hard professors continued to support the code, despite all odds.

Consulting engineers have also contributed to the development, some having selflessly donated code, and others under contract, notably CDM (Camp Dresser and McKee). Their participation is especially important for keeping the development tied to practical problems.

For each major issue of the documentation, the evolving code has been bundled up and given a new version number. Software archaeologists should take care not to infer that the historic versions of the documentation reflect the then contemporary SWMM code; available documentation clearly lags code development by several years.

1.2 USEPA SWMM

The four volumes of the 1971 STORMWATER MANAGEMENT MODEL report marks an early starting point. Two volumes related to the theoretical background, and to code validation. Volume three comprised the first user's manual, and volume four the original source code listing. That program was very limited by comparison with to-day's versions. Version 1 did not include a COMBINE block. Figure 1.1 is a schematic for version 2 and Figure 1.2 shows the engineering design relationship between its modules. Figure 1.2 is essentially unchanged from that for version 1.

The user's manual for version 2 was published in March 1975. The program still incorporated a RECEIVing block, but no EXTRAN block. The schematic is shown in Figure 1.1.

In November 1977, Interim SWMM documentation was published (still version 2 officially), and it included snowmelt, better continuous simulation, and EXTRAN documentation. The new schematic is presented in Figure 1.3. In November 1981 the user's manual for version 3 was published. The RECEIVing block was deleted, and a separate Addendum for EXTRAN was published under the authorship of CDM of Annandale, Virginia. The program was still big-batch-mainframe oriented, and distributed on 9-track tape. Later, version 3.3 was distributed by the EPA for use on microcomputer. Figure 1.4 is a schematic for version 3. Note that this version did not in fact include a RECEIVing block.

Starting in 1977, the author and his group ported the code to minicomputer and later to microcomputer systems. Free-format input, 80-column output, screen-oriented graphics, error-checking and transparent file handling was added. The documentation was rewritten in user-friendly style, the code stored on diskettes, and the package distributed by CHI as PCSWMM. PCSWMM3.2 was last formally distributed in 1987. A schematic is shown in Figure 1.5, depicting how the program was structured on the 5¼ inch diskettes.

Version 4 was the second EPA version that could run on microcomputers, and appeared with published documentation in August 1988. It was an USEPA response to PCSWMM which was a commercial product, and version 4 included most of the attributes of PCSWMM as well as a wide range of enhancements. The new EPA documentation reflected many of the contributions of PCSWMM, particularly the tables of input data requirements. Figure 1.6 is the schematic for version 4. It was configured for operation on hard disk, but retained much of the big-batch main

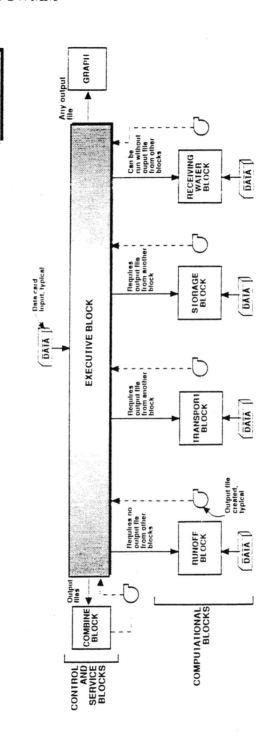

Figure 1.1: Schematic for SWMM version 2 (March 1975). Source: Huber, W.C. et al. SWMM User's Manual Version 2, EPA-670/2-75-017 (NTIS PB-257809) EPA Cincinnati OH March 75.

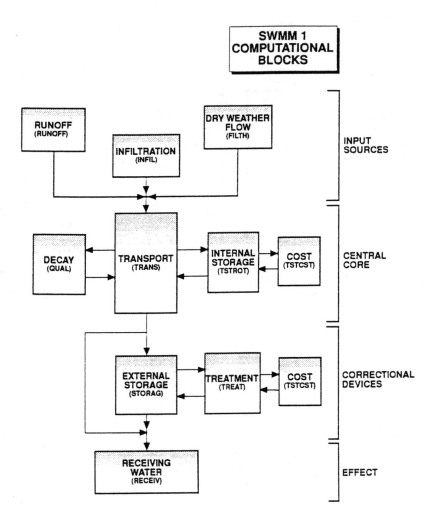

Figure 1.2:   Engineering design relationship between SWMM modules (and submodules). Source: Huber, W.C. et al., SWMM User's Manual Version 2, EPA-670/2-75-017 (NTIS PB-257809) EPA Cincinnati, OH March 75.

frame architecture. It was now much better configured for continuous simulation and time-series management, incorporating a RAIN block and a temperature block.

1.3 Derivatives of SWMM

In the early 90s a number of derivative codes became available in the private domain, of which the outstanding examples are EXTRAN-XP and SWMM-XP, developed and distributed by WP Software of Canberra, Australia. The XP codes, originating on a MacIntosh platform, use a graphics-based environment and are the most user-friendly of all the urban stormwater system design codes. The codes now include significant modification of the USEPA code, eliminating the grief involved in getting SWMM running smoothly.

1.4 Serial Literature

In this chapter we deal with the specialised serial literature only; textbooks, manuals, research reports, and other one-off documentation are not covered. Serials here are taken to be the result of more-or-less regular technical meetings, as opposed to (say) a series of unrelated design codes. The Australian meetings and conferences on Urban Stormwater Quality Modelling are not listed here.

1.5 Conference Proceedings

1.5.1. Sanctioned ASCE Urban Water Resources Research Council Meetings

As SWMM was evolving, the literature was growing. Several textbooks appeared, and the University of Kentucky ran an annual

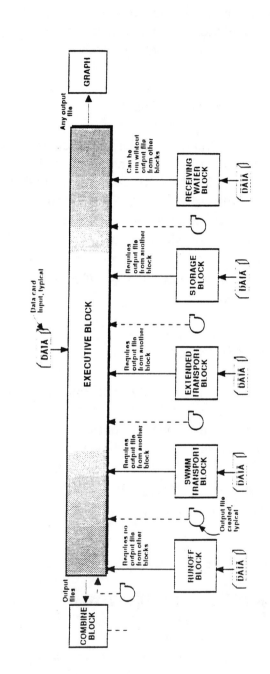

Figure 1.3:   Schematic for SWMM version 2 (Nov. 1977). Source: Huber, W.C. et al., Release of EPA SWMM, Nov. 77. Draft Interim Documentation, University of Florida, Dept. of Environmental Engineering Sciences, Gainesville, FL Nov. 1977.

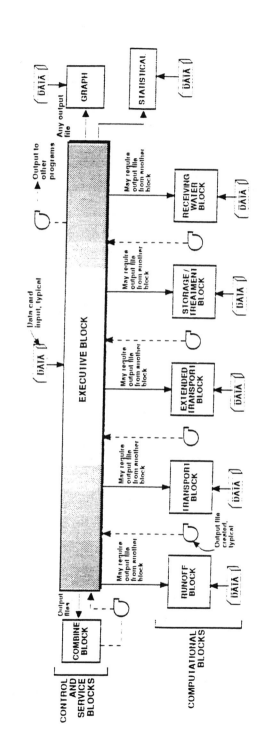

Figure 1.4: Schematic for SWMM version 3 (Nov 1981). Source: Huber W.C. et al., SWMM User's Manual Version 3 EPA-600/2-84-109a (NTIS PB84-198423), Environmental Protection Agency, Cincinnati, OH (4th printing October 1982).

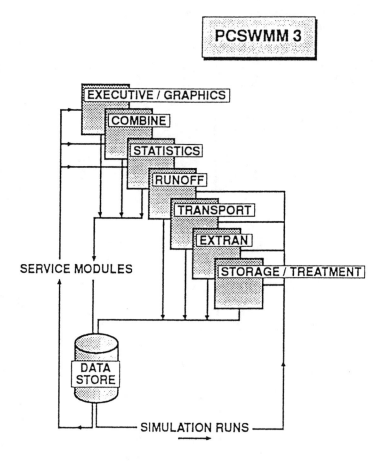

Figure 1.5: Schematic for PCSWMM3.2.

conference on urban hydrology, hydraulics and sediment transport every July from 1976 to 1985.

The Urban Water Resources Research Council (UWRRC) of the ASCE sponsored a number of special meetings, as well as sessions at various ASCE Division Specialty conferences. Every few years a conference was arranged under the close support of the UWRRC by the Engineering Foundation. The proceedings for all these conferences, as well as a number of related manuals, were published by the ASCE. A list of some relevant, focused conferences in the period 1971-89 whose proceedings are

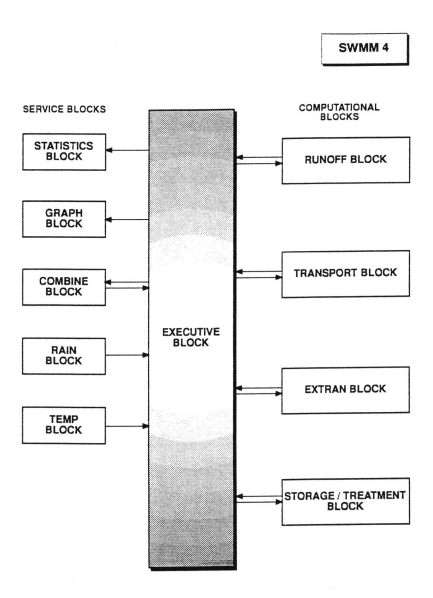

Figure 1.6:   Schematic for SWMM version 4 (August 1988).
              Source: Huber, W.C. and Dickinson, R.E.
              SWMM Version 4 User's Manual. EPA/600/3-
              88/001a (NTIS PB88-236641) EPA Athens, GA
              (August 88).

available from the ASCE is provided below. The list is not exhaustive, does not cover conferences with broader coverage than urban stormwater management, and does not include other international conferences, such as those of the Canadian Society of Civil Engineering (CSCE). Copies of the books of proceedings are available from the ASCE at 345 East 47th St, NY, NY 10017.

ASCE SANCTIONED CONFERENCES

DATE	PLACE	TITLE
1989	Davos, Switz.	Urban Stormwater Quality Enhancement - Source Control, Retrofitting, and Combined Sewer Tech.
1988	Potosi, Missouri	Design of Urban Runoff Quality Controls
1986	Hennicker, N.H.	Urban Runoff Quality - Impact & Quality Enhancement Technology
1983	Baltimore, M.D.	Urban Hydrology
1983	Niagara-on-the-Lake, Ont.	Emerging Computer Techniques in Stormwater and Flood Management
1982	Hennicker, N.H.	Stormwater Detention Facilities
1980	Blacksburg, Vir.	Urban Stormwater Management in Coastal Areas
1978	Hennicker, N.H.	Water Problems of Urbanizing Areas
1976	Easton, M.D.	Guide for Prediction Analysis and Use of Urban Stormwater Data
1974	Rindge, N.H.	Urban Runoff Quantity and Quality
1971	Deerfield, Mass.	Urban Water Resources Management

1.5.2. International Meetings Involving IAHR and/or IAWPRC

International meetings were also organised by a joint committee of the International Association for Hydraulic Research and the International Association for Water Pollution Research and Control. Venues were Southampton (UK), Urbana (IL), Gothenburg (Sweden), Lausanne (Switzerland), Osaka (Japan) and (in 1993) Niagara Falls (Ontario). Other smaller, more specialised meetings were held in Dubrovnik, Davos, Montpelier, Wageningen, Duisberg, and elsewhere. The UWRRC was instrumental in arranging participation by US agencies and academics. The next table lists a selection of those meetings and their titles.

SOME INTERNATIONAL CONFERENCES

DATE PLACE	TITLE	Number of Volumes
Jul 93 Niagara Falls, Ontario	Proc. 6th Int. Conf. on Urban Storm Drainage	(to be published)
Jul 90 Tokyo, Japan	Proc. 5th Int. Conf. on Urban Storm Drainage	3 volumes
Apr 88 Duisburg, West Germany	Hydrological Processes and Water Management in Urban Areas	1 volume
Aug 87 Lausanne, Switzerland	Topics in Urban Stormwater Quality, Planning and Management: Proc. 4th Int. Conf. on Urban Storm Drainage	2 volumes
Apr 86 Dubrovnik, Yugoslavia	Urban Drainage Modelling - Symposium on Comparison of Urban Drainage Models with Real Catchment Data	3 volumes
Aug 85 Montpellier, France	NATO Advanced Science Institute: Urban Runoff Pollution	1 volume
Jun 84 Goteborg, Sweden	Proc. 3rd Int. Conf. on Urban Storm Drainage	4 volumes
Jun 81 Urbana, Illinois	Urban Stormwater Quality Management and Planning: Proc. 2nd Int. Conf. on Urban Storm Drainage	2 volumes
Apr 78 Southampton, United Kingdom	Proc. 1st Int. Conf. on Urban Storm Drainage	1 volume

1.5.3. So-called SWMM User Group Meetings

Early co-operation was established between groups in Canada and the US, and the proceedings of several user group conferences were published in "The Storm Water Management Modeller - a bulletin to the users of the EPA's SWMM". The writer holds copies for meetings held in Toronto, April 1st 1976; in Toronto October 26th 1976; Gainesville (FL) April 4-5 1977; and Milwaukee Nov 3-4 1977. In addition, the Ontario Ministry of the Environment published proceedings, of which the writer holds the following: Toronto, Oct 19-21 1976; and Toronto March 28-30 1977.

Subsequent conferences were then organised by various users, sponsored by the USEPA inter alia, and held approximately twice a year, alternating between Canada and the U.S. The title of the conference became "Stormwater and Water Quality Management Modelling" or similar. The proceedings of the conferences, and sources, as known to the writer, are as follows (in reverse chronology):

Sources for the proceeding USEPA sanctioned conferences:

Nat Tech Info Services (NTIS)   Dr. Paul Wisner
5285 Port Royal Road   Dept of Civil Engineering
Springfield, VA 22161   University of Ottawa
                                 Ottawa, ON K1N 9B4

Dr. Wm. James
School of Engineering   Charles Howard & Assoc. Ltd.
University of Guelph   300-1144 Fort Street
Guelph, ON N1G 2W1   Victoria, B.C. V8V 3K8

Ont. Min. of the Environment
Pollution Control Branch
135 St. Clair Avenue, West
Toronto, ON M4V 1P5

USEPA SANCTIONED
STORMWATER AND WATER QUALITY MANAGEMENT MODELLING CONFERENCES

DATE	PP	REPORT NUMBER	NTIS NUMBER	SOURCE	PLACE
Feb 92	c600	CHI/U of Guelph		W. James	Toronto, ON
Apr 90	XX	unknown		T.Najarian	Eatontown, NJ
Oct 88	233	EPA-600/9-89-001	PB89 195002/AS	NTIS	Denver,CO
Oct 87	175	Howard Assocs		C. Howard	Victoria, BC
Mar 87	249	EPA-600/9-87-016	PB88-125430	Try NTIS	Denver,CO
Sep 86	463	U of Ottawa		P.Wisner	Toronto, ON
Mar 86	334	EPA-600/9-86-023	PB87-117438	Try NTIS	Orlando,FL
Dec 85	426	CHI-R149		W.James	Toronto, ON
Jan 85	268	EPA-600/9-85-016		NTIS	Gainesville,FL
Sep 84	261	CHI-R128	PB85-228302	W. James	Burlington, ON
Apr 84	298	EPA-600/9-85-003	PB85-168003	Try NTIS	Detroit, MI
Sep 83	274	GREMU 83/02		P. Beron	Montreal, PQ
Jan 83	282	EPA-600/9-83-015	PB84-118454	NTIS	Gainesville,FL
Oct 82	274	U of Ottawa		P. Wisner	Ottawa, ON
Mar 82	298	EPA-600/9-82-015	PB83 145540	NTIS	Washington,DC
Sep 81	551	CHI-R81		W. James	NiagaraFalls,ON
Jan 81	257	CHI-R83		W. James	Austin, TX
Jun 80	238	EPA-600/9-80-064	PB81 173858	NTIS	Toronto, ON
Jan 80	329	EPA-600/9-80-017	PB80 177876	NTIS	Gainesville,FL
May 79	413	EPA-600/9-79-026	PB80 105663	NTIS	Montreal, PQ
Nov 78	238	EPA-600/9-79-003	PB290 742/6BE	NTIS	Annapolis,MD
May 78	244	EPA-600/9-78-019	PB285 993/2BE	NTIS	Ottawa, ON
Mar 77	389	Conf. Proc. No.5		MOE	Toronto, ON
Oct 76	334	Conf. Proc. No.4		MOE	Toronto, ON

1.5.4. SWMM User Group Meeting Papers

Finally, this chapter concludes with a very simple classification of papers of the above irregular water quality model users meetings held in Canada and the US. In the lists following the index tables, each paper two pages or longer is given a "koechel" (an arbitrary pseudo-chronological) number. The coarse indexes are presented first.

Readers who need copies of any papers should contact the above sources, and if difficulties arise, FAX the author at 1-519-836-0227 or 1-519-767-2770.

GROUP TOPICS

GROUP 1: Emerging SWMM environment
1.1 Model complexity
1.2 Limitations of wet hydrology
1.3 Impacts of urbanization
1.4 Best management practices

GROUP 2: Stormwater management modelling
2.1 History of stormwater management
2.2 Brief introduction to various programs
2.21 HSPF
2.22 Receiving water models
2.23 All other models
2.3 Managing the modelling: accuracy, disaggregation, synthesis, sensitivity, calibration, validation, interpretation

GROUP 3: The Stormwater Management Model
3.1 History
3.2 General concepts

GROUP 4: Data management
4.1 Time series management
4.2 Spatial data management
4.3 Statistics

GROUP 5: Urban hydrology
5.1 Meteorology
5.2 Surface runoff
5.3 Groundwater and infiltration
5.4 Pollutant buildup and washoff

GROUP 6: Urban hydraulics
6.1 Storm sewer networks
6.2 Sanitary sewer hydraulics
6.3 Pumps, weirs, gates, orifices, diversion structures, outfalls
6.4 Surcharged flows

GROUP 7: Pollution control
7.1 Detention basins
7.2 Pollutant removal

GROUP 8: Continuous modelling

GROUP 9: The near future of stormwater management modelling

For the list of papers in the above groups, see the following index.

INDEX

STORMWATER AND WATER QUALITY MANAGEMENT MODEL USER MEETING PAPERS

9   September 1981:

901 Uncertainty analysis in stormwater and water quality modelling. R.H. Kummler, J.G. Frith, C.S. Liang & J.A. Anderson, pp 1-54

902 Implications of storm dynamics on design storm inputs. W.James, & Z.Shtifter, pp 55-78

903 A review of NCASI's mathematical water quality modelling program. R.C. Whittemore, J.S. Hovis & J.J. McKeown, pp 79-94

904 Stream modelling analysis with field measurements of reaeration rate constant (central New York). C.C.K. Liu, pp 95-110

905 Desk top estuary water quality modelling - a case study (Kennebunkport River). A.K. Deb & D. McCall, pp 111-124

906 Justification of advanced waste treatment at Mansfield, Ohio. A.S. French & S. Amragy, pp 125-158

907 Dynamic modelling used to develop waste load allocations (Savannah, GA). D.R. Bingham & R.A. Moore, pp 159-174

908 Calibration of the RECEIV model in a well mixed tidal estuary using the equilibrium procedure. I.B. Chou, pp 175-194

909 TEMSTAT - a model for statistically evaluating the variability of river temperatures. D.J. Murray, & D.R. Schregardus, pp 195-204

910 Pollutant removal in stormwater detention basins. R.A. Ferrara, P. Witkowski, & A. Hildick-Smith, pp 205-226

911 Cost analysis of a runoff detention policy. P.F. Lemieux & M. Codere, pp 227-242

912 Calibration of pesticide behaviour on a Georgia agricultural watershed using HSP-F. D.A. Woodruff, D.R. Gaboury, R.J. Hughto, & G.K. Young, pp 243-256

913 Master drainageway planning study, Boxelder Creek, CO. J.R. Kimzey, R.A. Burns & W.S. Liang, pp 257-270

914 Application of urban runoff models to cities in Saudi Arabia. A.M.Ishaq, pp 271-282

915 A comprehensive watershed management plan for Tinker's Creek - a case history in Prince George's County, MD. S. Udhiri, M. Vasa, M. Wallace, & K.D. Nambudripad, pp 283-294

916 Preliminary analysis of CSO impact on lower Rahway river water quality. T.O. Najarian, & V.K. Gunawardana, pp 295-314

917 An application of a closed form approximation for frequency of urban runoff volumes to a unique storm drainage problem. R.Bishop & M.Cosburn, pp 315-332

918 Modelling applications using QUAL-II and STORM. A.C. Rowney, J.C. Anderson, & A.R. Perks, pp 333-344

919 Application of SUBHYD to the Hamilton test catchment. L. Thompson & K. Dennison, pp 345-370

920 TRANQUAL - two dimensional modelling of transport water quality processes. R.B. Taylor & J.R. Pagenkopf, pp 371-398

921 An integrated modelling approach for evaluating CSO treatment alternatives. A.Dee, T. McConville, & C. C-S Song, pp 399-426

922 Evaluation of relief alternatives for combined sewer systems. A.Ashmala & M. Ahmad, pp 427-454

923 Persistence of toxic substances in stormwater - a case study. J.Wong & J. Marsalek, pp 455-468

924 Continuous SWMM quality modelling using Env. Can. data files. M.Robinson & W. James, pp 469-492

925 Investigation of management alternatives for relieving CSOs for the city of Hamilton. D.Henry & W. James, pp 493-512

926 Agricultural non-point source pollution and control in the Grand River basin. S.N. Singer & S.K. So, pp 513-544

---

10   March 1982:

1001 A study of the selection, calibration and verification of mathematical water quality model. R.C. Whittemore, J.S. Hovis, & J.J. McKeown, pp 1-35

1002 An assessment of the measurement uncertainty in the estimation of stream reaeration rate coefficients using direct tracer techniques. J.S. Hovis, R.C. Whittemore, L.C. Brown, & J.J. McKeown, pp 36-53

1003 Calibration of hydrology and sediment transport on small agricultural watersheds using HSPF. D.E.Schafer, D.A. Woodruff, R.J. Hughto, & G.K. Young, pp 54-68

1004 Hydrologic modelling studies of pollutant loadings and transport in large river basins. A.Cavacas, J.P.Hartigan, E.Southerland, & J.A.Friedman, pp 69-89

1005 Continuous DO response predicted using CSPSS is verified for Springfield, MI. J.E. Scholl & R.L. Wycoff, pp 90-100

11   October 1982:

12   January 1983:

13 November 1983:

14 April 1984:

1913 Real time control of combined sewer systems: operational alternatives and optimization techniques. W. Schilling, B. Morse, D. Consuegra & P. Wisner, pp 3.1-3.11

1914 Preliminary considerations in the modelling of storage facilities with small release rates. R.H. Pankratz, pp 4.1-4.13

1915 The development of an erosion index model for the planning level design and operation of stormwater management facilities. C. MacRae, pp 5.1-5.17

1916 A comparison of modelled and measured flows for McKenney Creek drainage basin. J.M.G. Bryck, H. Kelly, & D. Bowins, pp 6.1-6.13

1917 Comparison of the IMPRAM model (improved rational method) with other models. C. Rampersad, J.F. Sabourin, & P. Cheung, pp 7.1-7.23

1918 Town branch drainage study. A.K. Umble & L. Salcedo, pp 8.1-8.38

1919 Interface of lumped and detailed modelling in subdivision design. A. Lam & C. Rampersad, pp 9.1-9.15

1920 Experience with master drainage planning in western Quebec. R.S. Cebryk & J.G. Ouelette, pp 10.1-10.22

1921 Detention basin le Corbusier Sus. P. Lamarre, L. Gagnon, & L. Remillard, pp 11.1-11.15

1922 Models and decision making in urban drainage. P. Wisner, pp 12.1-12.22/A2

1923 Calibration of an urban hydrology model by trial and error and automatic calibration. D. Consuegra, P. Wisner, & A. El-Bahrawy, pp 13.1-13.20

1924 Application of the OTTSWMM model for a relief sewer study in Laval, Quebec. R. Roussel, J.C. Pigeon, & J.R. Noiseux, pp 14.1-14.13

---

20 March 1987:

2001 Storm sewer design by UDSEWER model. C.-Y. Guo & B. Urbonas, pp 1-9

2002 Microcomputers - the computer stormwater modelling future. G.R. Thompson & B.C. Phillips, pp 10-20

2003 Enhancing SWMM3 for combined sanitary sewers. W. James & T. W. Green, pp 21-41

2004 SEWERCADD. M.H. Jackson & J.L. Lambert, pp 42-50

2005 Current trends in Australian stormwater management. A.G. Goyen, pp 51-69

2006 A new groundwater subroutine in SWMM. B.A. Cunningham, W.C. Huber, & V.A. Gagliardo, pp 70-104

2007 SWMM applications for municipal stormwater management: the experience of Virginia Beach. J.A. Aldrich & J.E. Fowler, pp 105-109

2008 The effect of subwatershed basin characteristics on downstream storm-runoff quality and quantity. R.G. Brown, pp 110-118

2009 Some thoughts on the selection of design rainfall inputs for urban drainage systems. I.Muzak, pp119-124

2010 Field measurement and mathematical modelling of combined sewer overflows to Flushing Bay. G.Apicella, D.Distante, M.J. Skelly, & L.Kloman, pp 125-148

2011 Accounting for tidal flooding in developing urban stormwater management master plans. S. Dendrou & K. A. Cave, pp 149-170

2012 Wasteload allocation for conservative substances. M. Hutcheson, pp 171-179

2013 The use of detailed cost estimation for drainage design parameter analysis on spreadsheets. S.W. Miles, T.G. Potter, & J.P. Heaney, pp 180-193

2014 Corrective phosphorus removal for urban storm runoff at a residential development in the town of Parker, Colorado. W.C. Taggart & M.S. Wu, pp 194-204

2015 Evaluation of sediment erosion and pollutant associations for urban areas. K. Irvine, W. James, J. Drake, I. Droppo, & S. Vernette, pp 205-216

2016 Uncertainty in hydrologic models: a review of the literature. T.V. Hromadka II, pp 217-227

2017 Uncertainty in flood control design. T.V. Hromadka II, pp 228-247

---

21 October 1987:

2101 Urban Drainage Design Using MICRO-ILLUDAS. M.L. Terstriep & D.C. Noel, pp 1-6

2102 Design Comparisons for Peak and Volume Prediction. B.A. Cunningham, W.C. Huber, pp 7-22

2103 A Multi-Model Approach to Water Pollution Control Planning. P.W. Cheung, D.I. Smith & F.C. Moir, pp 23-44

2104 The Henson Creek Watershed Study. D.J. Motta & M.-S. Cheng, pp 45-66

2105 A Comprehensive Approach to Urban Stormwater Management. C.J. Edmonds, K.B. Lee, N.W. Schmidtke, & R. Ferguson, pp 67-82

Chapter 2

Developing an Ecosystem Context for the Management of Water and Water Systems

J. G. Imhof
  Fisheries Policy Branch, Ontario Ministry of Natural Resources, P.O. Box 5000, Maple, Ontario L6A 1S9

W.K. Annable
  Department of Earth Sciences, University of Waterloo, Waterloo, Ontario N2L 3G1

2.1 Introduction

The complexity of issues dealing with the surface waters within an urban area often overwhelm the planners, engineers, biologists and ultimately the decision-makers. Faced with a depressing array of seemingly contradictory or multi-directional objectives such as flood protection, drainage, water quality, health, odour, safety, recreation, domestic water demands, and natural environmental concerns, agencies regulate and each discipline responds to these issues by attempting to simplify the issues within their own field of experience.

By simplification within each discipline the overall problems become partitioned into a set of sub-issues which allow a feeling

0-87371-898-4/93/$0.00 + $.50

of satisfaction within each discipline. The results are that engineers consider water as a waste product and manage it accordingly, hydrogeologists recommend prevention of water infiltration in urban areas and then also express concern for dwindling water supplies, planners design communities and cannot understand why environmental concerns cannot be addressed after the fact and biologists manage in-channel habitat along small portions of stream oblivious to alterations in watershed pathways and processes that will negate their efforts.

Agencies in turn set regulatory standards for variables in the environment, based upon the minimum standard necessary to meet their bureaucratic, technical or environmental objectives. This process of regulation in turn leads to development designs that use the minimum standards as targets. This regulatory approach by minimum standard has fostered the creation of minimum environment.

A change in the philosophy and approach to modelling water and water systems is necessary. This paper discusses the need to move from managing the component parts of a system to model development, application and integration into a systems framework based upon a logical, interactive ecosystem unit which, in the case of water, is likely to be the watershed. This paper discusses the need to move from a regulatory basis for system management to a process based management system.

2.2 Overview and Perspective

2.2.1 Past Approach

Man has modified water-based environments throughout the world. Most major catchments in North America and Europe have been altered significantly (Minshall 1988; Bacalbasa-Dobrovici 1989; Backiel and Penczak 1989; Brousseau and Goodchild 1989; Hesse et al. 1989; Lelek 1989; Mann 1989; Sedell et al. 1989; Ward and Stanford 1989). Modifications of temperate North American rivers, their catchments and water budgets has occurred over a relatively short period (i.e. 200 years) compared to European rivers (i.e. >1000 years)(Bacalbasa-Dobrovici 1989; Lelek 1989; Mann 1989).

Degrading processes vary from catchment to catchment and from continent to continent but include changes in water quality through discharge of surface run-off and groundwaters from agricultural and urban landscapes, direct human and animal wastes, industrial wastes and sediment loading through poor landuse practices. Alterations in the morphometry of river channels and their floodplains occur through the increases in surficial run-off, flow volumes, increased flow energies and sediments from the catchment as well as damming, draining, channelizing, dredging, dyking, and filling. Water abstraction for municipal and industrial purposes reduce water supply from groundwater sources while landuse practices that increase the imperviousness of the catchment reduce groundwater replenishment. Acting as a whole, these alterations have created problems for the management of not only natural environments and ecosystems but also for the management of human dominated environments. Conventional urbanization with its two primary components of residential and industrial development appears to be the endpoint in the structural degradation of rivers and their basins (Steedman 1988; Regier et al. 1989; Imhof et al. 1991).

Alterations of the water budget of a catchment and the modifications to physical, chemical and biological pathways and processes can be classified as a syndrome of conventional urbanization. The results of conventional urbanization syndrome is characterized by reductions in tributary density, alteration or barriers to migration of fish and other animals, increases in frequency and magnitude of storm events and peak discharges. These modifications result in a concurrent reduction in baseflows, increased sediment loads, reduction in channel and floodplain complexity, and impaired water quality (Klein 1979; Steedman 1987, 1988; Imhof et al. 1991).

The structural degradation of rivers and their catchments alters water inputs into the ground and acts to simplify, modify or eliminate physical habitat features required by aquatic ecosystems (lakes, streams and wetlands) and their plant and animal communities (Regier et al. 1989). Klein (1979) demonstrates that water quantity (discharge), water quality and fish species diversity are directly linked to the percent imperviousness of a catchment. Changes in percent imperviousness of a catchment are linked to urbanization; downward infiltration of water is impeded and lateral surface run-off is expedited; tributary density is reduced. The water budget is altered. Run-off which in a forested watershed is an artifact becomes the dominant component of an urban/urbanizing watershed. Steedman (1988) demonstrated that urbanization affected the hydrology, hydraulics, water quality and fish community of streams flowing through urban and urbanizing areas.

Water quality degradation also occurs through the discharge of human and animal wastes resulting in either a toxic reaction (e.g. toxic levels of ammonia) or a chronic change (e.g. depleted oxygen due to high BOD). These effects increase costs of treatment for water supply managers and in natural environments often create chemical barriers which isolate fish and other animals from various portions of the river system that may be necessary for their life cycle. Industrial processes also discharge hazardous material into rivers and into saturated and unsaturated groundwater zones (e.g. halogenated chemicals; heavy metals; cyanide; arsenic; etc) increasing the costs associated with environmental management.

Water is viewed as a resource and as a waste product. This is the contradictory dichotomy under which managers presently operate. Within any watershed, the state of the streams and rivers are an expression of the results of this dichotomy and the management processes that have been put in place to use and dispose of water (Hynes 1975).

As an example, watershed management has operated within the engineering context that natural resources of a watershed such as its rivers must serve an absolute, direct human need in order to be of use. Therefore many rivers are viewed as sources of cheap energy, water supply, hydraulic waste, floodwater disposal and at its worst, convenient conduits of human waste. Hydrologic modelling of the watershed in its existing state is often conducted to determine the engineering specifications for drainage, the ability of the channel to handle the *waste water* and the implications on erosion and flooding. Once all these physical factors are satisfied, attempts are then made to accommodate the ecological needs of the river. In this manner, planning and management allows for a simplistic drainage template to be overlaid upon the river and its watershed and then attempt to *fit* sound environmental management within this context. This approach is environmentally and economically unsustainable.

In many jurisdictions, including Ontario, Canada, most hydrologic engineering in urban watersheds embraces the paradigm that water is a waste product or a hazard and must be moved off the land as fast as *humanly possible*. Modelling is usually done on the design storm basis using 1:2 year, 1:5 year, 1:25 year or 1:100 year recurring events for hydrologic and hydraulic modelling and design. Since rivers modify their channels during the annual storm which has an annual return rate of 1:1 to 1.5:1 (Leopold et al. 1964; Leopold 1968), this time scale discrepancy in the two approaches causes fundamental changes to the river: its' morphology; recharge capability and ultimately the water quality; productivity and biodiversity. The assumption that the existing stream morphology and flow regime is the correct system is a tacit approval of the historical record of human pioneering, settlement and development which has resulted in these existing degraded conditions. For example, the lack of fluvial geomorphological input in river channel management, despite the well-established science of the discipline is testimony to the compartmentalized engineering view of rivers as simply conduits of waste water. This traditional hydrologic engineering approach has been challenged in Germany over the last several years and has resulted in a multi-million dollar program to *de-engineer* and re-naturalize their rivers (Arnold et al. 1989).

2.2.2 Present Practices

In Ontario, there have been major changes in the last 10 years in the process and management of urban water quality and quantity. Some of these changes in the last several years are the result of importing technology refined in the State of Maryland.

In the late 1970's the Province of Ontario began to request a comprehensive approach to stormwater management. This was a result of concerns identified by various government agencies over changes in flood, erosion and water quality resulting from urban drainage designs and management. The focus for stormwater quantity and quality control shifted from the specific small development to a larger scale of analysis which encompassed portions of drainage systems in the sub-basins of larger river systems. This was an attempt to predict and manage for cumulative impacts of stormwater discharge on the river channel downstream of the newly developed lands within the watershed and to ensure proper capacity of channels, erosion control, flooding, etc. These plans were called *Master Drainage Plans* (MDP).

Despite this attempt to manage water and water supply in a more integrative fashion, the emphasis was flooding, focusing on surficial concerns for water on the surface of the land and its fate upon entering a natural channel (often ignoring headwater/-recharge regions). The objective was to convey water off the land and where necessary detain it so that discharge characteristics in the modified landscape would emulate *pre-development* flow conditions based upon certain design rainfall events (e.g. 1:2 yr; 1:5 yr; 1:25 yr). Often the context for *pre-development* was simply the degraded agricultural state. Once the major requirements for drainage were met, consideration was then given to attempting to *fit in* environmental concerns such as water quality and fisheries. In some of the models and analyses, consideration was given to shallow groundwater flow, but only as input to the hydrologic and hydraulic modelling components.

Although the MDP approach was a major step forward in the management of surface water quantity and later quality, it was still a two-dimensional approach, occurring as a supporting process to the landuse planning process. The landuse planning process determines the preferred use of the land based upon economic and sociological considerations. The MDP process is then used to ensure that the designs to be placed upon the landscape will function to remove water from the area and not overly affect river conditions and water quality downstream. Only recently in Ontario have MDP's examined the related issues of water supply quantity and quality. In order to do this, other water budget parameters such as infiltration, throughflow, and evapotranspiration would have to be considered and the unit of investigation expanded to encompass an integrative, physical/-ecological unit of land.

The result of this planning and modelling approach is that many of the environmental alterations resulting from changes in the water budget including physical, chemical and biological processes within the study basin are not predictable. These changes include modifications to channel form, dimensions, stability, ability to efficiently transport water and bedload at all major design stages; modification to water quality in the receiving channel; interference, reduction or contamination of well water supplies; destruction of fish habitat; alteration of nutrient cycling in streams; etc.

There is no opportunity in the present MDP process to set quantifiable objectives, targets or standards that are designed to: avoid environmental damage; achieve reasonable development within an area that does not compromise the integrity of local ecosystems; and improve the condition of the local environment as development proceeds.

The present process of model development and planning for stormwater management and water quality is not being applied within an ecosystem context. Without this integrative focus there is no way to examine within a common physical/chemical/-biological context the landbase and potential alterations to environmental features and processes that society requires, desires, or wishes to avoid (eg. fish, wildlife, water supply and quality, avoid damaging floods, contamination of wells, erosion, etc.). As well the ecosystem context is important in order to create the *one window* approach to the examination of water and water resource management in a way that allows all issues and opportunities to be examined at the same time. This information would also be extremely useful as objective input into the cause:effect relationships of potential landuse planning decisions in municipalities residing in the ecosystem unit.

2.2.3 Present Processes

The present landuse planning process has not succeeded in resolving the interactive and oft-times seemingly conflicting issues revolving around landuse, water supply and management. Presently, landuse decisions are made prior to more technical analyses of the condition and functioning of the physical environment. It is only once development is proposed for a portion of land that the full rigour of the environmental analyses and reviews occur. The approach apparently taken by most States and Provinces has been that environmental concerns are addressed once the land is actually slated for development.

Most jurisdictions in North America have some form of environmental legislation and planning in place to minimize destruction or degradation of the environment. Ontario has one of the most complex and seemingly sophisticated systems of any jurisdiction.

Government initiated landuse proposals are subject to the Ontario Environmental Assessment Act which requires environmental assessment for any public undertaking. Depending upon the size, complexity or potential impact of an undertaking, each proposal is subject to some level of environmental scrutiny ranging from a full environmental assessment for high impact proposals (e.g. hydroelectric projects) down to relatively simple straightforward conditions of development (e.g. dock construction). The requirements of the Act can be applied to private sector undertakings, although this rarely occurs.

Privately initiated proposals are generally regulated under the provisions of the Ontario Planning Act. The Act regulates landuse planning in the Province and provides for the subdivision of land and development controls through a specific process of landuse designation, enforcement, modification and approval. The Act is implemented at the regional and municipal level. To ensure a sound consistent approach to land management at the municipal level, the Planning Act requires the development of Municipal Official Plans (OP). The OP is a strategic plan that establishes long-term landuse management goals and objectives for the municipality. It designates a present and proposed landuse designation on all lands in the municipality and ensures a consistent mechanism for amendments to the plan, development proposals, drainage and servicing studies and plans for subdivision, etc. Although Provincial approval agencies (e.g. environmental agencies) may recommend modifications to the OP, the designation of land is driven by economic, development and political interests.

On a day-to-day basis within a municipality, repercussions to the natural environment and the sustainability of water and water based resources are usually addressed by environmental agencies at the plan for subdivision level when detailed plans are submitted. At this point, the developer has already received approval to develop his land from various Provincial, Regional and local government agencies and has invested substantial monies in the design and engineering of his undertaking. Comments by environmental managers and biologists on the need to modify design at this point are usually viewed with some hostility leading to confrontation. Environmental reviewers also have difficulty in determining the cumulative impacts of a series of small developments on the physical, chemical and biological interactions within the watershed. This planning process leads to further degradation of the watershed environment and its physical and biological resources. This situation has lead to the realization that ecosystems cannot be managed on a plan-for-subdivision level.

Presently, regulations are used to ensure that the negative effects of landuse development are mitigated or minimized. This is usually done through a regulatory process implemented at the site planning level. This is somewhat akin to enforcing restrictions of a certain nature on a site when the first question that should have been asked is whether this type of development or landuse is appropriate at all. This approach has continued to result in incremental deterioration of water quantity and quality despite the best efforts of technologists, in part because it appears that management by minimum standards becomes a target for design which in turn ensures a minimum environment. Perhaps it is time to view technology as a tool and develop an integrative systemic or ecosystemic management and planning process that directs the best use of the land for the entire ecosystem (people, plants, fish and animals). The process would also improve resolution of the most appropriate tool or tools (eg. soft or hard BMPs) to use for the land to be altered. The process ultimately enables managers to *design out* potential problems before they occur.

2.2.4 Developing an Ecosystem Approach and Context

There have been discussions on using an Ecosystem Approach for managing resources and landuse (I.J.C. 1978; Likens 1984). An Ecosystem Approach attempts to examine and identify the inter-relationships among biophysical, chemical and human elements of the ecological system. It recognizes the dynamic nature of the ecosystem, incorporates concepts of carrying capacity, resilience and sustainability. The approach strives to develop management targets based upon the potential of the ecosystem in accordance with a balance between the needs of the natural system and human requirements. Although this concept has been proposed in the past, the difficulty has always been in determining how to apply the approach, how to integrate the disciplines and the modelling efforts that would examine the various components of the ecosystem and finally how to implement the approach through established approval systems.

An Ecosystem Approach requires a number of key elements:

a logical geographic/ecological unit; an analytical process that allows for a view of the interactions between human and natural components of the ecosystem; and use of a physical pathway that interacts and integrates physical, chemical and biological processes (Imhof et al. 1991). The approach must have the ability to develop targets, standards and guidelines for landuse management, development, protection and rehabilitation, and there must be a planning process used that merges the ecological targets and preferred ecological management scenarios into the municipal, regional and state/provincial planning processes.

For water and water resource systems, we suggest that investigators and managers use the hydrologic cycle encompassed within certain physical ecosystem boundaries (Figure 2.1). For most urbanizing riverine based areas in North America, we suggest a process entitled *Watershed Management Planning* (WMP) be considered as one application of the Ecosystem Approach to management. In this process, a watershed is delineated as the geographical unit that encompasses the 4-dimensional characteristics of an ecosystem composed of aquatic environments and up-slope terrestrial environments linked by a common element, water (Figure 2.2). This then allows us to study the land:water interactions in a clearly defined geographic setting. The use of the hydrologic cycle provides a clearly measurable pathway in which the characteristics of the land can be modelled and the implications of water movement over and through the watershed can be analyzed for its implications on the biotic and human dominated environments. The hydrologic cycle also allows for the development of targets for environmental and landuse management that are quantifiable.

2.3 Framework for Model Development and Application

Two processes are required in order to employ an ecosystem approach effectively. One is a scientific/technical framework and

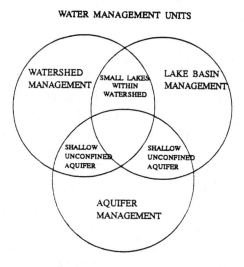

Figure 2.1: Geographically based hydrologic units.

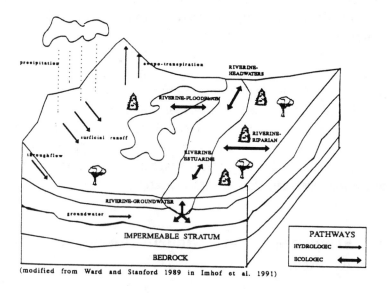

(modified from Ward and Stanford 1989 in Imhof et al. 1991)

Figure 2.2:　Hydrologic and ecologic pathways within a watershed.

process for model development, calibration and validation based upon a logical, ecosystem unit. The other component is a planning framework and process that incorporates the information from the ecosystem model into the various resource, environmental and landuse planning streams. For the purpose of this discussion we will examine these components using the watershed/subwatershed as the ecosystem unit.

2.3.1 Creation of a Scientific/Technical Modelling Process

Modelling is used extensively in the field of water and water resource management. In generic terms, any reproduction of a system is a model. Models that make use of algorithms in an attempt to reproduce a system are termed mathematical models.

What is common to all mathematical models is the balance of some observable quantities or variables which conceptualize the object to be simulated. The equations defining balances usually imply the formulation of constitutive laws. For example, in the case of various water discharge models these would include the hydrologic transport, generation and decay processes. Presently models are used to fit an endpoint objective such as a desire to move a certain amount of water down a channel without considering the implications to the ideal functioning of the channel. The final results often have limited consistency because the models of conveyance transport ignore the local micro-scale equations (eg. fluvial geomorphology) which actually define full system function. Because of the scale of the system to be modelled, simplifying assumptions must be used, this approach often produces results in which computed values do not compare well with measured values. In these instances the modeller will adapt the natural system into the confines of the generic model being employed at the time. This approach leads to inconsistencies between the analytical and empirical data with the result that the modeller will attempt to fit the natural system parameters into the parameters of the analytical model.

The forecasting capabilities of mathematical models are tied to statistics because of the nature of the systems that models try to reproduce. In this spirit statistical mechanics and its two dimensional thermodynamic applications have inspired various ecological models. In spite of some other well-established similarities between traditional statistical and thermodynamic models and ecological models, the linkages between the two fields are not so clearly understood.

Previously, in the field of ecological modelling the deterministic character of the basic equations have made them very attractive for forecasting. This characteristic has been generally useful in decision-making. Presently, these expert systems develop a life of their own because of the experimental method by which they are derived: i) screen the quality of information contained in the measured values, ii) yield results affected by uncertainty, and iii) offer options to the decision-maker. As a result, the form and function predicted by this modelling approach often bares no resemblance to the natural functioning system and as a result the options provided the decision-makers are invalid.

Presently, in ecological studies, process conceptualization for model development is carried out in an intuitive manner. Assumed in this process is that the link between the original system and the conceptual model can be defined through a functional transformation of macroscopic variables. However, environmental experimentation is strongly affected by the scale of observation for both sampling protocols and the significance of the data. Present ecological models are also affected by the choice of time and spatial scales. These scales define the parameters and determine the suitable algorithms for the model. As a result of this approach the modeller is placing very narrow contrived mathematical boundaries from generic models into heterogeneous domains. As well the time scales may not be interactive between the spatial, temporal, physical, biological, chemical, and geological processes and scales. Therefore the effects of cross-linking submodels from two or more disciplines might obscure or alter the expected events because time averages will differ from ensemble averages. The present method to resolve these anomalies is to incorporate fuzzy partitioning algorithms at the cross-over position to quantify the cross-disciplinary linkages. Consequently, the utilization of fuzzy parameters in present inter-disciplinary modelling methods is a convenient means of qualifying the model, leading to a highly acceptable yet low precision representation of the ecosystem (Figure 2.3, left hand side).

Networked ascending integration is an alternative method which foregoes the problematic factoring of fuzzy parameters into the modelling infrastructure. This approach generates a self-derived system which defines the models representative of the unique watershed system. This approach provides greater reproducibility and precision in definition of the watershed. In essence the *real life* physical parameters dictate the appropriate models rather than using standardized, convenient, generic models.

A vertical integration procedure is used for selection of models (Figure 2.4) which have to ultimately supply information to several inter-disciplinary levels of the networked model. This selection criteria process reduces the dependency on fuzzy parameters, thereby improving integrated modelling precision and accuracy. The iterative steps are model building blocks, (eg. geology, climate, hydrology, hydrogeology, and biology) taken in a methodical non-biased sequence to formulate the definition of the conceptual watershed model. Each discipline is responsible for its submodel selection procedure. Each submodel should be mathematically suited to the physical conditions while providing input/output pathways to the consolidated network infrastructure.

Therefore, the first step in a watershed study is the determination of the initial disciplines for the study. Once there is a level of confidence in the disciplines selected, the first iteration of the watershed model is the selection of submodel components. The submodels are networked to validate the watershed model against empirical information. If the watershed model is not valid, the suite of submodels are reviewed based upon the incongruities found within the watershed model. Based upon this review, redefinition of the disciplines and the submodels

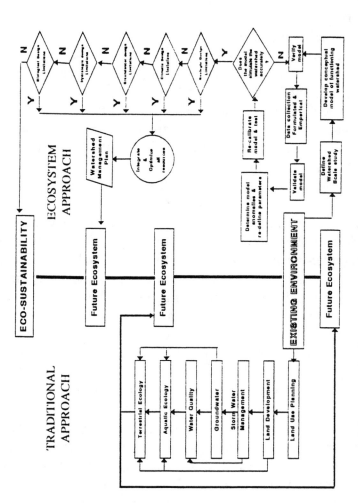

Figure 2.3: A conceptual framework showing the results of the traditional and ecosystem approach to watershed modelling.

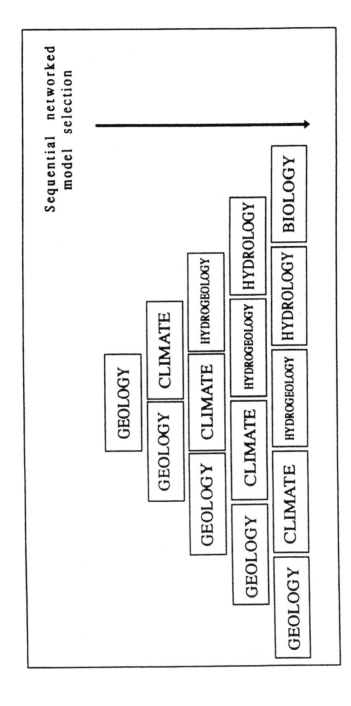

Figure 2.4: A vertical integration procedure for the selection of watershed sub-models.

would be reevaluated. The above process is repeated until such time as the watershed model as a whole is validated. For the abiotic portions of the model, the fundamental means by which validation is quantified and qualified is by means of a watershed mass balance (Figure 2.3, right hand side). For example, the submodel for groundwater must reflect accurately the functions of the regional and multi-surficial groundwater systems, the locations of discharge and recharge so that they may be integrated in the hydrologic and biological system analyses while maintaining hydrologic and chemical mass balances within own modelling domain and the domains of the other submodels. Biotic components in turn would be validated by their correlations with the abiotic components.

By our definition, Watershed Management Planning is a holistic, integrative approach to the management of land:water on a watershed basis. It strives to identify the functional attributes of each area of the watershed and recognize the inter-dependencies of components within the watershed. Based upon this understanding, the process then determines optimal, feasible ecosystem targets within which the human social fabric can be interwoven.

The use of the watershed as an ecosystem unit is contingent upon setting the appropriate scale and resolution for planning (e.g. full basin; sub-basin; single tributary system). The selection of the appropriate scale and its corresponding resolution will be based, at least in the short-term, upon local and regional planning issues. Although sub-basin watershed plans may be adequate to address ecological and human considerations in the sub-basin, by foregoing the full basin study prior to the sub-basin study, there will be a loss in efficiency and flexibility of managing the entire system and its supporting sub-basins as an integrated unit.

Since WMPs will require an interdisciplinary team made up of a variety of disciplines (including and not limited to hydrogeology, hydrology, hydraulics, geomorphology, aquatic and terrestrial ecology, engineering, environmental and municipal planning, water quality and toxicology and data management), there is an explicit need to ensure that the study process incorporates an interdisciplinary and data integration mechanism so that pathway and process modelling of the watershed or sub-basin can be developed, tested and used to develop and assess various target scenarios for the watershed.

2.3.2 Creation of a Ecosystem Planning Process

The model development process and its ecosystemic application are deterministic. The ecosystem being studied (eg. a watershed) has formed and evolved based upon the geology and climate of the region, modified by biotic processes and additional abiotic processes. In turn the ecosystem itself, if examined carefully, defines its functions and form and these in turn provide the information for determining how best to manage the system to ensure both a thriving, functional natural environment and human dominated environment (see Figure 2.3).

Once the study has developed a view of the form and function of the watershed system, environmental/ecological targets can be determined that will ensure that the physical pathways and processes remain intact and functional. Once these targets have been set, various landuse/resource management options can be examined in order to determine opportunities for development or resource management. Analyses of cause:effect relationships within the watershed can then be determined for the various desired uses and ecological/economic implications of any proposed management decision can be identified.

To ensure the proposed WMP management scenarios are implemented, the WMP process must be integrated into the state or provincial planning process. In Ontario, planning legislation such as Planning Act and the Official Planning (OP) process are two of the keys to implementing the ecological targets and management scenarios developed by the WMP process. The WMP process would complement the OP process by providing a scientific and objective process for the determining opportunities and constraints for the use of land in the watershed. In order for WMP to be used as an effective, objective tool, WMP's must occur prior to or early into the resource, environmental or landuse planning process. If WMP's precede planning decisions, then quantifiable, achievable, environmental objectives can be identified and used to make informed planning decisions. Potential cumulative impacts on natural environments can be avoided, designed out of a project or at worst, mitigated by using WMP to set ecological goals, targets and landuse recommendations as the input into municipal planning. People can have their homes, their livelihood, and healthy, clean rivers and healthy fish communities.

2.4 Conclusions

The management of water and water-based resources is in fundamental transition. Never before has there been a greater demand across North America for better water conservation, water quality, healthier aquatic systems, and benign development. People now want to maintain and rehabilitate their streams, rivers and lakes so that these systems are once more *drinkable, swimmable, fishable and enjoyable.*

The old tools, standards, regulations and models do not appear to be achieving the results that many desire. Concerns range from the loss of streams, lakes and wetlands to depletion of well fields, water quality degradation, etc. There are several possible reasons for this: the tools, models and processes are compartmentized and non-integrative; they are being applied at the wrong scale; models are being applied outside their domain; there is no common context in which to examine, analyze and model the water system within an ecological unit. Although an ecosystem, such as a watershed, may be of interest for a variety of reasons, the various planning processes (resource; environmental; municipal) all plan in isolation of each other. This is especially true when portions of the watershed are slated for modification for any reason. This has resulted in an inability to develop environmental objectives, targets and standards to guide and design developments in order to maintain the natural water-based biophysical systems that society wishes to protect and rehabilitate.

What is needed is a new process that manages the land:water base within an ecological unit in which the needs of natural components of the ecosystem are met or exceeded while at the same time accommodating reasonable human development and growth.

Application of an ecosystem approach and context requires a fundamental change in philosophy. The shift is from a regulatory based process for environmental management addressed through a site-specific, compartmentized, application of environmental standards to a process based planning and analytical/modelling system integrated within a definable ecosystem unit (eg. watershed or sub-basin). The process creates the integrative context for the determination of ecosystem form and function, application and networking of appropriate models. This can lead to the determination of ecological targets as the standards for use of the land and its water resources.

For the management of water and water systems, use of a physical pathway such as the hydrologic cycle, bound within a logical, geographic ecosystem unit, such as a watershed and using an appropriate functioning time scale, appears to offer great efficiencies. Efficiencies not only in costs and benefits but also as valuable, objective information for the development of human uses of the water and water resources that will not compromise the sustainability of the ecosystem and its physical resources.

Ecosystems are complex systems, driven in part by physical and chemical pathways and processes, but also modified by biotic communities and processes. Ecosystems are extremely resilient because of the built-in feed-back loops, as well as their adaptability and flexibility to change. The individual who approaches ecological modelling as a science might assume that, as computing capabilities progress, system dynamics can be determined with a prefixed level of accuracy. This deterministic view of natural systems derives from classical mechanics which asserts that if one knew all the physical laws and the initial state of the universe, one could predict the future. Unfortunately (or, fortunately) it is impossible to analytically solve all the equations needed even for a small subsystem. In the present compartmentized approach to modelling, this has lead to the overuse of fuzzy parameters that act as qualitative and quantitative connectors between the models used by each discipline. The use of fuzzy parameters is in part a result of cross-discipline uncertainties.

As an alternative, for consideration, is the process illustrated in Figure 2.3 (right side). Whereby the natural system itself is used by the team to determine the structure of the study, the data collection requirements, submodels used and the linkages between them. This will not avoid the use of some fuzzy parameters but will eliminate the cross-disciplinary uncertainties and minimize intra-disciplinary data deficiencies. This ecological model development process will produce a deterministic watershed model of higher precision which has the fundamental scientific requirement of being reproducible and representative of the study domain: in this case the natural watershed system.

If this process is integrated with a landuse planning process, implications of proposed landuse alterations upon natural environments, water conservation (surficial and groundwater) and human needs could be determined. Management scenarios could be proposed that would recommend landuse designations, development targets and standards designed to protect and improve natural environments and improve the efficiency of water management (e.g. flood and erosion control, aquifer management, servicing, stormwater management, fish habitat protection, etc.) as development proceeded within the watershed.

2.5 References

Arnold, V, J. Hottges, G. Rourie, M. Einsele, S. Englich, V. Ernst, H. Herberg, and F.K. Holtmeier. 1989. Removing strait-jackets from rivers and streams. *German Research*, Reports of the DFG 1989 (1): 22-24.

Bacalbasa-Dobrovici, N. 1989. The Danube River and its fisheries. Proceedings of the International Large Rivers Symposium. *Canadian Special Publication of Fisheries and Aquatic Sciences* 106: 455-468.

Backiel, T., and T. Penczak. 1989. The fish and fisheries in the Vistula River and its tributary, the Pilica River. Proceedings of the International Large Rivers Symposium. *Canadian Special Publication of Fisheries and Aquatic Sciences* 106: 488-503.

Brousseau C.S., and G.A. Goodchild. 1989. Fisheries and yields in the Moose River basin, Ontario. Proceedings of the International Large Rivers Symposium. *Canadian Special Publication of Fisheries and Aquatic Sciences* 106: 145-158.

Hesse, L.W., J.C. Schmulbach, J.M. Carr, K.D. Keenlyne, D.G. Unkenholz, J.W. Robinson, and G.E. Mestl. 1989. Missouri River fishery resources in relation to past, present, and future stresses. Proceedings of the International Large Rivers Symposium. *Canadian Special Publication of Fisheries and Aquatic Sciences* 106: 352-371.

Hynes, H.B.N. 1975. The stream and its valley. Edgardo Baldi Memorial Lecture. *Verhin International Verein. Limnology* 19: 1-15.

Imhof, J.G., R.J. Planck, F.M. Johnson, and L.C. Halyk. 1991. Watershed urbanization and managing stream habitat for fish. *Transactions of the 56th North American Wildlife & Natural Resources Conference* 56: 269-283.

International Joint Commission. 1978. The ecosystem approach. Great Lakes Research Advisory Board. 47 p.

Klein, R.D. 1979. Urbanization and stream quality impairment. *Water Resources Bulletin* 15: 948-963.

Lelek, A. 1989. The Rhine River and some of its tributaries under human impact in the last two centuries. Proceedings of the International Large Rivers Symposium. *Canadian Special Publication of Fisheries and Aquatic Sciences* 106: 469-487.

Leopold, L.B. 1968. Hydrology for urban land planning - a guidebook on the hydrologic effects of urban land use. *United States Geological Surveys, Circular* 554. 18 p.

Leopold, L.B., W.G. Wolman, and J.P. Miller. 1964. *Fluvial Processes in Geomorphology.* Freeman and Sons, San Francisco, CA. 522 p.

Likens, G.E. 1984. Edgardo Baldi Memorial Lecture. Beyond the shoreline: A watershed-ecosystem approach. *Verhin International Verein. Limnology* 22: 1-22.

Mann, R.H.K. 1989. The management problems and fisheries of three major British rivers: the Thames, Trent and Wye. Proceedings of the International Large Rivers Symposium. *Canadian Special Publication of Fisheries and Aquatic Sciences* 106: 444-454.

Minshall, G.W. 1988. Stream ecosystem theory: a global perspective. *Journal of the North American Benthological Society* 7: 263-288.

Regier, H.A., R.L. Welcomme, R.J. Steedman, and H.F. Henderson. 1989. Rehabilitation of degraded river ecosystems. Proceedings of the International Large Rivers Symposium. *Canadian Special Publication of Fisheries and Aquatic Sciences* 106: 86-97.

Sedell, J.R., J.E. Richey, and F.J. Swanson. 1989. The river continuum concept: A basis for the expected ecosystem behaviour of very large rivers? Proceedings of the International Large Rivers Symposium. *Canadian Special Publication of Fisheries and Aquatic Sciences* 106: 49-55.

Steedman, R.J. 1987. *Comparative Analysis of Stream Degradation and Rehabilitation in the Toronto Area.* Doctoral dissertation, University of Toronto, Toronto. Canada.

Steedman, R.J. 1988. Modification and assessment of an index of biotic integrity to quantify stream quality in southern Ontario. *Canadian Journal of Fisheries and Aquatic Sciences* 45: 492-501.

Ward J.V., and J.A. Stanford. 1989. Riverine ecosystems: The influence of man catchment dynamics and fish ecology. Proceedings of the International Large Rivers Symposium. *Canadian Special Publication of Fisheries and Aquatic Sciences* 106: 56-64.

Chapter 3

A Comprehensive Stream Study

T.W. Mahood and G. Zukovs
CH2M Hill Engineering Ltd.
2000 Argentia Rd. Plaza 3 Suite 100
Mississauga, Ontario. L5N 1V9

The continued deterioration of water quality, the destruction of wildlife habitats and the limited potential for recreational use are three of the major adverse effects of urban development and past watershed management practices evident in a number of small streams in Southern Ontario.

One such stream, the Little River in Windsor, flows northward from the rural areas of the Township of Sandwich South through the eastern end of the City of Windsor where it discharges to the Detroit River. The Little River has been the subject of a study initiated by the City of Windsor and the Ministry of the Environment (MOE) to quantify the various point and nonpoint pollutant discharges to the Little River, to assess the relative impacts of pollutant sources on the river's ecosystem and to develop a strategy for mitigating the effects of these pollutant sources and improving the *health* of the stream.

0-87371-898-4/93/$0.00 + $.50
© 1993 by Lewis Publishers

55

3.1 Introduction

The rehabilitation and protection of urban streams has become a major focus of attention in a number of municipalities in Southern Ontario. Increasing levels of urbanization within the drainage basins of these streams has been a contributing factor in the deterioration of stream quality and in the impairment of aquatic habitats.

Stream management practices within urban watersheds have historically focused on reducing the potential of flooding during extreme hydrological events. This practice has led to the channelization and dyking of many streams in Southern Ontario cities. In many instances, channelization and dyking practices have damaged or even eliminated aquatic habitats that had already been severely stressed by increased pollutant loads. More recently, many municipalities have initiated comprehensive management programs that include enhancement and protection of aquatic habitats, terrestrial habitats and recreation potential as fundamental components.

In 1989, the City of Windsor and the Ontario Ministry of the Environment (MOE) initiated a comprehensive stream study to develop a management program for the Little River in Windsor. The Little River Comprehensive Stream Study (LRCSS) was carried out by the City of Windsor, the MOE, the Great Lakes Institute (GLI) of the University of Windsor and the Consulting firms of Lafontaine, Cowie, Buratto and Associates Ltd., and CH2M HILL ENGINEERING LTD.

This paper discusses the approach and describes the findings of the integrated study team who conducted the LRCSS as part of an overall Pollution Control Plan (PCP) for the City of Windsor.

3.2 Study Area

Key study area features are shown in Figure 3.1.

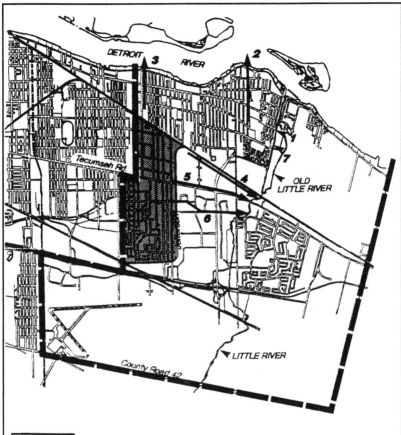

 COMBINED SEWER AREA

1. WYANDOTTE SANITARY SEWER OVERFLOW
2. WATSON/WYANDOTTE SANITARY
   SEWER OVERFLOW
3. FORD/SOUTH NATIONAL CSO
4. JEFFERSON/SOUTH NATIONAL CSO
5. TECUMSEH CSO
6. ROSE CSO
7. LITTLE RIVER POLLUTION CONTROL PLANT

Figure 3.1: Study area.

The study area encompasses the entire section of the Little River basin within the City of Windsor. This area is bounded by the Detroit River to the north, the Township of Sandwich South to the south at County Road 42, the Village of Tecumseh in the east and Pillette Road in the west.

The section of the Little River within the study area from County Road 42 to the Detroit River is approximately 8 km long. The lower reaches of the River downstream of Tecumseh Road have been channelized and straightened to provide efficient and safe conveyance of storm water from the urban portions of the study area to the Detroit River.

The total sewershed within the study area is approximately 4,320 hectares in size and serves a population of about 55,460 people. Most development is concentrated in the lower reaches of the river. Land use in the upper reaches is predominantly agricultural. There are 150 industries within the study area although none presently directly discharge wastewater effluents to the Little River. Until the fall of 1990, Wickes Manufacturing, a car bumper plating facility, discharged treated effluent into the Little River immediately north of the river's intersection with Tecumseh Road.

Much of the City of Windsor within the LRCSS area is serviced by separate sanitary and storm sewer systems. There is however, a significant area serviced by only combined sewers. Overflow discharges from the sewer network to either the Little River or the Detroit River occur at six separate locations. The Little River Pollution Control Plant (LRPCP) discharges to the Little River approximately 1 km upstream of the river mouth.

The section of the Little River that existed prior to channelization is known as the *Old Little River*. The stream bed of the Old Little River essentially has no flow and is separated by a dyke system from the main channel of the Little River. A wetlands habitat area exists in the Old Little River channel.

The Little River:

1. **has no viable sport fishery** (habitats are not sufficient for the establishment of an indigenous sport fish population);

2. **is not used for swimming** (water quality is not presently suitable in most areas); and

3. **is used as a boating access to some homes** (stream is navigable to the CP Railway Line).

3.3 Study Issues and Objectives

Two major issues facing the City of Windsor were the focus of the LRCSS. These issues were:

1. levels of pollutant loadings to the Little River, and

2. impacts of pollutant loads and stream physical characteristics on the Little River ecosystem.

Specific objectives were developed as part of the LRCSS to address the concerns of the City of Windsor. These objectives included:

1. determination of the relative contribution of contaminants from each of the sources contributing flows to the Little River;

2. assessment of the impacts of pollution and stream physical characteristics within the Little River watershed; and

3. development of a comprehensive remedial action plan for the Little River Watershed.

Study Approach

The approach was divided into two separate major components. The first component, a contaminant loading inventory was conducted to determine the relative pollutant loadings to the Little River from a number of sources. These sources included:

1. upstream boundary stream flow,
2. stormwater runoff,
3. combined sewer overflows,
4. LRPCP final effluent,
5. LRPCP emergency bypass,
6. industrial discharges, and
7. dry weather outfall seepage.

The second component, an ecological assessment was carried out to determine the present condition of the Little River ecosystem. The assessment included:

1. benthic invertebrate surveys,
2. biomonitoring with freshwater clams,
3. sediment contaminants survey, and
4. in-stream water quality analysis.

3.4 Contaminant Loading Inventory

The sources of inputs to the Little River were classified as either dry weather sources or wet weather sources.

Dry Weather Sources

Loading estimates for dry weather sources included:

1. upstream boundary flow,
2. LRPCP final effluent,
3. industrial discharges, and
4. dry weather outfall seepage.

Dry weather flow volumes and contaminant concentration data were collected during a field survey program carried out during the summer and fall of 1990.

Wet Weather Sources

Loading estimates for wet weather sources included:

1. upstream boundary flow,
2. LRPCP emergency bypass,
3. stormwater runoff, and
4. combined sewer overflows.

Samples for wet weather contaminant concentration analysis were collected by automatic samplers at key locations in the study area.

The estimated annual discharge volumes from each source are presented in Table 3.1.

Figure 3.2 presents a piechart of the total volume distribution of inputs to the Little River, and Figure 3.3 presents a piechart of the wet weather volume distributions.

Table 3.1: Current annual discharge volumes.

Source	Flow Volume (m^3/yr)
Dry Weather Sources	
· Upstream Boundary	7,385,000
· Industrial Effluents	120,000
· Outfall Seepage	305,000
· LRPCP Effluent	14,167,000
Total Dry Weather	**21,977,000**
Wet Weather Sources	
· Upstream Boundary	1,760,000
· Stormwater	1,415,000
· Combined Sewer Overflow (CSO)	375,000
· LRPCP Bypass	374,000
Total Wet Weather	**3,924,000**
Overall Total Flow	**25,901,000**

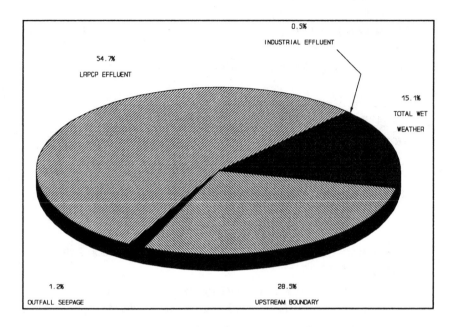

Figure 3.2: Total flow input distribution.

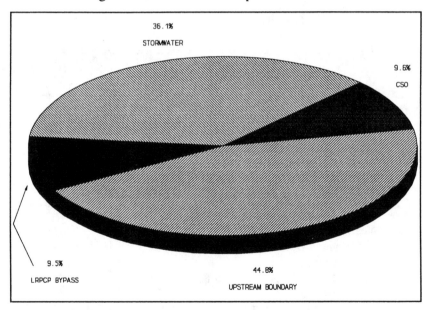

Figure 3.3: Wet weather flow input distribution.

A modified STORM (storage, treatment, overflow, runoff model) model was selected to estimate the discharge volumes from these wet weather sources.

The original version of STORM (Hydrologic Engineering Centre, 1977) is a single catchment model without the capability to analyze a complicated interconnected and overflowing sewer network. The modified version (Cheung et. al. 1987) allows the continuous simulation of stormwater runoff and CSO discharges over a year long record in a complex network of combined and separated sanitary sewers.

For any single catchment, the modified storm model reads the continuous inflow generated at the catchment immediately upstream in the network, adds it to the locally generated flow and calculates the throughflow of the catchment, the overflow volume and frequency of overflow. The throughflow then becomes the upstream inflow of the next catchment in the network.

Although STORM has been used in a similar fashion in other studies (CH2M HILL, 1990), (CH2M HILL, 1991), the sanitary sewer system modelled in Windsor was a particularly complex system of combined and partially separated sewers.

To provide STORM with the necessary input data, thus enabling the model to accurately assess the performance of the sewer network, a number of field surveys and flow monitoring programs were carried out. A typical year of rainfall data was then selected for input to the model by analyzing rainfall data collected at the Windsor airport (Environment Canada, 1991).

Bacterial, nutrient/conventional and metals loadings were calculated for wet weather and dry weather sources using the concentration data and flow data collected. The distribution of loadings to the Little River from the three highest contributors for selected parameters on a percent of annual loadings basis is presented in Tables 3.2, 3.3, and 3.4.

The following observations can be made in regard to the loadings information:

1. The majority of the loadings to the Little River are from:

   · stormwater,
   · upstream boundary flow,
   · LRPCP, and
   · CSO.

2. Most bacterial loading to the river are the result of stormwater runoff and combined sewer overflow discharges.

Table 3.2: Bacteriological loading distributions.

Loading Source	Percent of Total Annual Loading to Little River (%)	
	E.C.[1]	F.C.[1]
Dry Weather		
Upstream Boundary	<0.1	0.1
Outfall Seepage	0.9	0.3
Industrial Effluents	0	0
LRPCP Effluent	0.1	0.1
Wet Weather		
Upstream Boundary	2.6	3.2
Stormwater	55.0	57.1
CSO	41.0	28.3
Bypass	0.3	11.0

Notes:
1. E.C. -   *Escherichia coli*
    F.C. -   Fecal coliform

3. Despite low concentration levels, a substantial fraction of the metals loadings, to the Little River is a result of LRPCP final effluent discharges. The LRPCP metals loadings are a result of the large volume of discharge in relation to most other sources, not the concentrations.

4. Metals contributions from the upstream boundary were

significant during both dry and wet weather.

5. Most suspended solids loadings are from wet weather sources. In terms of percent of annual loadings, the upstream boundary contribution was the most significant.

6. Although effluent criteria of 15 mg/L for BOD and 1 mg/L for TP are consistently met by the LRPCP, it still contributes a significant fraction of the total phosphorus and BOD to the River because of it's large volume of discharge.

Table 3.3: Metals loading distribution.

Loading Source	Percent of Total Annual Loading to the Little River (%)		
	Cr[1]	Ni[1]	Zn[1]
Dry Weather			
Upstream Boundary	22.2	27.8	14.8
Outfall Seepage	1.1	1.4	0.7
Industrial Effluents	0.6	0.3	0.4
LRPCP Effluent	50.9	38.5	60.1
Wet Weather			
Upstream Boundary	9.6	24.9	8.6
Stormwater	10.2	4.6	9.0
CSO	3.0	1.4	2.8
Bypass	2.4	1.1	3.6

Notes:
1. Cr -   Chromium
    Ni -   Nickel
    Zn -   Zinc

3.5 Ecological Assessment

Significant areas of increased sediment contamination were identified on the Little River. Depositional samples taken on a number of sampling occasions indicated increased metals and

Table 3.4: Nutrient and conventionals loading distribution.

Loading Source	Percent of Total Annual Loading to the Little River (%)		
	SS[1]	BOD[1]	TP[1]
Dry Weather			
Upstream Boundary	17.4	5.3	15.1
Outfall Seepage	0.4	0.6	0.5
Industrial Effluents	0.1	0.3	0.1
LRPCP Effluent	21.2	42.3	50.8
Wet Weather			
Upstream Boundary	31.2	14.2	9.9
Stormwater	10.3	5.1	3.8
CSO	6.3	8.8	6.7
Bypass	13.1	23.5	12.9

Notes:
1. BOD- Five-day biochemical oxygen demand
    TP-   Total phosphorus
    SS-   Suspended solids

nutrient concentrations in the sediments below Tecumseh Road (Station 17) and again below the LRPCP (Station 8) to the mouth of the river (Station 4). Levels of zinc, chromium, nickel and phosphorus from the summer 1990 sampling expedition are summarized and included here in Figures 3.4 and 3.5. The levels of these parameters are representative of the general trends in the sediment data.

Figures 3.6 to 3.11 show the trends in benthic community composition from upstream sample locations to the lower reaches of the river. Station #25 was at the upstream boundary of the LRCSS at County Road 42. Station #21 was between County Road 42 & Tecumseh Road. Station #17 was at Tecumseh Road. Stations #12, #9 and #6 were downstream of Tecumseh Road.

The benthic invertebrate sampling results indicate:

1. the composition of invertebrates at stations below Tecumseh Road were indicative of highly degraded conditions;

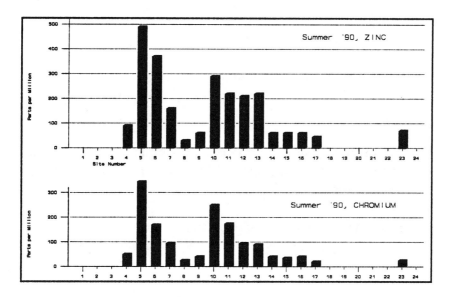

Figure 3.4: Summer 1990 zinc & chromium sediment concentrations in the Little River.

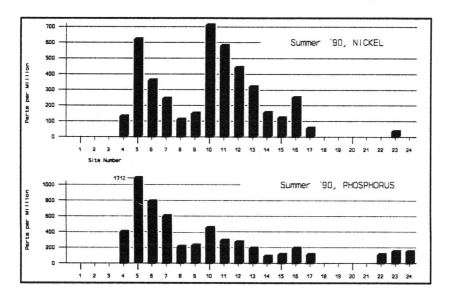

Figure 3.5: Summer 1990 nickel & phosphorus sediment concentrations in the Little River.

2. significant habitat degradation occurs as far upstream as Tecumseh Road;

3. invertebrate communities above Tecumseh Road were more rich in crustacean, molluscan and insect taxa than downstream sites;

4. the transition from arthropod/mollusc dominated communities to annelid dominated communities cannot be attributed to any single factor;

5. channelization and dyking in the lower reaches can be expected to have an important impact on benthic composition; and

6. impaired environmental quality conditions are found along the entire length of the Little River.

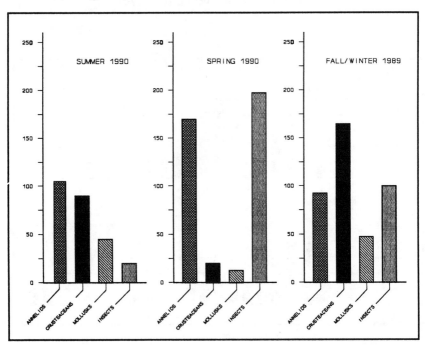

Figure 3.6: Benthic sampling results from station #25.

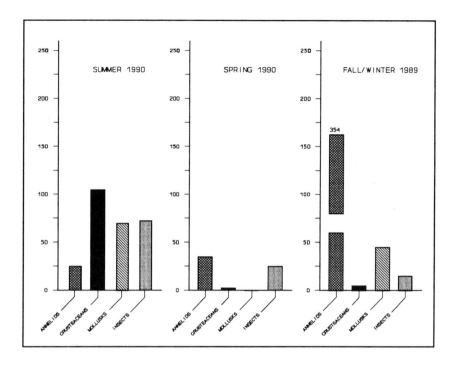

Figure 3.7: Benthic sampling results from station #21.

3.6 Remediation Alternatives

Ecological conditions in the Little River may be impaired by a number of contaminant loadings and by the lack of available habitat. Alternatives for stream enhancement should therefore focus on feasible options for the control of the major contaminant loads and enhancing stream habitat conditions in areas where the most benefits can be expected.

3.6.1 Loading Controls

Stormwater

Urban stormwater runoff becomes contaminated principally as a result of contact with various pollutants deposited on the catchment surface. Reduction of stormwater runoff contaminant loadings may be achieved through a number of measures. These measures may include:

1. Source quality controls which may include:

    · enhanced street sweeping,
    · catchbasin cleaning, and
    · anti-litter regulations.

2. Interception and dispersion of flows to other receiving locations.

3. Interception and treatment of flows prior to discharge including:

    · solids removal,
    · metals precipitation, and
    · disinfection.

Stormwater controls can be expected to have the most significant effect on bacterial loadings to the Little River.

Boundary Flow

Contaminants in the Little River enter the City of Windsor and the LRCSS Study area from the Township of Sandwich South at County Road 42. The upstream boundary flow was found to contribute a significant fraction of the metals and suspended solids loadings to the Little River.

The sections of the Little River in Sandwich South were not however included in the LRCSS and are beyond the City of Windsor's municipal boundary. A separate program of upstream source identification would be necessary to address this issue.

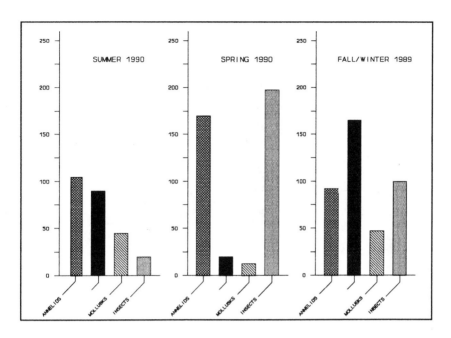

Figure 3.8: Benthic sampling results from station #6.

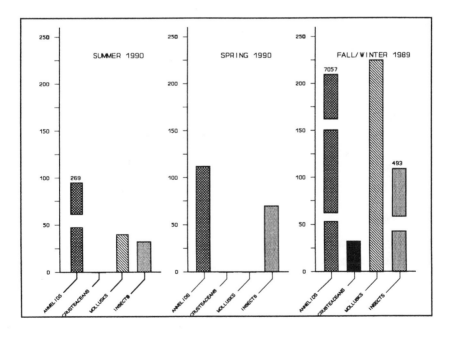

Figure 3.9: Benthic sampling results from station #17.

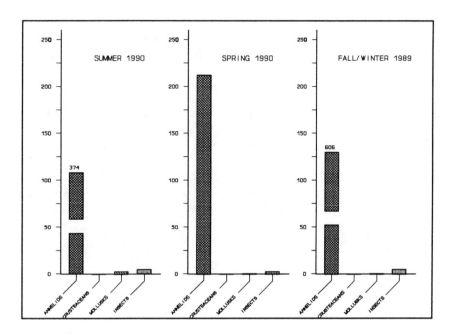

Figure 3.10: Benthic sampling results from station #12.

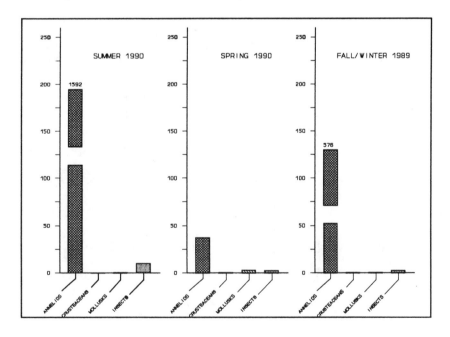

Figure 3.11: Benthic sampling results from station #9.

Combined Sewer Overflows

Alternative CSO control measures that were investigated as part of the LRCSS include:

1. **Separation** - Various combinations and degrees of road and on-lot drainage separation can be considered. In practice, however, sewer separation is costly and never 100% effective (CH2M HILL, 1991). It has been difficult if not impossible in most jurisdictions to achieve separation of on-lot drainage particularly footing drains. Moreover, numerous combined sewer overflow studies have shown that separation alone is not a cost-effective means of controlling overflows. Sewer separation also has the disadvantage of increasing stormwater discharge to receiving waters. Depending on the ratio of combined sewage pollutant concentrations to stormwater pollutant concentrations, separation may not provide any significant benefits in terms of protecting stream habitats (MOE, 1991).

2. **Tank Storage and Treatment** - Storage of CSO volumes and LRPCP emergency bypass flows could be accomplished using two separate off-line storage facilities.

    The storage tanks would be situated at critical locations to capture and store the overflow volumes from the CSOs and the LRPCP bypass.

    Two locations were chosen, one at the LRPCP to store bypass and separated sanitary sewer overflows and another in the combined sewer area to store CSO discharges.

    The volumes of storage calculated using the modified STORM model and the percent reductions in discharge volumes expected at each of the two sites for control levels of ten, four and one overflow per year are listed in Table 3.5.

    Treatment could be accomplished by dewatering the storage tanks to the LRPCP.

3. **Tunnel Storage and Treatment** - The storage of CSO discharge volumes and LRPCP emergency bypass flows could be accomplished using an in-line storage tunnel.

   Tunnel storage would provide the same percent overflow volume reductions at ten, four or one overflow event per year as tank storage. The estimated tunnel size and the reductions in discharge expected for each control option are shown in Table 3.6. Storage volumes were calculated using the modified STORM model.

   Treatment of stored volumes could be accomplished by dewatering to the LRPCP.

Table 3.5:   Estimated storage tank volumes and percent overflow volume reductions.

Tank Location	Number of Overflows Per Year		
	10	4	1
Combined Sewer Area	10,500 m³	16,300 m³	25,000 m³
LRPCP	14,000 m³	25,500 m³	31,300 m³
Totals	24,500 m³	41,800 m³	56,300 m³
Percent Overflow Volume Reduction	80%	95%	99%

4. **LRPCP** - The Little River Pollution Control Plant has recently (1990) undergone a major expansion that has increased its treatment capacity from 38,000 m³/d to 64,000 m³/d. The plant currently operates with final effluent discharges of total phosphorus, BOD and suspended solids below levels stipulated in the MOE Certificate of Approval.

   Control alternatives for LRPCP effluent discharges could include upgrading to a tertiary treatment facility or moving the final effluent discharge location.

Table 3.6: Estimated tunnel size and overflow volume reductions.

Control Option (Overflows per year)	Tunnel Volume (m³)	Tunnel Diameter (m)	Tunnel Length	Percent Overflow Volume Reductions
10	24,600	3.0	3.3 km	80%
4	41,800	4.0	3.3 km	95%
1	56,296	5.0	3.3 km	99%

3.6.2 Habitat Enhancement

Channelization

Stream channelization greatly influences the ecosystem of a stream (Brooks, 1988). Channelization in the Little River below Tecumseh Road has increased the hydraulic efficiency of the channel but has also led to a homogeneous substrate which undergoes frequent occurrences of scouring.

Any alteration of the stream flow characteristics with the intent of improving aquatic habitats would conflict with the use of the stream as a stormwater conveyance channel. The effects of such alterations could possibly cause an increase in flooding frequency in the areas adjacent to the Little River.

Sediments

Sediments in the Little River upstream of Tecumseh Road were of generally good quality. Downstream of Tecumseh Road elevated concentrations of nickel, copper, iron, lead and zinc were found that exceeded draft provincial guidelines for open water, unrestricted or restricted land use disposal. Particularly high concentrations of metals were found immediately downstream of the former Wickes Bumper Plating Plant. Sediment quality in this section of the river is expected to improve with the closing of the plant.

The density and diversity of benthic communities has also been shown to be greatly reduced below Tecumseh Road. It is not clear however to what degree the level of sediment contamination affects the benthic communities because of the cumulative effects of stream scouring and the lack of substrate variability.

Riparian Vegetation

In the upper reaches of the Little River upstream of Tecumseh Road, the stream still meanders through its original channel. The establishment of Riparian vegetation has been shown in a number of similar instances to improve aquatic habitats.

Riparian vegetation adjacent to the Little River could:

1. control extremes in water temperature;

2. provide nutrients to aquatic species;

3. reduce contaminants in run-off water entering the stream;

4. control erosion;

5. control runoff volumes entering stream and thus limit stream flow velocities; and

6. provide wildlife habitats.

Strips of riparian vegetation (buffer zones) grown alongside watercourses are defined in a number of municipal and conservation authority policy documents (Buchinger, 1990). These buffer zone policies vary for the most part in the depth of the zones from the water course or from a stable bank slope line. The requirements and policies with respect to vegetative buffer zones on the Little River are still to be determined.

3.7 Conclusions

The most distinctive feature of the Little River that places a constraint on the effectiveness of remedial measures is the channelization and dyking of the River below Tecumseh Road. Channelization reduces sediment variability and stability and thus limits the improvements in aquatic habitat that might otherwise be expected from a reduction in contaminant loadings.

There are a number of stormwater quality and quantity control options available. The feasibility and effectiveness of these controls must be assessed in relation to the existing stormwater management system which includes the Little River.

Tank storage/treatment and tunnel storage/treatment options can provide the most effective level of combined sewer overflow control from 80% (at 10 overflow events per year) to 99% (at one overflow per year). Storage alternatives substantially reduce the discharge of sanitary waste to the Little River.

Habitat impairment is less severe in the upper reaches of the Little River. The natural stream bed above Tecumseh Road also provides more opportunity for habitat improvement. The establishment of vegetative buffer zones and associated riparian canopies adjacent to the Little River could be expected to improve the aquatic habitats in the stream.

3.8 References

Buchinger, M., *Review of Buffer Zone Policies*, Proceedings, P.C.A.O. Seminar on Integrated Watershed Management, December 2, 1991.

Cheung, P.W., Smith, D.I. and Moir, F.C. *A Multi-Model Approach to Water Pollution Control Planning*, Proceedings, Stormwater and Water Quality Model Users Group Meeting, October 15-16, 1987, pp 23-24.

CH2M HILL, *St. Catharines and Area Pollution Control Planning Study*, April 1990.

CH2M HILL, *City of Kingston Pollution Control Planning Study*, January 1992.

Environment Canada, Atmospheric Environment Service, *Canadian Climate Normals*, 1951-1980.

Hydrologic Engineering Centre, *Storage, Treatment, Overflow, Runoff Model - STORM*, Computer Program 723-S8-L7520, U.S. Army Corps of Engineers, Davis, California, August 1977.

Ontario Ministry of the Environment, *Analysis of Combined Sewer Overflow Control Technologies*, Water Resources Branch, Toronto, September 1991.

Chapter 4

Tools for Evaluating Environmental Quality, Water Quality and Water Quantity Issues

W. Snodgrass[1], D.E. Maunder[2],
K. Schiefer[1] and K.C. Whistler[1]
Beak Consultants [1] and Aquafor Engineering[2]
14 Abacus Road, Brampton, Ontario L6T 5B7

4.1 Introduction

The objective of this paper is to present an overview of an integrated approach for using modelling tools, monitoring data, and planning studies for evaluating management options in Watershed Management Studies. Emphasis is placed upon the impacts of urbanization.

The various methods presented have been used to address issues pertaining to ecosystem health (environmental health), water quality and water quantity. The watersheds which are considered include the Rouge River, the Don River, and Duffins Creek. By evaluating the alternative approaches, the reader can gain some insight on the benefits, limitations and appropriateness of each approach for their potential study.

0-87371-898-4/93/$0.00 + $.50
© 1993 by Lewis Publishers

The methods presented in this paper are illustrative. A comprehensive evaluation of the applicability of these methods to water quality management modelling is beyond the scope of this paper.

4.2 Selection of Modelling Approach

The selection of the modelling approach should follow a predefined set of steps. Several steps that we have found useful include:

1. Define the goals and objectives of the model. That is, questions that are addressed by the model.

2. Evaluate the database available or to be gathered during the study. Detailed models require extensive sets of data, either as input data, for calibration purposes, or for testing purposes (ie. comparison of modelling calculations with observed data).

3. Consider monetary constraints. This involves evaluating budget available, and allocating it for additional field studies (monitoring), modelling studies, planning studies and public involvement/institutional review/hearings.

4. Examine uses of the model after urbanization has occurred. Modelling together with monitoring may be required to audit the success of proposed environmental management plans. Models should be selected which are useful both for planning studies and for auditing purposes.

5. Select the modelling approach.

4.3   Environmental Frameworks For Defining Issues

A variety of frameworks are currently in use in the Province of Ontario for defining the environmental issues to be addressed in watershed/subwatershed studies. The way that the issues are defined, and the issues considered in the framework, provide the basis for integration of study results. Two frameworks used in recent studies in Ontario are given in Tables 4.1 and 4.2. Goal statements and management objectives were established for each of the fourteen divisions given in Table 4.1 for the Rouge River watershed. The framework emphasized:

1. the linkage between the riverine watershed and the Great Lakes water, and the quality of human life derived from these ecosystems,
2. water quality, public health and aesthetics,
3. public safety, and
4. fisheries, riparian and terrestrial habitats.

The framework initially used in a subwatershed study, the Hanlon Creek study (Table 4.2) emphasizes:

1. public safety and natural floodplain hydrologic functions,
2. water quality and associated aquatic resources and water supplies, and
3. amenities of rural and urban stream corridors.

Both frameworks emphasize management of water, the lands directly connected with aquatic systems and the associated biological communities. As pointed out by Jack Imhoff (1992), watershed management must emphasize both land and water.

We have found it useful to use the following environmental issues as the framework for managing environmental quality in watersheds impacted by urbanization (page 84):

Table 4.1   Conceptual divisions for ecosystem-based management plan, used in Rouge River study.

I. Quality of Life Within Great Lakes Ecosystem

1. linkage to Great Lakes ecosystem
2. pride in Great Lakes and Rouge River ecosystem
3. balance of economic and environmental value
4. quality of life and land ownership

II. Water Quality, Public Health and Aesthetics

5. contact, non-contact recreation
6. drinking water
7. fish consumption
8. aesthetics

III. Public Safety

9. erosion and flood protection
10. risk to life in valley lands

IV. Fisheries, Riparian and Terrestrial Habitats

11. river beds as fish habitat
12. angling
13. enjoyment of plants and wildlife
14. wildlife and waterfowl and their habitats

Table 4.2   Mission statement and goals and objectives of the Hanlon Creek watershed plan.

Mission Statement

To develop a watershed plan that allows sustainable development aimed at maximizing benefits to the natural and human environments on a watershed basis.

Goals and Objectives

1) Goal
    • To minimize the threat to life and the destruction of property and natural resources from flooding, and preserve (or re-establish) natural flood plain hydrologic functions.

Objectives
- *To ensure that runoff from developing and urbanizing areas is controlled such as it does not unnecessarily increase the frequency and intensity of flooding at the risk of threatening life and property.*

- *To adopt appropriate land use controls and performance standards for controlling development of flood plains.*

2) Goal
- **To restore, protect, and enhance water quality and associated aquatic resources and water supplies.**

Objectives
- *To minimize erosion and prevent sedimentation of waterways.*

- *To prevent the accelerated enrichment of streams and contamination of watershed from runoff containing nutrients, pathogenic organisms, organic substances, and heavy metals and toxic substances.*

- *To maintain or restore a natural vegetative canopy along streams where required to ensure the mid-summer stream temperatures do no exceed tolerance limits of desirable aquatic organisms.*

- *To maintain the stream or waterway free from litter, trash, and other debris.*

- *To minimize the disturbance of streambed and prevent streambank erosion and, where practical, to restore eroding streambanks to a natural or stable condition.*

- *To restore, rehabilitate, or enhance water quality and associated resources through the implementation of appropriate Best Management Practices on the land.*

- *To take full advantage of stream baseflow enhancement opportunities.*

- *To enhance the fishery habitat. Specifically to increase the quantity and quality of Brook Trout in the headwaters are and to extend their range downstream of the Hanlon Expressway to the Speed River.*

- *To maintain or enhance the buffer provided by wetlands.*

- *To minimize disturbance of wetlands, preserving or enhancing the habitat they provide.*

- *Provide buffers to wetlands to maintain or enhance their biological health.*

3) Goal
- **To restore, protect develop, and enhance the historic, cultural, recreational, and visual amenities of rural and urban stream corridors.**

Objectives
- *To ensure the that environmental resources constraints are fully considered in establishing land use patterns in the watershed.*

- *To retain and preserve open space and visual amenities in urban and rural areas by establishing and maintaining greenbelts along stream corridors and adjacent natural areas.*

- *To ensure that development in down stream corridor is consistent with the historical and cultural character of the surroundings and fully reflects the need to protect visual amenities.*

- *To ensure that the recreational and fisheries potential of a stream corridor are developed to the fullest extent practicable.*

- *To maximize the use of creative and imaginative resources to rehabilitate and transform urban stream corridors, which through neglect may represent a source or urban decay and blight, into attractive community assets consistent with historical or other cultural amenities.*

- flooding,
- erosion,
- surface water quality,
- groundwater (quality and quantity),
- natural features (wetlands, ESAs, ANSIs),
- aquatic communities,
- recreation,
- aesthetics (water, valleyland),
- terrestrial (wildlife, woodlots), and
- ultimate receiving body (Great Lakes).

This list is not ranked. Rather, it reflects the historical evolution of watershed management within Ontario.

4.4 Examples of Environmental Issues

This section discusses a few of the key environmental issues for which water quality/environmental quality modelling may be required as a tool in developing a watershed or subwatershed management plan.

4.4.1 Water Balance

The change in the water balance due to urbanization is one of the most profound effects of urbanization. Urbanization increases flood flows, erosion potential (through the increased frequency of full-bank flow), and destroys fish habitat by widening channels and making them more shallow. Mitigating measures can ameliorate these effects to some degree.

Estimated changes in the water balance as a forested watershed urbanizes are given in Table 4.3. The example is used as a reference point in the Duffins Creek study. Depression storage and infiltration are reduced by 33% and 50% respectively, but the volume of runoff increases approximately an order of magnitude after urbanization. Implementation of infiltration

Table 4.3:   Distribution of May to November Rainfall for Forested and Urban Areas.

| Item | Forested Areas | | Urban Areas with 40% Impervious Land | | | |
| | | | No Infiltration | | With Infiltration | |
	Depth (mm)	% Total Depth	Depth (mm)	% Total Depth	Depth (mm)	% Total Depth
May to November Rainfall	515	100	515	100	515	100
Interception Storage and Depression Storage on Impervious Areas	342	66.5	235	45	235	45
Infiltration	155	30	100	20	200	40
Runoff	18	3.5	180	35	80	15

devices can approximately double the degree of infiltration in urban areas. However, the volume of runoff is still approximately 400% larger compared to the forested condition.

4.4.2 Surface Water Quality

The two major sources of contaminated discharges to surface water quality from urban areas are water pollution control plants (WPCPs) and urban runoff. Typical concentrations of the parameters from the sources given in Table 4.4 often exceed water quality objectives. If natural stream processes do not significantly cause decreases in these discharged concentrations, urbanization requires mitigation for streams which fully drain urban watersheds, if instream water quality is to meet PWQO's.

Many water quality parameters may have to be considered. Traditional concerns have included biological oxygen demand, suspended solids, pathogenic bacteria, and nutrients.

Table 4.4:   Comparison of Pollutant Concentrations to Provincial Water Quality Objectives.

Water Quality Parameter	Urban Runoff Concentrations (mg/L)	Sewage Treatment Plant Concentration (mg/L)	Provincial Water Quality Objective (mg/L)
Total Phosphorus	0.35	0.33	0.03
Total Copper	0.035	0.014	0.005
Lead	0.14	0.02	0.05
Fecal[1] Coliforms	25,000[1]	250	100

Notes: [1] Fecal Coliform measurement is x counts/100 mL.

Recent concerns have increased the number of parameters to include:

• trace metals, and
• polynuclear aromatic hydrocarbons and other organics (pesticides, herbicides).

In addition, other effects of urbanization are not found in typical parameter lists. For example, spills of petroleum products are not normally detected in monitoring programs.

Tools for modelling a few water quality parameters are described below.

4.4.3 Wetlands

Factors to be considered include:

• buffer zone;
• alteration in hydrologic cycle (baseflow, fluctuation);
• alteration in groundwater cycle; and
• water quality.

4.4.4 Aquatic Communities

As a part of past watershed management studies, we have developed a conceptual model of the environmental requirements of indicator species of fish within the riverine watershed. Specific requirements for several habitats were synthesized from Habitat Suitability Index models and the general literature for a variety of aquatic communities.

Ecological objectives and the habitat requirements were developed for each of the following aquatic communities:

i)   a native cold-water fishery,
ii)   a self-sustaining cold-water fishery,
iii)   a non self-sustaining cold-water fishery,
iv)   a cool-water fishery,
v)   a warm water fishery, and
vi)   community composed of tolerant species.

These ecological objectives are stated in terms of a fishery but represent an ecosystem at the top of which a quality fishery is a key ecological niche.

An example of habitat factors and their general requirements for a native cold water fishery are the following:

Factor	Requirement
1. Stream order	1 - 3
2. Riparian canopy	> 80%
3. Groundwater	> 35%
4. Peak flows	Maintain historical peaks, volumes
5. Instream cover	Maintain historical cover
6. Stream morphology	Pools/riffles
7. Water quality	Turbidity (clear), dissolved oxygen (> 5 mg/L), and no spills
8. Stream temperature	< 20 degrees Celsius

4.4.5 Groundwater

Key factors to be considered at a watershed and subwatershed scale are:

- water quantity (supply rate for drinking water, water levels for wetlands); and
- water quality (for example, chlorides, nitrates).

At a site scale which is often the focus for public awareness of groundwater problems, the key factors are still both water quantity and water quality. But water quality parameters become site-specific, dependent upon the source of contamination (eg., landfills, industrial operations, etc.)

4.5 Watershed Analysis

This section presents examples of analyses at a watershed scale. The examples consider the impacts of urbanization upon water quality and water quantity.

4.5.1 Data Analysis

Prior to establishing the water quality model calibration, the PWQMN network data in the three watersheds have been analyzed using a locally-weighted time series technique. A typical graph is given in Figure 4.1 for total phosphorus. The dotted line evaluates seasonal effects, while the solid line represents the long-term trend. The advantage of the technique is that it can detect long-term trends in noisy data which are not spaced at equal time points. The results show that there has been a substantive improvement in water quality, due to previous phosphorus control efforts (removal of P from detergents, chemical precipitation in the WPCP). Such results can be used to document the success of past efforts where trends are detected and presented appropriately to the public. In addition the

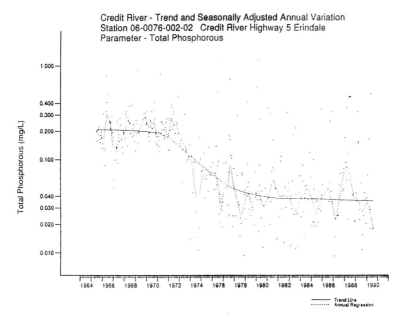

Figure 4.1:   Time and trend analysis for total phosphorous in
               Credit River.

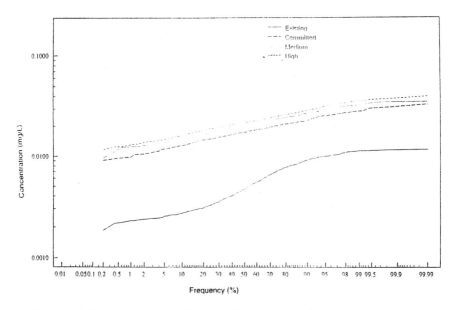

Figure 4.2:   Impact of urbanization on Copper using Monte
               Carlo simulation techniques.

analysis was used to establish a time period where the database is essentially stationary, and hence useful for calibrating a mass balance water quality model.

4.5.2   Mass Balance Model for Evaluating Water Quality and Water Quantity

The database usually available at a watershed scale are stream gauging stations from the Water Survey of Canada (with daily estimates of flow) and grab samples at approximately monthly intervals for water quality at stations in the Provincial Water Quality Monitoring Network (PWQMN). Inputs of water into the watershed in the form of precipitation are usually measured within the watershed or at a nearby location. Loadings of contaminants into surface waters and groundwaters can be estimated from all sources, but usually have not been measured for all sources.

Accordingly, the database available for calibrating a watershed model is much more extensive for flow-related issues than water quality-related issues. This is a major factor in defining the spatial resolution to be used in a model.

Simplified mass balance and water balance models have been used in watershed studies such as the Rouge River, the Don River and the Credit River recently.

For the Don River, the change in flow and loadings due to different size of events were used to estimate riverine concentrations for different flow conditions. Then the reductions in loadings due to different control measures were established and the resultant water quality calculated. An example of this approach is given in Table 4.5 for the whole Don River watershed for dry weather conditions and a 20 mm event. The water quality calculated for five control options were evaluated, and compared to "existing" water quality and "target" values. The loadings model was calibrated on a time-weighted basis with data from the PWQMN network to establish average annual conditions.

Table 4.5:   Estimated response in Don River water quality for alternative control options.

Option	DRY CONDITIONS			20 mm EVENT		
	SS (mg/L)	Copper (mg/L)	FC (#/100 mL)	SS (mg/L)	Copper (mg/L)	FC (#/100 mL)
1. STP Upgrade	6	0.024	2000	235	0.033	22000
2. CSO 90% Removal	6	0.024	2000	231	0.030	10000
3. Urban Offline Wet Ponds	6	0.024	2000	62	0.021	20000
4. Urban Flow Control	6	0.024	2000	170	0.024	19000
5. All (2-4) with STP Removal	1	0.006	1000	33	0.010	5000
Existing	6	0.024	2000	235	0.033	22000
Target Value	200	0.005	100	200	0.005	100

The values for "existing" water quality given in Table 4.5 are those computed by the model for the respective meteorological conditions.

The calculations were used to assist in prioritizing the control options for developing a medium term (2-10 year) and long term (50 year) strategy for improving water quality in the Don River. The calculations were presented in a comparative sense because, except for bacteria, internal sources or sinks were not considered on the model. Hence, for example, instream erosion at high flows is not considered in the estimates of instream concentrations.

4.5.3 Uncertainty Analyses in Water Quality Models

After water quality data is determined, flow effects removed, and a stationary data string selected, there is still substantial variability in the dataset. To address such uncertainty, a Monte Carlo simulation technique was used in the Rouge River watershed study. An example of the results are given in Figure 4.2 for the effects of different degrees of urbanization on a particular subwatershed within the Rouge River.

The graph for existing conditions indicates that close to an order of magnitude range in water quality is computed. The effects of urbanization suggest that urban development of committed lands will have the largest effect upon copper concentrations, with only small additional increases due to further infilling of the subwatershed.

4.6   Watershed or Subwatershed Scale Analysis: Evaluation of Factors in Fish Habitat Models

This is an example which can be applied at a watershed or subwatershed scale.

More detailed models were used to evaluate the effects of urbanization upon flow and temperature in the Rouge River watershed, two key factors in maintaining a quality fishery.

The habitat suitability index model was used to evaluate the effects of urbanization on the fishery. It is a semi-quantitative tool for assessing the relative importance of factors influencing fish habitat. All factors could be good, except one which is limiting. Key factors include:

- flow parameters,
- bottom characteristics (cobbles, etc.),
- temperature, and
- water quality (toxicity, dissolved oxygen).

The downstream reach of the Main Rouge River below the Milne Reservoir is a candidate stream for an ecological objective of rainbow trout. Three key habitat factors are water velocity, water depth, and water temperature (see Figure 4.3). Each factor is represented in the HSI model on a 0 to 1 scale, with 0 representing poor conditions, and 1 representing good conditions. Hence the graphs indicate the following optimum conditions: stream velocity between 0.5 and 2 fps; a water depth greater than 1.5 ft and a temperature between 55 and 70 °F. Since the HSI model is published in the United States, the graphs are not in SI units.

Calculated frequency curves (see Figure 4.4) indicate that small increases in the velocity are expected due to urbanization

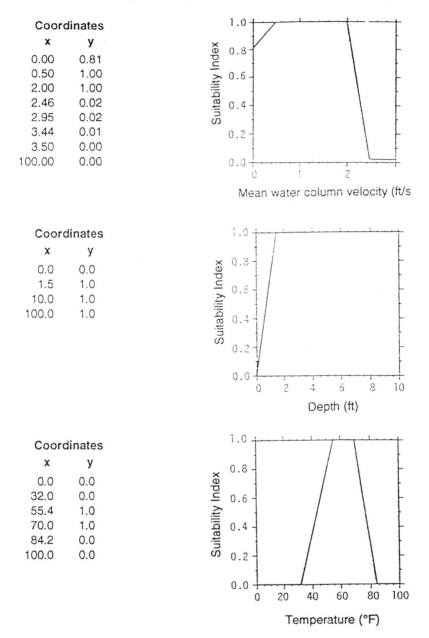

Coordinates	
x	y
0.00	0.81
0.50	1.00
2.00	1.00
2.46	0.02
2.95	0.02
3.44	0.01
3.50	0.00
100.00	0.00

Coordinates	
x	y
0.0	0.0
1.5	1.0
10.0	1.0
100.0	1.0

Coordinates	
x	y
0.0	0.0
32.0	0.0
55.4	1.0
70.0	1.0
84.2	0.0
100.0	0.0

Figure 4.3: SI curves for rainbow trout adult for water velocity, water depth and temperature.

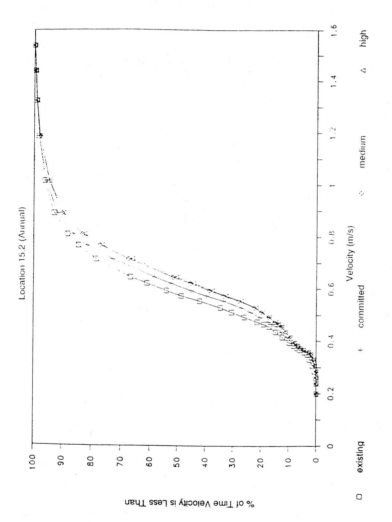

Figure 4.4: Velocity-duration curve for stream velocity in Rouge River basin.

for many conditions on an annual basis.

A river temperature model was used to compute temperatures downstream from the reservoir as a function of time of day. The calculations (see Figure 4.5) suggested that a daily temperature range from below 20°C to above 26°C was expected 4 km below the reservoir. This temperature range agreed with diurnal stream measurements during some of the hottest days in the summer providing some veracity to the model's calculations.

The effects of urbanization in the river section are shown as an imposed stream flow at distance 5 km and at 3 km. With the flows evaluated and assumed to enter at 23°C, the inflow streams have a small effect relative to the diurnal range. If it were assumed that the streams contained water at 30°C, there would be a larger effect added to the diurnal perturbation.

Hence these calculations show that the stream temperature in the section of the Rouge River accessible to rainbow trout from Lake Ontario are affected mainly by thermal conditions in the upstream reservoir and the daily heating cycle. Urbanization in upstream reaches does not affect temperatures entering the river section for the range of flows evaluated due to the thermal inertia of the reservoir. It compounds the effect in downstream reaches if water enters at much higher temperatures (say 30°C). In addition, velocity variations would not become a critical factor for the fishery relative to temperature.

Hence, the present stream temperatures preclude maintenance of a resident population of rainbow trout in the main stem of the river during summer months. Stream temperatures will be appropriate during spring or fall conditions - the stream temperatures found for a migratory population of rainbow trout - the current use made of this river section by these species.

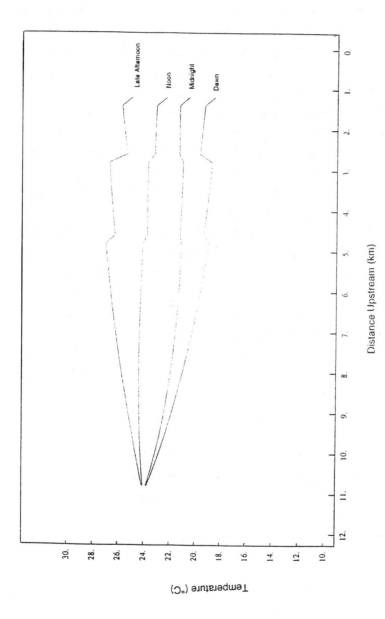

Figure 4.5: Calculation of temperature vs. distance curves as a function of time of day.

4.7   Evaluation of Baseflow, Peak Flow and Water Quality in a Subwatershed

For a subwatershed analysis, an integrated approach was adopted using GIS linked to a hydrological model - water quality model. One key question of ultimate concern in the study was to evaluate whether a resident brook trout fishery could be maintained if urbanization with necessary protective measures occurred.

Hence, the hydrological water quality model requirements included:

- integrate groundwater flow with surface water flow where aquifers have a 10-20 years residence time;
- evaluate the annual water balance;
- evaluate peak flow characteristics;
- evaluate stream temperature and effects of canopy destruction; and
- evaluate water quality constituents.

The tools selected were:

- GIS system for display and linking data bases to hydrological model; and
- a model such as HSPF to evaluate baseflow, high flow, water balance, and water temperature.

The GIS system was used to establish the relative key factors affecting the cold water fishery. It showed that discharge from an aquifer intersected with the upstream end of brook trout habitat. This habitat was maintained downstream by the presence of a dense cedar canopy which protected the cold stream water from heating. It was then used to develop the data base for modelling present conditions and to provide a guide to additional monitoring needs for the model.

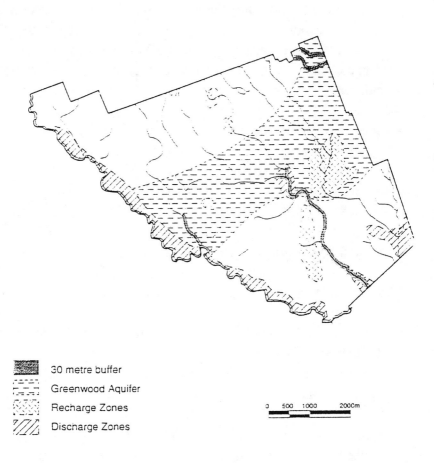

30 metre buffer
Greenwood Aquifer
Recharge Zones
Discharge Zones

0   500   1000   2000m

Figure 4.6:   Application of GIS system to show relative
location of aquifers; recharge and discharge zones,
and brook trout habitat.

From this subwatershed scale study, the pertinent points include:

- more detailed analysis requires more quantitative tools;
- more monitoring data may be available or is required;
- more money may be spent;
- but still need to have general knowledge and descriptive field data, e.g., location of recharge areas.

4.8 Summary

The major area of application of this paper has been toward:

- integrated management planning; and
- assessment of urban impacts upon environmental quality at the watershed scale and subwatershed scale.

Three specific items emphasized are:

i)   Monitoring impacts modelling - ie. the amount and type of data effects the model and modelling approach selected.

ii)   Modelling is only a tool. Since a model is an abstraction of reality, its assumptions and calculations of the response of environmental quality to management actions must be interpreted and integrated into planning processes.

iii)   Complexity of the model. The complexity of the model selected will depend upon whether it is to be used for planning purposes in a relative sense, or for flood-flow forecasting purposes. For planning, relative answers are often sufficient, whereas for flood-flow forecasting, much more accuracy may be required.

These studies suggest the following considerations in terms of selecting tools and directions for future study.

1. Future directions in watershed and subwatershed studies include:

    - environmental quality monitoring and assessment together with integrated studies will be a major focus; and
    - monetary constraints are directing studies towards simplified approaches for overall watershed studies.

2. Monitoring improvements are needed to provide the data base for conducting modelling studies. This data base should maintain PWQMN in some form and create enhanced programs such as the following to address specific questions:

    - event monitoring;
    - additional parameters (PAHs in sediments, stormwater, etc);
    - soluble versus totals for metals;
    - toxicity testing; and
    - benthic community evaluation.

3. Auditing of watershed plans will probably use both monitoring and modelling to establish the effectiveness of plans in protecting environmental quality because the cost of monitoring to establish the effectiveness may be too onerous given environmental variability observed.

4. At best, modelling is only a tool in watershed management.

4.9 References

Imhoff, J. and W.K. Annable (1992). Developing an Ecosystem Context for the Management of Water and Water Systems. Paper published in this volume.

BEAK. 1988. Water Quality Study - Rouge River Water Management Plan. Report for Metropolitan Toronto Region Conservation Authority.

Chapter 5

A Decision Support System for Highway Runoff

Eugene D. Driscoll
HydroQual Inc.
1 Lethbridge Plaza,
Mahwah, New Jersey 07436

5.1 Introduction

In a study for the Federal Highway Administration (FHWA) completed several years ago, my co-workers and I assembled and analyzed pollutant runoff data for highway sites (Driscoll, Shelley, and Strecker, 1990). The data were obtained in a number of independent studies at locations throughout the USA. The studies were conducted between 1978 and 1984, and the monitoring periods were generally one or two years in duration. In addition to characterizing the quantity and quality of highway stormwater, another project objective was to develop a procedure by which a highway engineer or planner could perform a screening analysis to determine whether a receiving water quality problem was likely to be created by stormwater discharges from a particular highway site. In the event a problem condition was projected, it was desired to identify the extent of runoff control that would be required to mitigate the indicated water quality problem.

0-87371-898-4/93/$0.00 + $.50
© 1993 by Lewis Publishers

We developed an analysis procedure and described it in a guidance document (Driscoll, Shelley, and Strecker, 1990). The procedure was structured in the form of spreadsheet-based fill-in tables to convert raw input data into values for the parameters used in the receiving water impact computation. Tables and maps were provided to assist selection of values for raw inputs. Other tables were developed which listed separately-computed impact results for a range of input values, from which a user would "look up" the answer, interpolating as necessary. General guidance was provided on how independently-determined stormwater pollutant control effects could be incorporated into the analysis.

FHWA agreed with our suggestion that, given the intended audience and use, more effective use of the impact assessment procedure would result from the availability of a user-friendly computer program which would guide a user through the procedure. This program was organized in the form of a decision support system, the structure and operation of which is described in this chapter. The Highway Runoff Decision Support System (HWY DSS) that was developed expanded on the scope of the method manual to permit a user to examine the effect of selected stormwater control measures on the mitigation of a projected adverse water quality impact. Both the types of pollutant control measures appropriate for highway stormwater, and the performance estimates that are incorporated in the DSS, are based on a separate FHWA research study (Versar & Camp, Dresser, and McKee, 1988).

5.2 Program Design

The target audience was considered to be State Department of Transportation (DOT) or other engineers or planners who would have responsibility for providing environmental input to overall planning and design of new or reconstructed highway segments. These individuals would be expected to address water quality issues as one element of a broad spectrum of concern, and were anticipated to require use of the procedure only intermittently, perhaps a few times a year. They were presumed to be generally knowlegeable on the subject and issues, but given their range of duties and lack of sustained effort on water quality analysis, it was considered that they would benefit most from an analysis tool structured in the form of a decision support system.

The conceptual basis of the decision support system is illustrated schematically in Figure 5.1.

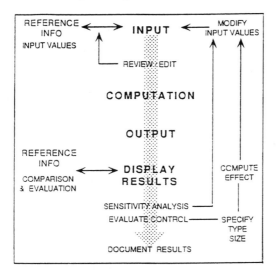

Figure 5.1: Schematic outline of HWY DSS.

Features that were considered important to incorporate include the following:

1. Lead the user step by step through the selection of values for the input parameters required for the analysis.

2. Provide ready access to reference information to assist the user's selection of appropriate values.

3. Avoid providing *automatic default* values so that the user is forced to choose, thereby carrying through a sense of the degree of uncertainty in the results that are generated.

4. Make it convenient to perform sensitivity analyses, to

determine the significance of alternative estimates for uncertain input values, on the final decision that would be indicated by the analysis.

5. Select a computation scheme that generates results of a type and in a form that relates to the decision to be made.

6. Select an analysis procedure that lends itself to use for screening analyses, that is, one that is sufficiently simple in execution and in data requirements to permit convenient analysis of an appreciable number of different alternatives. Sufficient accuracy for the type of decision to be made is presumed, as is the understanding that highly sensitive or complex situations may require something beyond a screening analysis.

7. Make it easy to test a variety of alternatives, keep track of and compare results for different conditions, and where appropriate (as with control evaluation), to evaluate combinations of types as well as individual measures.

8. Build in shortcuts to permit more rapid operation by users familiar with inputs required and the assignment of values for them, but maintain a base procedure that effectively serves a new or occasional user.

9. Display results in a manner that facilitates interpretation and evaluation. Favour graphical displays wherever feasible and useful.

10. Provide for the generation of hard copy or disk files for analyses that the user elects to document, but allow exploratory analysis that need not be retained to be ignored.

5.3 Impact Analysis Method

The general objective of the analysis was the assessment of the water quality problem potential in a specific stream segment receiving the stormwater runoff discharge from a section of highway. In selecting the impact analysis method to be utilized, we narrowed the focus to the concentrations of runoff pollutants in the receiving stream. Given the normal variation of rainfall and stream flows, the impact would be expressed as the probability distribution of the concentrations. For decision purposes, the computed probability distribution would be evaluated to determine the magnitude and frequency at which an appropriate target concentration would be exceeded. Heavy metals are among the more significant pollutants present in highway stormwater runoff. At sufficiently high concentrations, they are toxic to aquatic life, the support and protection of which was presumed to be the most common beneficial use for the stream segments that will receive highway runoff. The US EPA's toxicity criteria for the protection of aquatic life are presented as concentrations which should be exceeded no more frequently than once in three years.

The decision guidance structure incorporated into the method presents a comparison which indicates the degree to which target concentration exceedence frequencies are higher or lower than the three year return period specified by the EPA toxic criteria. The methodology also permits a similar analysis for pollutants other than metals. When, as in the case of TSS, there is no comparable water quality criteria or standard for comparison, the user may apply a target concentration based on judgement or on other standards that may apply.

The receiving water impact computation method used is the Probabilistic Dilution Model (PDM), developed under EPA's NURP program, and described by DiToro (1984). Figure 5.2 presents a schematic outline of the procedure, which applies an analytical solution to compute the distribution of stream concentrations from inputs consisting of the statistics (mean and coefficient of variation) of lognormal distributions of stream and highway runoff flows, and stream and highway runoff concentrations.

The information needed to perform the analysis includes:

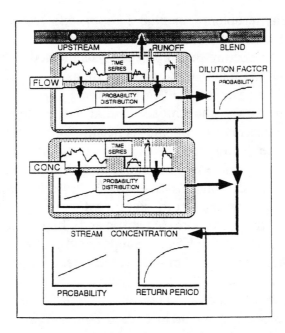

Figure 5.2: PDM stream impact analysis.

1. Site area and impervious fraction and rainfall characteristics representative of the area, from which runoff flow characteristics are computed. Reference information on independently-developed rain-event statistics is provided.

2. Stream flow statistics for daily average flows is required. An independently developed summary of mean flow on a unit area basis (cfs/sq mi) is provided for user reference. This is combined with the user determined upstream drainage area to determine the mean stream flow at the site. A procedure for estimating the coefficient of variation of stream flows is provided.

3. Runoff pollutant concentrations for highway sites are based on the results of the FHWA study of which the DSS development was a part. Estimates of runoff quality statistics are called up on user prompt, with provision for adjustment by the user and reference information to guide selections.

4. Upstream concentrations of the pollutant being analyzed are not considered in the analysis incorporated in the DSS. Although the analysis method provides for it's use, an independently-developed generalized summary was not available for use in providing guidance, as was the case for rainfall and streamflow. With upstream concentration tacitly assumed to be zero, the results generated represent the downstream concentration of a pollutant due *solely* to the highway discharge.

5.4 Operation of the Decision Support System

On startup, the program presents the user with several options for providing the computation component with the required input data. The *Normal Track* proceeds stepwise through the required inputs, presenting user prompts and direct access to commentary and reference material pertaining to the particular type of input being addressed. A *Fast Track* takes an experienced user directly to a series of fill-in edit fields where pertinent input parameter values may be entered. A third option, *Data File*, produces a display listing previously-stored input-data sets. Selecting the desired file-name loads that set of input data.

The user is next asked to indicate whether the receiving water to be analyzed is a stream or a lake. The discussion which follows describes program operation using the normal track routine, with a stream selected as the receiving water. Lake impacts are assessed in a parallel computation which determines phosphorus loading and lake concentrations, and compares projected levels to levels associated with different trophic states.

Assigning Input Values

Input data is required to describe the pertinent characteristics of the highway site, and for three of four parameters illustrated in the schematic presented in Figure 5.2. As noted earlier, upstream concentrations of the pollutant being analyzed are not considered, and the analysis will reflect the impact produced exclusively by the highway stormwater.

Site Characteristics

Site characteristics are the first set of input parameters to be entered. Data entries are prompted for the watershed drainage area, and for the area of the highway paved surface and total right-of-way (and thus the percent impervious). Figure 5.3 illustrates the site data prompts presented by the program. From the percentage of impervious area, a runoff coefficient (Rv) is estimated which is used to convert rainfall to runoff flows and volumes.

Rainfall

Statistics based on a long term record are used to characterize storm event parameters that are used in the analysis. The mean and coefficient of variation of event volumes, intensities, durations and intervals are the required parameters. The program provides the necessary reference data and allows the user to call up rainfall statistics for gages in individual cities, or for preliminary estimates, typical regional values can be selected from a map. Figure 5.4 illustrates the program displays. The intention is that users will develop files for gages pertinent to their area of concern, and the program allows for the addition of rain files.

The required rainfall statistics can be developed using the statistical rainfall analysis program SYNOP, which analyzes an hourly US Weather Service rainfall record to produce the storm event statistics. Under work assignments for FHWA and EPA, we made a number of enhancements to the SYNOP program (now designated SYNOP II) and developed versions for the IBM PC and Apple Macintosh (Driscoll et al. 1990 and Driscoll et al. 1989).

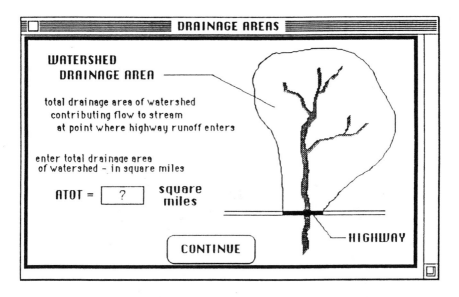

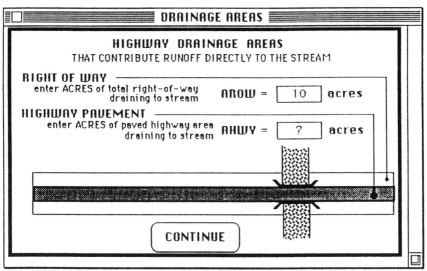

Figure 5.3: Prompts for site characteristics inputs.

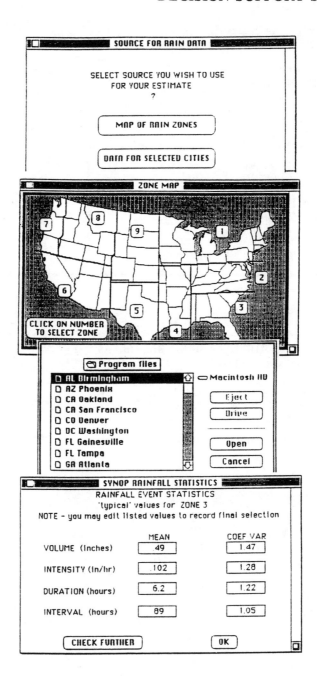

Figure 5.4: Rainfall data input displays.

Streamflow

The analysis method requires input values for mean and the coefficient of variation of streamflows at the point of highway runoff discharge. The program provides a map of mean runoff per square mile as well as access to a data file containing records for specific gages. Estimates of the coefficient of variation of daily stream flows are based on a separately-developed empirical relationship between the variability and the ratio of the 7Q10 to the average flow. Figure 5.5 illustrates the stream flow support material that can be accessed by the user. When users develop stream flow statistics for streams in their area of concern, results may be stored for future access in the program's reference data file.

Pollutant Concentrations

Input data required for highway runoff concentrations includes the site median concentration of the constituent being analyzed and the coefficient of variation of the storm event concentrations. The data on site median concentrations for urban and rural sites that was developed in the FHWA study is stored for reference by the program. Figure 5.6 illustrates the user prompts for selection of the pollutant type, highway type, and the site percentile to use to select from the range of observed values. For metals, the toxicity is assumed to be produced by the soluble fraction, which the user is prompted to supply, with help windows available to guide the selection. Impact assessment based on the total concentration (particulate plus dissolved fraction) can be done by assigning 100 percent as the soluble fraction.

Target Concentrations

EPA's water quality criteria are written such that effective concentrations (acute and chronic) should not be exceeded more than once every three years. For most metals, the concentrations that produce toxic effects increase with total hardness of the

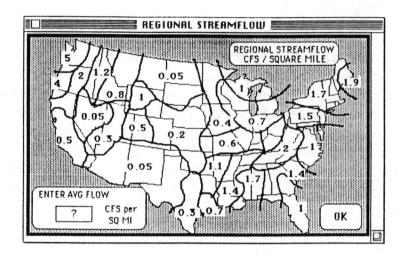

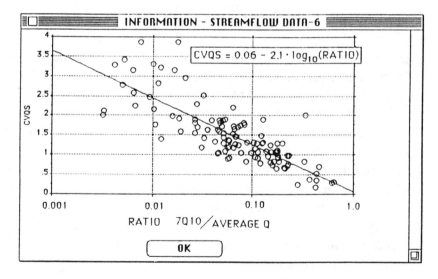

Figure 5.5: Stream flow reference data displays.

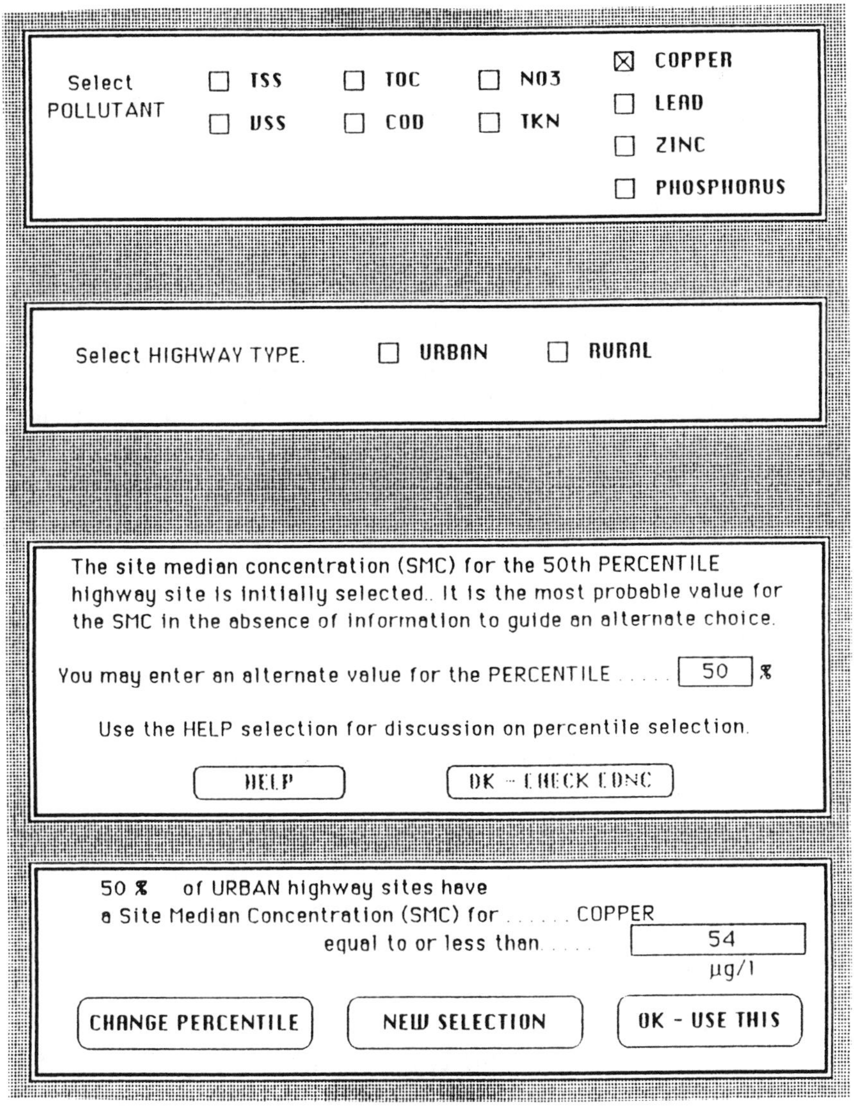

Figure 5.6: Pollutant selection displays.

water. The EPA criteria for toxicity to aquatic life, including the hardness adjustment are incorporated in the program. Figure 5.7 indicates prompts for user selection of water hardness and the resulting target concentration that is computed and displayed. The user may modify this target value if other standards apply in the case being examined.

Input Summary

Once input data entry is complete, the program displays a summary of the values assigned for all of the input parameters, and the identifying name assigned to the input set. This display is illustrated in Figure 5.8. The user is provided the opportunity to edit any of the assigned values before proceeding further.

Display of Impact Analysis Results

When the above information has been entered, the program then calculates a predicted probability distribution of stream concentrations below the point of entry of the highway runoff, after complete mixing with streamflow. The probability computed defines the percentage of storm events that will produce stream concentrations equal or greater than a particular concentration level. When this percentile is combined with the average number of storms per year, the return period (MRI = Mean Recurrence Interval) at which the target concentration is exceeded can be calculated and displayed. Figure 5.9 shows an example of a histogram generated by the program. In the example, the predicted frequency of target exceedence (MRI = 116 days) indicates that some type of control is required to meet the assigned water quality criteria. The user can then choose to investigate the addition of a control measure or measures. In the design workbook, the user calculates the MRI and then proceeds independently to control measure evaluation (if necessary).

The display in Figure 5.9, which shows the relationship of an appropriate target level to the projected magnitude and frequency

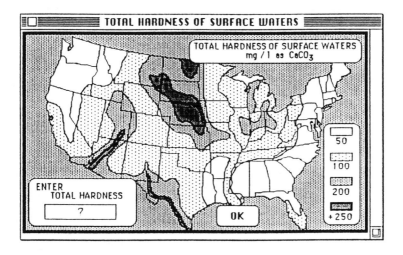

Figure 5.7: Displays for selection of Target concentration.

```
SUMMARY OF INPUTS for TEST CONDITION :    BASE CASE

Total Drainage Area =  10.00 sq.mi.      HWY Paved Area  =     8.00  acres
TH of Surface Water =    50 mg/l     HWY Right-of-Way =  12.00  acres

RAIN EVENT     VOL(in)    INTENS(in/hr)    DURAT(hr)    INTERV(hr)
    MEAN        0.49         0.102          6.20         89.00
    CofV        1.47          1.28          1.22          1.05

STREAMFLOW              1.00 CFS/sq mi
             MEAN =    10.00 CFS         Coef of Var =   1.50

RUNOFF FLOW       MEAN =  0.70 CFS          Coef of Var =   1.28

    POLLUTANT = COPPER           Target Concentration =    9 µg/l
Site Median Concentration =    54 µg/l
  Coef of Var of Site EMC's = 0.71           Soluble Fraction  =  40 %
      [  GO BACK and EDIT  ]                  [  OK - CONTINUE  ]
```

Figure 5.8: Display for summary of input data.

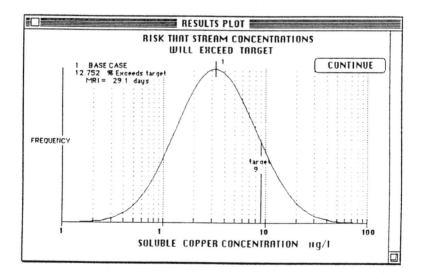

Figure 5.9: Impact analysis result display.

of concentrations produced by highway stormwater, provides a concise display of the information that can guide a determination of whether or not the impact is significant, and a decision on the need for stormwater control.

The EPA toxic criteria have no legal or regulatory force; they provide guidance on water quality conditions that have been determined to be protective of a use. In many cases, the criteria are translated directly into legal water quality standards. However, a decision on a need to mitigate could, in the case of a sensitive water body for example, be based on an arbitrarily lower target level, or on a longer return period for exceedence of the selected target.

Output Processing Options

When the user elects to continue, the program displays a text summary of the result, and presents a set of options for further action. As indicated by Figure 5.10, user options at this point are:

PRINT - produces a hard copy summary of the input parameters assigned and the impact analysis results, including an optional printout of the histogram plot. A prompt allows the entry of brief comments to elaborate on the conditions that apply to this analysis set.

SAVE INPUT - creates a data file of the input values selected for this analysis. This file may be selected on startup of a future session to avoid stepwise entry of input values.

INVESTIGATE...TREATMENT - takes the user to a program element that provides for exploring the effect of the design size of several control types first on the reduction of pollutant discharges, and then on the extent to which the stream impact is modified. Program action following this selection is discussed further below.

INVESTIGATE SENSITIVITY - steps the user through a

tabulation of the parameter input values that were selected, with an opportunity to modify any of them for comparing the extent to which the output (and the decision that would be based on it) is influenced by uncertainty in any of the input values. A new output histogram display is produced with the modified input parameter values.

TERMINATE - ends the session.

For the input conditions assigned to... BASE CASE
The STREAM impact analysis indicates that the 9 µg/1 TARGET
concentration will be exceeded during 12.752 % of storm events
or ... on an average of once every .... 29.1 days

Once in 3 years- the concentration will exceed 37 µg/1

YOU MAY NOW DO ANY OF THE FOLLOWING
(one at a time)

☐   PRINT a summary of inputs and results for this case

☐   Save input values for this condidtion in a DATA FILE.

☐   Investigate the effect of runoff TREATMENT.

☐   Investigate SENSITIVITY to estimates of input values.

☐   TERMINATE this analysis.

Figure 5.10: Display of output options.

Any or all can of these options can be implemented during a session, with the program returning to this menu at the completion of each option. Program operation is terminated from this display.

Control Measure Evaluation

When the model user elects to investigate control measures, the program provides a choice of four types: grass channels, detention basins, overland flow, and infiltration devices. These controls can be applied individually or in series (such as a grass swale leading to a detention basin). The predicted efficiencies of the grass channels and overland flow devices are based upon the draft FHWA study by Versar et al. (1988). The detention basin and infiltration devices methodology uses EPA's detention basin analysis approach developed by Driscoll (1986).

Figure 5.11 illustrates the displays presented following selection of a grass channel and a detention basin respectively. Original performance estimates for grass channels (which have been elaborated since the publication on which the DSS procedure is based) related pollutant removal to channel length. For this control, the user manually assigns a removal efficiency based on visual inspection of the displayed relationship. In the case of the detention basin shown, and the other control types, removal efficiency is computed using the site rainfall and drainage area characteristics, and device size information entered by the user. The user is permitted to examine the removal efficiency produced by various size selections before selecting the one that will be carried through for the repeat of the impact analysis.

Once the removal efficiency of the control has been projected or assigned, and is accepted as the estimate to use, the program displays the revised set of inputs and recomputes the stream concentration frequency-distribution. In the new display, illustrated by Figure 5.12, the result for the current alternative is compared with that for the base case. Continuing the program operation displays the output options window (Figure 5.9).

The user may continue to examine different control alternatives, cycling through this part of the program, and assigning a descriptive name to each alternative, ignoring inappropriate choices and documenting those worth recording.

After the first pass through the foregoing procedure, all subsequent selections of the *Investigate Treatment* option will present the display shown at the top of Figure 5.12. The removals determined may be applied to untreated stormwater to evaluate the selection as an alternate to others investigated. The new control may also be applied to the residual pollutants in the

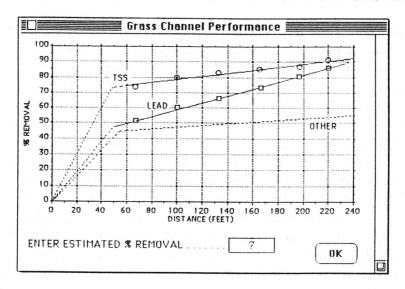

Figure 5.11: Display for control evaluation.

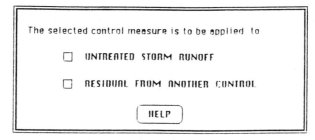

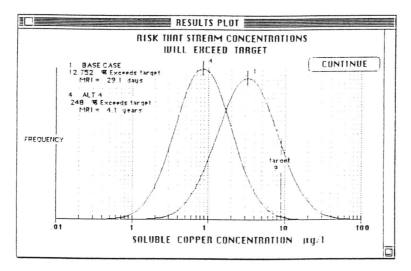

Figure 5.12: Comparison display for alternative assessment.

discharge from another control, which permits the examination of combinations of controls arranged in series (e.g., a swale followed by a detention basin). When this option is selected, the user is presented a list of the previous control alternatives examined, and permitted to select the one the current choice will be linked with.

5.5 Summary

The computer program provides a method for assessing the impacts of highway runoff on receiving waters, structured in a decision support system format. The technique could be applied to similar type analyses for assessing stormwater impacts from urban sources in general, with the substitution of appropriate reference data.

The decision support system structure is one that could be applied quite effectively to a variety of analyses. In such cases, an appropriate computation element for the situation being addressed would be required, in addition to the input and decision reference data.

The reports that were cited, and floppy disks with the program (in Macintosh or IBM-PC format) are available from FHWA thru the National Technical Information Service (NTIS).

5.6 References

Di Toro, D.M. , Probability Model of Stream Quality Due to Runoff, ASCE, Journal of Environmental Engineering, Vol 110, No. 3, June 1984, pp 607-627.

Driscoll, E.D. , P.E. Shelley, and E.W. Strecker; Pollutant Loadings and Impacts from Highway Stormwater Runoff - Volume III: Analytical Investigation and Research Report. FHWA/RD/88-008, April 1990

Driscoll, E.D. , P.E. Shelley, and E.W. Strecker, Pollutant Loadings and Impacts from Highway Stormwater Runoff - Volume I: Design Procedures, FHWA/RD/88-006, April 1990

Driscoll, E.D. , P.E. Shelley, and E.W. Strecker, Pollutant Loadings and Impacts from Highway Stormwater Runoff - Volume II: Users Guide for Interactive Computer Implementation of Design Procedure, FHWA/RD/88-007, April 1990

Driscoll, E.D. , G.E. Palhegyi, E.W. Strecker and P. E. Shelley; Analysis of Storm Event Characteristics for Selected Rainfall Gages Throughout the United States, US EPA, Washington DC, November 1989.

Driscoll, E.D., Methodology for Analysis of Detention Basins for Control of Urban Runoff Quality, US EPA Office of Water, Washington DC, September 1986.

Versar & Camp, Dresser, and McKee, Retention, Detention and Overland Flow for Pollutant Removal from Highway Stormwater Runoff. FHWA/RD/87-056, 1988.

Chapter 6

Modelling Changes in Surface Water Quality for Watersheds Undergoing Urbanization

Donald G. Weatherbe
Donald G. Weatherbe Associates Inc.
1352 Safeway Crescent, Mississauga, Ont. L4X 1H7

Harold O. Schroeter
Schroeter & Associates
126 Scottsdale Drive, Guelph, Ont. N1G 2K8

Dennis W. Draper
D.W. Draper & Associates Ltd.,
3-1750 Queensway, Suite 1368, Etobicoke, Ont. M9C 5H5

Hugh R. Whiteley
School of Engineering, University of Guelph
Guelph, Ontario. N1G 2W1

6.1 Introduction

In this paper we describe a linked cascade of modelling procedures assembled specifically to assess changes created by urbanization in the quantity and quality of streamflow. This

0-87371-898-4/93/$0.00 + $.50
© 1993 by Lewis Publishers

123

assessment is a key to successful integrated watershed management, a concept which has received widespread support in Ontario recently (PCAO, 1991; Queens Printer, 1991; Charlton and Tufgar, 1991).

Subwatershed studies are needed as part of integrated watershed management to determine the impact of development that exists or is anticipated, to evaluate the effectiveness of measures to prevent or remediate damage, and to assist decisions on acceptable levels of urbanization. Urbanization changes both the flow rates of water within various hydrological processes and the physical, chemical and biological attributes associated with these processes. Techniques used for assessment of the environmental impacts of urbanization must account for all these changes.

The modelling approach that is summarized here has been created as part of the Laurel Creek Watershed Study (Charlton and Tufgar,1991). This study is being carried out primarily by Triton Engineering Services Ltd. of Kitchener and Ecological Services for Planning Ltd. of Guelph, for the Grand River Conservation Authority, the Regional Municipality of Waterloo and the cities of Kitchener and Waterloo in Ontario. The first three authors of this paper are subconsultants for this study.

The model components we assembled and used have been carefully selected to provide the assessments required to meet the objectives of the Laurel Creek Study. The objectives of the study are to establish acceptable levels of urbanization, and to recommend measures to protect, rehabilitate and enhance water quality, associated water resources, fisheries, wetlands and environmentally sensitive areas.

In the following sections of this chapter we describe component modelling procedures and the links among them that were used to assess changes in water quality from urbanization. Typical results from Laurel Creek illustrate the range of impacts that can be assessed. The chapter concludes with comments on the effectiveness and applicability of the methods.

6.1.1 Laurel Creek Watershed Study

Laurel Creek is a tributary of the Grand River located within Waterloo Region in southwestern Ontario. The watershed area of approximately 77 km^2 is currently half urban and half rural land use. There is both a strong demand for additional urban development in the rural area, and a desire expressed by citizens and planners in Waterloo Region to maintain natural features in the watershed. A study funded by the Cities of Waterloo and Kitchener, the Regional Municipality of Waterloo, and the Grand River Conservation Authority was initiated in 1991. The objectives of the study are to:

*   establish acceptable levels of urbanization,
*   protect and rehabilitate fisheries,
*   protect wetlands and environmentally sensitive areas, and
*   protect and enhance water quality including associated water resources.

The assessment technique had to meet these objectives, and account for the impact on water resources of land use changes, and preventative and remedial measures. The study is being carried out primarily by the Consulting team of Triton Engineering Services Ltd, Kitchener, Ontario and Ecological Services for Planning Ltd., Guelph, Ontario. Additional details of the study may be found in Charlton and Tufgar, 1991.

6.2 Modelling Framework

In the Laurel Creek Study results from modelling of surface water quality were needed to assess impacts on various uses. There were specific objectives for streamflow, baseflow, protection of groundwater, water quality, fish species and habitat, wildlife habitat, natural resource areas and recreation. Assessment of possible impacts of urbanization on fisheries and stream habitat is particularly challenging because of the multiplicity of factors involved. This type of impact assessment is emphasized in the following discussion of modelling procedures. The main impacts and their associated evaluative approaches are:

1. IMPACT: Increased overland runoff, increased peak flows in channels, and reduced baseflow.

    EVALUATION: A watershed model which accounts for
    REQUIREMENT imperviousness and channel characteristics in rural and urban land uses.

    APPROACH: The GAWSER model linked to a GIS indicates these effects.

2. IMPACT: Physical changes caused by changes in flow, including higher peak velocity in runoff events, reduced depth during low flows, and increased channel erosion due to increased velocity.

    EVALUATION: Average channel velocity and depth as a
    REQUIREMENT function of flow.

    APPROACH: The GAWSER model calculates velocity-depth-flow relationships which are used to calculate stream power.

3. IMPACT: Reduced water quality due to runoff from urban and rural areas. The main contaminants of concern are sediment, nitrogen compounds, phosphorous, dissolved oxygen and temperature.

    EVALUATION: Water quality contaminant levels to
    REQUIREMENT compare with stream reach based targets for fisheries.

APPROACH: The DOMECOL Model package is used to calculate plant biomass and dissolved oxygen. Suspended solids concentrations are generated by the GAWSER model. Temperatures are calculated by an empirical approach based on observed conditions (with expected improvements in approach prior to project completion). The HSI suitability model is used to convert contaminant levels and physical factors to fish species suitability.

Figure 6.1 shows how the models interrelate. The details of the approaches are provided in succeeding sections.

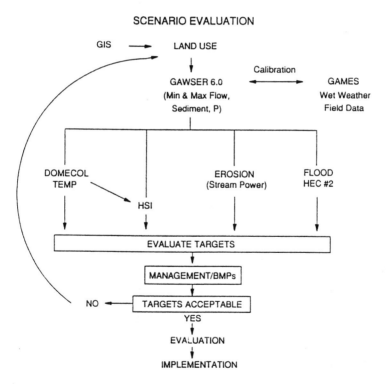

Figure 6.1: Interrelationship of models used in evaluation.

6.3 Watershed Hydrologic Model: GAWSER

An overall watershed model was employed to simulate the hydrological conditions and amounts of pollutant washoff, and to provide a framework from which to operate the DOMECOL model package.

The Guelph All-Weather Sequential-Events Runoff (GAWSER) model is a deterministic hydrologic model, based on the HYMO format (Ghate and Whiteley, 1982; OMNR, 1989), that is applied widely in Ontario for planning, design, real-time flood forecasting, and evaluating the effects of physical changes in the drainage basin (e.g. GRCA, 1988; Triton 1990; Schroeter et al., 1992). GAWSER is used to simulate the total streamflow resulting from precipitation inputs defined in terms of rainfall, snowmelt or a combination of both. For simulation, drainage basins can be divided into a series of linked elements representing watersheds, channels and reservoirs (see Figure 6.2). The physical effects of each element are simulated using efficient numerical algorithms representing tested hydrologic models that are described briefly below.

GAWSER was modified recently to operate in a continuous mode, and to predict pollutant accumulation, washoff and transport. These changes have resulted in version 6.0 of GAWSER.

6.3.1 Representation of Spatial Variation in Runoff Amounts

To account for the spatial variability in sequences of rainfall, snowfall, air temperature and wind speed, a watershed is divided into zones of uniform meteorology (ZUM) (see Schroeter et al, 1991). Each ZUM uses one set of meteorological measurements and covers one or more subwatersheds. Typically, a ZUM may be from 100 to 350 km^2 (Schroeter et al., 1991). Variations among spatially-distinct surfaces of infiltration, percolation rates within the soil, and overland runoff amounts are accounted for in each ZUM by separate calculations for one impervious and up to four pervious areas.

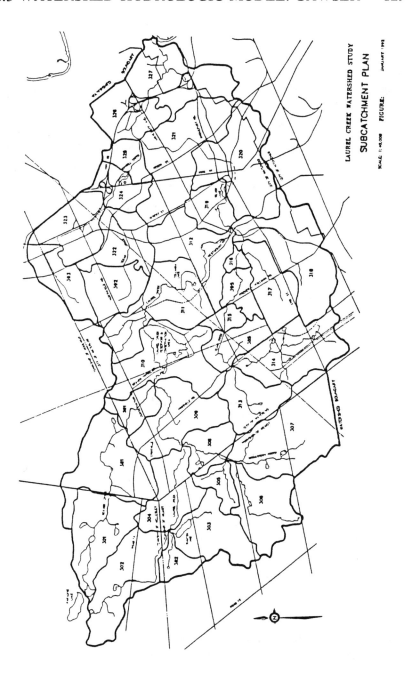

Figure 6.2: Subcatchment plan for Laurel Creek Watershed Study.

6.3.2 Model Inputs and Simulation of Hydrologic Processes

The model can operate at time steps as small as one minute or as long as 24 h, depending on the availability of meteorological inputs for the selected time interval. In most applications to date, GAWSER has been operated at a one hour time step to correspond with available rainfall rate inputs and streamflow comparison data. The program can access standard input data formats (e.g. Atmospheric Environment Service compressed hourly and daily files), and will compute the rainfall distribution for standard urban design storms directly (e.g. Chicago, SCS, Hydrotek).

GAWSER considers eight hydrological processes: snow accumulation and melt, infiltration, evapotranspiration, runoff estimates, overland flow routing, subsurface and baseflow routing, stream channel routing, and reservoir routing (with operations).

A flow chart for the runoff generation procedures is given in Figure 6.3. Each pervious area within a ZUM is modelled as two soil layers. Soil-water outflow from the second layer is distributed between subsurface flow and groundwater storage by a proportioning factor. The Green-Ampt equation is used in the infiltration calculations; the computations account for the recovery of infiltrability between events. Evaporation, or sublimation during periods of snow cover, is set at a fixed daily amount, and applied in time steps with no rain.

Overland runoff routing is accomplished by the area/time versus time method, in which computed rating curves based on cross-section measurements for main and off channels are used to define the computational parameters. Outflows from subsurface and groundwater storage (baseflow) are simulated using a single linear reservoir approach.

Channel routing is handled by the Muskingum-Cunge method (see Schroeter and Epp, 1988), and the reservoir routing is accomplished by Puls method in which controlled releases are allowed.

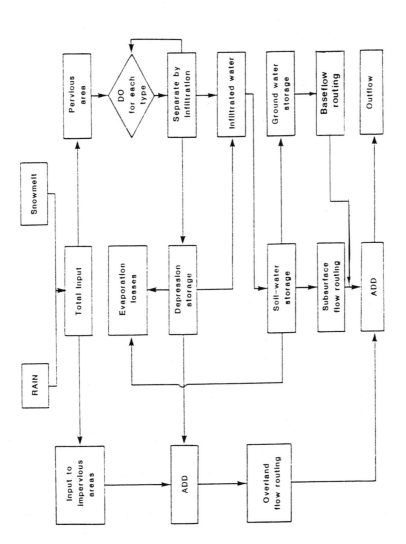

Figure 6.3: GAWSER model runoff generation procedure.

6.3.3 Pollutant Accumulation-Washoff and Transport

The approach for simulation of water quality selected for use in GAWSER V6.0 was based on sediment transport. It is generally recognized that many contaminants, especially organics, nutrients, toxics and heavy metals, are associated with sediment or solids transported by stormwater runoff (e.g. Dickinson and Green, 1988; Marsalek and Schroeter, 1989). The new water quality submodel in GAWSER is based on the Equivalent Solids Reservoir (ESR) algorithm developed and tested by Schroeter and Watt (1989) for use in the Queen's University Urban Runoff Model (Q'URM), but expanded to include a pollution accumulation feature for continuous simulation work.

The dry-weather accumulation of pollutants is modelled using the approach employed in STORM (U.S. Army Corps of Engineers, 1977). Here, dry-weather build-up between events is accounted for by specifying a daily rate of accumulation. The base daily rate is reduced by the number of dry-days since the last runoff event or street sweeping episode times a reduction constant.

The complex sediment transport process is represented by the mechanisms of translation and scour/deposition. First, the mass inflow of contaminants is lagged by the kinematic travel time of the water flow in each drainage element (for subwatershed elements - each impervious and pervious zone). Next, the deposition and resuspension (erosion/scour) processes are simulated using a fictitious equivalent solid reservoir located at the outlet of each drainage element. Required parameters and inputs are limited to two sediment characteristics (particle size and relative density), scour and deposition rate parameters, and initial sediment loadings. Sediment inputs from rainfall, surface and baseflow are considered by assigning fixed concentrations to these sources.

Up to four water quality parameters can be simulated in GAWSER. Where it is known that a given parameter is a function of another one already being simulated (e.g. total phosphorous as a function of suspended solids), the modelling is made simpler by specifying a constant ratio between the two parameters for washoff calculations.

6.3.4   Continuous Operation Mode and Linkage to Other Models

To allow operation of GAWSER in a continuous simulation mode a configuration (or batch) file system was installed, whereby the user specifies a series of event files that follow sequentially in time, and hence, are run by the program one after another. While executing a particular event file for a given calculation period (e.g. Nov. 1 to 25), GAWSER creates a file on disk to initialize the next event file in sequence. This end file is read by the program during the next calculation period to initialize the same computation variables.

Seasonal changes in model parameters (e.g. effective hydraulic conductivity, or snowmelt/refreeze factor) are directly specified in the sequential event files. At present, each individual event file can simulate a 25 day period when the computation time step is set at one hour.

To help transfer hydrograph and "qualograph" output from GAWSER in a format for input to other programs (e.g. ECOL, DOMOD7, and hydrograph plotting software), a special program called GAWSTRAN was created. At present, GAWSTRAN will read hourly temperature and solar radiation data from AES files and create input directly for the ECOL model. It will also create daily flow and water quality load summary tables that are similar to those printed in Water Survey of Canada's annual Surface Water Data publication.

6.4   Plant Biomass Model: ECOL

6.4.1   Overview

Attached algae and macrophytes (rooted aquatic plants) are the agents that predominantly determine dissolved oxygen levels in rivers of southern Ontario. Models must account for their effects.

The Grand River Simulation Model (GRSM) (Weatherbe, 1986) is a continuous simulation model that accounts for sewage treatment plant inputs, main channel inputs, tributaries, and urban runoff. Dissolved oxygen relationships include carbonaceous and nitrogenous oxygen demand, sediment oxygen demand, reaeration, and the impact of three species of aquatic plant and algae through oxygen production by photosynthesis and oxygen uptake by respiration. The plant growth relationships were incorporated into a subroutine called ECOL. A stand-alone version of the ecological subroutine was also programmed, called ECOL1. (Walker et al, 1982). A simple steady state version of the dissolved oxygen model was also prepared called DOMOD7. These three models were resident on mainframe computers at the Ontario Ministry of Government Services and the Grand River Conservation Authority. The two simpler models, ECOL1 and DOMOD7 were converted to run on IBM PC compatible computers for use in the Laurel Creek Study by Dr. Alan A. Smith, Burlington, Ont.

6.4.2 ECOL Model Descriptions

The ECOL model provides continuous simulation of the growth of aquatic plants and algae over a typical season, ie. April to November. Inputs of plant nutrients (phosphorous and nitrogen) along with sunlight are the main factors leading to growth. The predominant species observed in the Grand River Basin are accounted for.

Figure 6.4 shows the interrelationships in the model. The energy source for the plant growth is sunlight. The model accounts for sunlight variations in two-hour time-steps, and allows each species to grow with the amount of growth depending on its current biomass, available nutrients in the plant and water, available light, temperature, and species-specific growth coefficients. Available light is affected by incoming solar

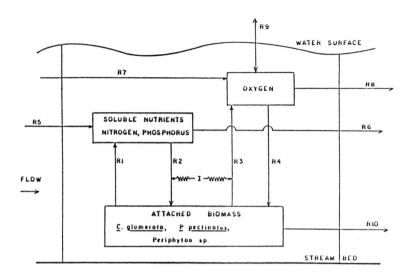

Figure 6.4: Conceptual view of biomass model - ECOL.

radiation, suspended solids or turbidity and the self shading effect of the present biomass. Each species also continuously respires, which reduces biomass and releases nutrients into the water. The model formulation is described in detail in Walker et al, (1982) and the draft model users guide, (MOE, 1982). The coefficients for plant growth were derived from extensive field work carried out in the Grand River in the 1980s as reported in Walker (1981). These coefficients were not varied for Laurel Creek. However other inputs such as flow, channel geometry, temperature, sediment and nutrient concentration were based on information on Laurel Creek.

6.5 Dissolved Oxygen Model: DOMOD7

6.5.1 Overview

The ECOL model is of direct value in evaluating watershed impacts which change conditions in the stream (temperature, incident sunlight, stream flow, concentrations of sediment and phosphorus) and result in changes in aquatic plant biomass. To determine if changed biomass level is of concern, the effect of the plant biomass on the dissolved oxygen (DO) regime, and particularly the minimum daily DO of the stream is required. This can be facilitated with the aid of the model, DOMOD7 (Dissolved Oxygen MODel- Version 7) (MOE, 1987).

DOMOD7 simulates dissolved oxygen concentration in the stream during a 24 hour day accounting for principal sources and sinks. Once calibrated using field data, the model can specify minimum DO under "design" conditions of critical stream flow, temperature and biomass level.

6.5.2 DOMOD7 Model Description

DOMOD7 simulates stream conditions for a 24-hour period on a 2-hour time step basis. Six water quality parameters are simulated: dissolved oxygen (DO), ultimate biochemical oxygen demand (BOD_u), nitrogenous oxygen demand (NOD), nitrite plus nitrate (NITR), suspended solids (SS) and total phosphorus (TP).

For modelling purposes, the river system is divided into sequential sections termed "reaches". The junction points of these reaches are called "nodes". DOMOD7 can handle up to two tributary streams or outfall discharges in each reach. Flow and water quality from these point sources are input on a two-hourly interval basis.

The model incorporates the principle oxygen sources and "sinks" associated with a flowing stream. Sources are atmospheric re-aeration, photosynthetic oxygen production of plants; sinks are sediment oxygen demand (SOD), exertion of nitrogenous and carbonaceous biochemical oxygen demand and plant respiration.

Plant respiration and photosynthetic production are modelled by a subroutine which is essentially a reduced version of ECOL.

First-order kinetics apply to the exertion of BOD and NOD and to the reaeration process. BOD and NOD decay rates generally have to be measured in the field. Re-aeration rates are more readily estimated by empirical formulae; for Laurel Creek, the rate was derived using the depth and velocity of the stream reach of interest. Velocity, travel time and depth were computed based on the stream flow rate of interest and the same channel hydraulic relationships used in the GAWSER model. Decay rates are not critical in Laurel Creek, since BOD and NOD are relatively low; therefore reasonable approximations to decay rates were made based on rates from other streams in the vicinity.

Sediment oxygen demand data have been measured in the study area and are converted from the measured areal rate to a volumetric rate by dividing by depth of flow.

Both the average daily temperature and daily range for the corresponding Julian Day are input into DOMOD7. The model applies a sinusoidal diurnal variation in water temperature, based on the mean temperature and temperature range input by the user, which allows for a temperature to be calculated at each time step. DO saturation concentrations are computed for each time step from the water temperature. All key rates in the model are adjusted for temperature.

Stream temperature was adjusted to simulate the effect of development according to a relationship suggested by Galli (1990). Typically, mean stream temperature increases of about 1.0° C. are seen with watershed imperviousness increases of 12%. Further development is required to fully integrate all major temperature effects into the models.

6.6 Sample Application of Modelling Approach

In this section, we give sample output from each of the models described above to demonstrate the capabilities of our modelling framework for surface water quality assessment. Our results are only preliminary as the final assessment procedures in the Laurel Creek study are still in progress and under review.

For modelling purposes, the watershed was divided into a

number of reaches, each draining the accumulative area from more than 25 subwatersheds as depicted in Figure 6.2. The watershed model included consideration of the Laurel Creek Dam, which controls outflows from a 39 km^2 area and is operated by the Grand River Conservation Authority (GRCA) for flood control and recreation.

6.6.1 The Typical Year Approach

Changes to the surface water resources within the watershed in response to anticipated development were assessed on the basis of a typical year approach. The typical year was selected using the historical streamflow records for the Laurel Creek gauge in downtown Waterloo. Here, a year was defined on a water year basis (Nov. to Oct.), because the long-terms simulations had to begin during a period prior to the accumulation of snow.

Our typical year was defined as being a mean flow year that had a distribution of monthly flows that corresponded to the most frequent pattern observed in the historical records. The period Nov. 1, 1968 to Oct. 31, 1969 had eight months with close to long-term mean flows, the greatest number of any year in the record, and hence was selected as our typical year.

6.6.2 Sample GAWSER V6.0 Output

This section illustrates the types of output from GAWSER V6.0 that are used in water quality assessments. The results presented are for 1968-69 typical year input with data for the existing (1991) watershed conditions and a few of the future development scenarios.

The hydrologic model parameters throughout the year were established from previous experience in applying GAWSER in forecast mode for the whole Grand River basin over the last four years (Schroeter et al., 1992). Due to the lack of event specific data on water quality for Laurel Creek, published values for the

ESR model parameters (Schroeter and Watt, 1989) were used to provide the initial settings for the GAWSER water quality submodel. Because the sediment washoff routine had not been applied previously in rural watersheds, the computed annual sediment loads for the rural areas (e.g. Reach 309) were compared with the reported annual sediment loads for selected Ontario streams (Dickinson and Green, 1988). There was considerable agreement in the computed sediment loads.

Figure 6.5 shows the simulated daily discharge hydrographs for existing and ultimate development conditions at the Laurel Creek Weber Street gauge for the period May 15 to July 15 in the typical year. Notice the program's ability to model the baseflow recession between runoff events. For the period shown, the discharges during events are slightly higher for the ultimate over the existing case, whereas the baseflows are slightly lower in the ultimate case.

A plot of monthly total sediment load for Laurel Creek Reach 309, upstream of the reservoir, is presented in Figure 6.6 for two future development scenarios. Notice that the ultimate development case (Fut 4A) produces a significant increase in sediment load for the November period.

The sediment transport algorithm produces useful information for fish habitat assessments. Figure 6.7 gives a plot of net sediment deposition in a number reaches in the upper Laurel Creek watershed. Such a plot can be used to show how changes in stream channel conditions (e.g. channelization, or naturalization) will influence the amount of sediment that deposits in a stream segment. From this plot, Reach 309 is generally free of sediment deposits.

6.6.3 ECOL Model Applications and Results

The model was set up and applied for several reaches with input provided by the GAWSER output, augmented by water temperature and sunlight data. The predicted biomass of each species is shown in Figure 6.8 for one set of temperature and

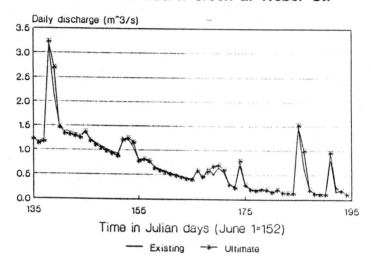

Figure 6.5: Sample GAWSER flow output.

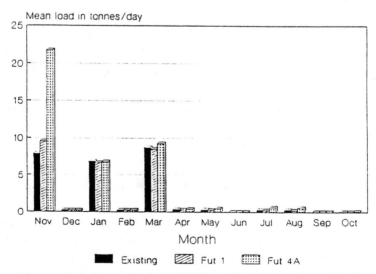

Figure 6.6: Sample GAWSER sediment load prediction.

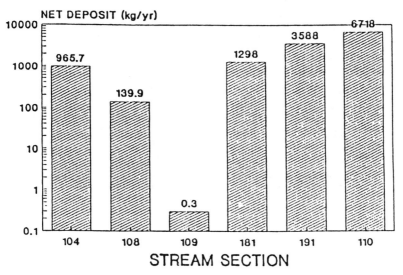

Figure 6.7: Sediment deposition predicted by GAWSER.

flow conditions, and total phosphorus concentrations generally below .01 mg/l. Note that the model defaults to a minimum biomass of $5gm/m^2$ for each species. Note also that plant biomass consists primarily of epiphytes and cladophora. This is consistent with the observed species in the Laurel Creek. Total biomass peaks around the end of May (day 140) as temperatures start to increase and spring flows are receding.

The model was also used to test the sensitivity to phosphorous input variations. Figure 6.9 shows the total predicted biomass for the same reach later in the year at various constant phosphorous input levels. (Note that 109 and 309 refer to the same reach).

6.6.4 Sample DOMOD7 Output

In the Laurel Creek application of DOMOD7, model rates as described above were initially adjusted according to reach-specific characteristics. The model was calibrated using the model's subroutine light and plant inhibition factors. Calibration success

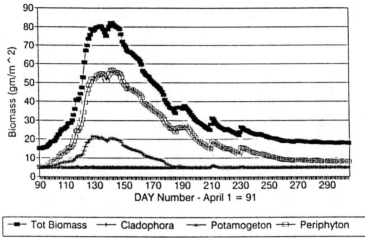

Figure 6.8:   Sample ECOL output - predicted biomass for
               different species.

was based on matching the daily DO maximum and DO
minimum with the survey data.

The calibrated model was then used for predicting DO under
various conditions of stream flow, temperature and aquatic plant
biomass as predicted by ECOL. To set up these runs, the ECOL
output for a specific development scenario was examined to
determine the peak biomass occurrence date and the stream flow
and water temperature associated with this peak growth. DOMO-
D7 was run under these conditions. An example DO predicted
with this high biomass condition is shown in Figure 6.10.

In our application of this model, plant respiration and
photosynthetic DO production, and re-aeration were the key
processes influencing the diurnal variation of dissolved oxygen.

Aquatic plant biomass as simulated by ECOL did not respond
dramatically to development; consequently no major shifts in
dissolved oxygen minimums simulated by DOMOD7 for specified
reaches and design conditions were noted. The processes
simulated by DOMOD7 are sensitive to temperature, however;

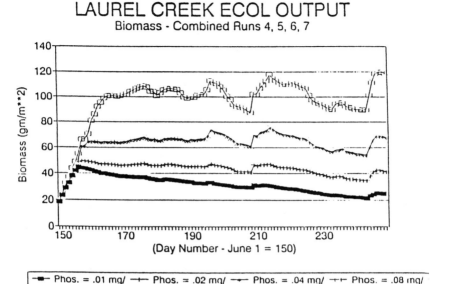

Figure 6.9:   Total biomass predicted by ECOL at different phosphorous input concentrations.

hence increases in temperature related to urban development must be taken into account when using the DOMOD7 model in predictive mode.

Outputs of daily minimum DO from DOMOD7 runs are plotted in Figure 6.11, showing the effects of increased biomass and temperature for one stream reach. Model runs were done with existing, and increased biomass levels, and with design stream temperatures increased as a result of two levels of development. These model runs were "worst case" scenarios which combined peak seasonal biomass predicted by ECOL with least favourable stream flow. Even with no increase in biomass, temperature effects associated with development can be seen to result in minimum daily DO being lowered by nearly 0.5 mg/l. In terms of HSI targets, this is a significant departure from the HSI minimum desirable DO concentration of 6.0 mg/L for trout.

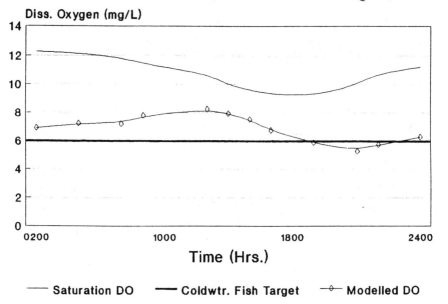

Figure 6.10:   Dissolved oxygen predicted by DOMOD& during peak biomass conditions for reach 109.

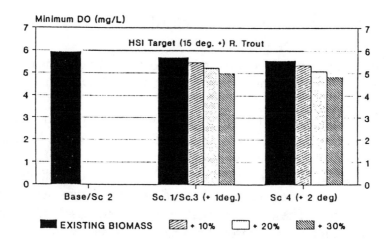

Figure 6.11:   Dissolved oxygen levels predicted at various land use scenarios and temperature levels.

6.6.5 Application of Output for Fish Habitat Assessment

In this study a number of water quality and quantity parameters related to fish habitat and other water uses were selected as key for assessment of the watershed impacts of different development scenarios. These parameters, or related outputs can be generated by GAWSER, ECOL and DOMOD7 and compared between scenarios or with criteria of interest.

The selected outputs are in part derived from the Habitat Suitability Indices (HSI) developed for use by the U.S. Fish and Wildlife Service (Raleigh *et al*, 1984; Edwards *et al*, 1983). This approach uses an extensive number of parameters concerning physical, chemical and biological characteristics of streams to develop indices ranging from 0 to 1 for fish species of interest.

For the present study, two fish species were of interest: smallmouth bass and rainbow trout, which are considered "warm water" and "cold water" fish species, respectively. From the extensive list of HSI variables, several may be selected about which information may be gained using the models. An example list as selected from the smallmouth bass HSI is shown in Table 6.1, also indicating the model(s) used to generate the required data.

Table 6.1 Selected Habitat Suitability Indices - Smallmouth Bass.

Variable	Desireable Range	Model Source
Spawning period Temp.	11 – 27 deg. C.	Temperature Model
May – Oct Water Temp.	Adult: 11 – 29 deg. C. Fry: 13 – 33 deg. C. Juvenile: 10 – 31 deg. C	Temperature Model " " " "
Monthly ave. Turbidity	0 – 70 JTU	GAWSER
Minimum DO through year	4 mg/L (6 mg/L is optimum)	DOMOD7

Note that Suitability Indices (SI's) have been selected from the overall array of HSI variables only because of model ability to generate values for comparison; a comprehensive habitat assessment would include the full array of HSI variables related to physical, other chemical and biological factors.

6.7 Conclusions

The models presented in this paper are useful for predicting the impact of urbanization for the major impacts anticipated in the watershed under study. At the time of presentation of the paper (March, 1992), additional work was still to be completed, to be reported in future. This is identified here, both to indicate the additional applications, and to clarify some limitations of the procedure.

Future work includes:

1. completion of the evaluation of different land-use scenarios;

2. addition of a temperature modelling routine to the procedure;

3. evaluation of reservoir impacts. The option of removing reservoirs or bypassing base flow is being considered in the watershed;

4. the model packages could be improved with enhanced integration, to make them more efficient; and

5. the models would benefit from additional verification.

6.8 References

Charlton, D.L. and R. Tufgar. 1991. Integrated watershed management approach for small southern Ontario rural/urban watersheds. Canadian Water Resources Journal, Vol. 16, No. 4, pp. 421-432.

Dickinson, W.T. and D.R. Green. 1988. Characteristics of sediment loads in Ontario streams. Canadian Journal of Civil Engineering, Vol. 15, No. 6, pp. 1067-1079.

Edwards, E. A., Gebhart, G., and Maughan, O. E., 1983. Habitat Suitability Information: Smallmouth Bass. C. 1984. Fish and Wildlife Service, U.S. Dept. of Interior. Report No. FWS/OBS-82/10.36.

Galli, J., 1990. Thermal Impacts Associated with Urbanization and Stormwater Management Best Management Practices. Publication 91701. Metropolitan Washington Council of Governments. Washington D.C.

Ghate, S.R. and H.R. Whiteley. 1982. GAWSER - A modified HYMO model incorporating area variable infiltration. Trans. ASAE, Vol. 25, No. 1, pp. 134-142

Grand River Conservation Authority (GRCA). 1988. Grand River Hydrology Study-Phase I: Final Technical Report (draft).

Marsalek, J. and H.O. Schroeter. 1989. Annual loadings of toxic contaminants in urban runoff from the Ontario Great Lakes Basin. Water Pollution Research Journal of Canada, Vol. 23, No. 3, pp. 360-378.

O'Connor, D.J., Ditoro, D.M. 1970. Photosynthesis and oxygen balance in streams. Jour. San. Eng. Div. ASCE. SA 2, April, 1970.

Ontario Ministry of the Environment (MOE). 1987. DOMOD7 User's Guide (Draft). MOE Water Resources Branch, Watershed Management Section, Toronto, Ontario.

Ontario Ministry of the Environment (MOE). 1980. Stream Water Quality Assessment Procedures Manual. MOE Water Resources Branch, Watershed Management Section, Toronto, Ontario.

Ontario Ministry of the Environment (MOE). 1982. ECOL1 Users Guide. (in draft).Water Resources Branch.

Ontario Ministry of Natural Resources (OMNR). 1989. Guelph All-Weather Storm-Event Runoff Model (GAWSER) Version 5.4: Training Guide and Reference Manual.

Pollution Control Association of Ontario (PCAO). 1991. Integrated watershed management. Proc. of a Seminar, Toronto, Ont. Dec. 2, 1991

Queens Printer for Ontario. 1991. Stormwater quality best management practices. Report prepared for the Ontario Ministry of the Environment. June, 1991.

Raleigh, R. F., Hickman, T., Solomon, R. C., and Nelson, P. C. 1984. Habitat Suitability Information: Rainbow Trout. Fish and Wildlife Service, U.S. Dept. of Interior. Report No. FWS/OBS-82/10.60.

Schroeter, H.O., D.K. Boyd, L.L. Minshall, and H.R. Whiteley. 1992. Grand River Integrated Flood Forecast System (GRIFFS): development and real-time application. Proc. of the 2nd Canadian Conference on Computing in Civil Engineering, CSCE, Carlton University, Ottawa, Ont. (in press).

Schroeter, H.O. and R.P. Epp. 1988. Muskingum-Cunge: a practical alternative to the HYMO VSC Method of channel routing. Canadian Water Resources Journal, Vol. 13, No. 4, pp. 68-89.

Schroeter, H.O. and W.E. Watt. 1989. Practical simulation of sediment transport in urban runoff. Canadian Journal of Civil Engineering, Vol. 16, No. 5, pp. 704-711.

Schroeter, H.O., D.K. Boyd and H.R. Whiteley. 1991. Areal Snow Accumulation-Ablation Model (ASAAM): experience of real-time use in southwestern Ontario. Proc. of the 48th Annual Meeting of Eastern Snow Conference, June 6-7, 1991, Guelph, Ont., pp.25-37.

Triton Engineering Services Ltd. 1990. Credit River Water Management Strategy-Phase I: Study Report. Submitted to the Credit Valley Conservation Authority, Meadowvale, Ontario.

U.S. Army Corp of Engineers. 1977. Computer program 723-S8-L7520, Storage, Treatment, Overflow, Runoff Model "STORM", User's Manual.

Walker, R., Willson, K.. 1981. Plant community assessment techniques, data collection and field procedures. 1981. Grand River Basin Water Management Study, Technical Report 15. Grand River Implementation Committee by the Water Resources Branch, Ontario Ministry of the Environment.

Walker, R., Weatherbe, D.G., Willson, K. 1982. Aquatic plant model - derivation and application. Grand River Basin Water Management Study, Technical Report 14. Grand River Implementation Committee by the Water Resources Branch, Ontario Ministry of the Environment

Weatherbe, D.G. 1986. Continuous simulation models to evaluate urban drainage impacts. Urban runoff pollution. Proc. of a NATO research workshop on urban runoff pollution, Montpelier, France, 1985. Springer-Verlag.

Chapter 7

Comprehensive Urban Runoff Quantity/Quality Management Modelling

James Y. Li
Lake Simcoe Region Conservation Authority
120 Bayview Parkway, Box 282,
Newmarket, Ontario, Canada   L3Y 4X1

Barry J. Adams
Department of Civil Engineering
University of Toronto
Toronto, Ontario, Canada   M5S 1A4

This study introduces an analysis methodology for the preliminary planning of urban runoff quantity and quality control systems. The methodology consists of five basic steps: rainfall analysis, evaluation of existing runoff conditions, prediction of runoff control system performance, determination of the least-cost combinations of control measures, and sensitivity analysis. An application is demonstrated by employing the methodology on the Barrington catchment in East York, Toronto.

7.1 Introduction

Traditionally, drainage systems have been designed primarily to control the quantity of urban runoff. As the deterioration of receiving water quality due to combined and/or storm sewer overflows is frequently observed in many urban centers, drainage systems are being designed or rehabilitated to provide control over both the quantity and quality of urban runoff.

Urban runoff control can be planned for both quantity and quality objectives. As a result, there is an acute need for a planning methodology which takes into consideration the performance and cost of quantity and quality control measures.

The following sections present an optimization methodology for the preliminary planning of urban runoff quantity and quality management systems. Both upstream control measures such as storage and infiltration facilities, and downstream control measures such as storage and treatment facilities can be analyzed by the proposed methodology.

7.2 Literature Review

Lager et al. (1976) developed a stormwater management planning model to evaluate the performance of storage-treatment facilities for controlling stormwater runoff. Although this model provides an extensive analysis of stormwater control, it can only be used to determine the performance of the various combinations of downstream storage and treatment alternatives. Moreover, the continuous simulation of the storage and treatment performance required is time-consuming for preliminary planning purposes.

Sullivan et al. (1978) incorporated cost and performance analyses in the assessment of the magnitude and significance of pollution loadings from urban runoff of Ontario cities. The STORM model (HEC, 1974) was employed for the development of performance isoquants (combinations of downstream storage and treatment measures that achieve the same degree of pollution control) while the least-cost combination of downstream storage-treatment measures that can achieve a certain level of pollution control was determined using the constrained cost minimization approach.

To overcome the burden of continuous simulation of control system performance, Flatt and Howard (1978) derived analytical probabilistic models for the prediction of control system performance.   The analytical models were compared with the simulation model STORM and it was found that the predicted performance was close to that obtained by the STORM model. The tradeoff between storage and treatment was analyzed by employing the production theory of microeconomics.

Nix et al. (1983) and Nix and Heaney (1988) employed the production theory of microeconomics to determine the least-cost combinations of storage-release strategies.   The control system performance was simulated by the SWMM model (Huber et al., 1982) and the design performance isoquant was developed from the simulation results.   The least-cost mix of storage-release strategies was determined using the constrained cost minimization approach.

The approaches of Sullivan et al., Flatt and Howard, and Nix et al. and Nix and Heaney are an improvement over that of Lager because they consider control system cost explicitly in the analysis for urban runoff planning.   However, the analytical probabilistic models proposed by Flatt and Howard can be improved and the least-cost analysis of Nix et al. can be extended to include upstream control measures in addition to downstream storage-treatment measures.

7.3   Urban   Runoff   Quantity/Quality   Control Methodology Overview

Layouts of a typical combined and storm sewer system are illustrated in Figure 7.1. Rainfall fills the depression storage of a catchment and part of the remaining rainfall becomes runoff which is finally collected at a potential downstream storage site. If the storage capacity and treatment rate available is less than the runoff volume, the excess runoff volume overflows uncontrolled to the receiving water. The proposed planning methodology is

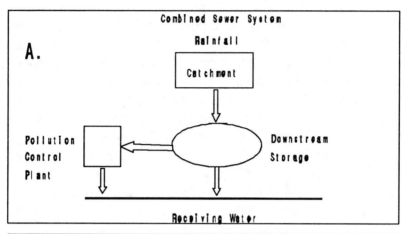

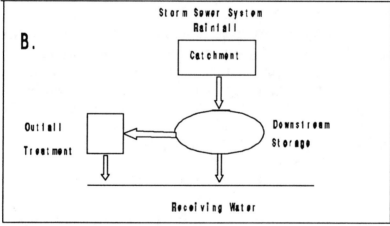

Figure 7.1:   Typical layout of combined and storm sewer systems.

developed to analyze this simplified rainfall-runoff-overflow process.

An overview of the planning procedure is illustrated in Figure 7.2. Quantity analysis follows the left path of the flowchart while quality analysis follows the right path of the flowchart. Integrated quantity and quality analysis follows both the left and right paths of the flowcharts.

The above quantity control planning process starts with input rainfall analysis in which the statistical properties of rainfall event characteristics are quantified. Measures of the existing runoff conditions such as the average annual or extreme-event runoff volume are then determined using the statistical models of the methodology. These statistical models are closed-form analytical transformations of the probability density functions (pdf's) of rainfall characteristics to the pdf's of runoff characteristics. If there are existing control measures, the control performance of those measures can be assessed using the derived analytical probabilistic models.

The quality control planning process also starts with rainfall analysis which is then followed by the runoff pollution load analyses. Quality analysis of runoff pollution load is based on a constant concentration approach in which pollution load is determined as the product of an average event concentration and event rainfall volume. The next step is to determine the pollution control performance of existing control measures using the derived analytical probabilistic models.

If existing control measures cannot achieve the required level of quantity and/or quality control, additional control measures must be provided. The least-cost analysis enables the planner to evaluate different combinations of upstream and downstream control measures and to determine the most cost-effective combinations of control measures that can achieve the required level of control.

The final step is to conduct sensitivity analyses of the model parameters where the impact of uncertainty in parameter estimation on runoff control analysis is assessed. From these analyses, upper and lower bounds of the least-cost mix of control

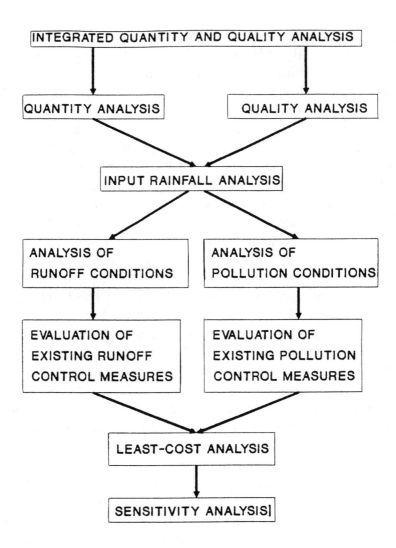

Figure 7.2: Urban runoff control planning methodology.

measures can be determined. Details of the methodology are discussed in the following sections.

7.4 Input Rainfall Analysis

Since a probabilistic approach is used to predict runoff and control system performance, the statistical properties of input rainfall events are required. Rainfall volume is recorded continuously at rainfall gauging stations across Canada. These series of continuous rainfall pulses can be separated into individual rainfall events with respect to an interevent time definition (IETD). The IETD is defined as the minimum elapsed time with no rainfall that distinguishes one independent rainfall event from another. Rainfall pulses separated by a time larger than the IETD are considered to be separate events. Typical IETD's for urban catchments range from 1 to 6 hours.

After a continuous rainfall series is separated into events, event rainfall characteristics such as rainfall volume (v), duration (t), average intensity (i), interevent time (b), and number of events per year can be computed. Statistics of these characteristics such as the mean, standard deviation, etc. are then determined.

The probabilistic models developed for the methodology are based on statistically independent rainfall characteristics (i.e., v, t, b) which can be described by exponential probability density function (pdf's) as follows:

1. Event rainfall volume, v (mm)

$$f(v) = z \, \exp(-zv) \quad ; \quad z = 1/E(v) \tag{7.1}$$

2. Event duration, t (hr)

$$f(t) = l \exp(-lt) \quad ; \quad l = 1/E(t) \qquad (7.2)$$

3. Event average intensity, i (mm/hr)

$$f(i) = w \exp(-wi) \quad ; \quad w = 1/E(i) \qquad (7.3)$$

4. Interevent time, b (hr)

$$f(b) = k \exp(-kb) \quad ; \quad k = 1/E(b) \qquad (7.4)$$

in which f(v), f(t), f(i), f(b) are the pdf's of v, t, i , b; E(v), E(t), E(i), and E(b) are the mean value of v, t, i, b ; and z, l, w, k are the pdf parameters.

It has been found by Adams et al. (1986) and others that the exponential distribution is suitable for describing the rainfall characteristics of many cities across Canada.

7.5 Analysis of Runoff Quantity/Quality

The rainfall-runoff process is modelled by a simple runoff coefficient method in which initial rainfall is stored in depression storage and a portion of subsequent rainfall becomes runoff from the catchment. The transformation of rainfall events to runoff events is given by:

$$Vr = 0 \qquad\qquad v \leq Sd \qquad (7.5)$$

$$Vr = PHI\ (v - Sd) \qquad v > Sd \qquad (7.6)$$

in which Vr is the runoff event volume, v is the rainfall event volume, Sd is the depression storage, and PHI is the runoff coefficient. With the pdf of rainfall event volume (Equation 7.1), the pdf of runoff event volume can be derived analytically using derived probability distribution theory (Benjamin and Connell, 1970). The expected value of runoff event volume and the average annual volume of runoff volume can then be determined (Adams and Bontje, 1983).

The runoff quality models are derived on the basis of rainfall events. The runoff event pollution load (Lr) is the product of runoff event volume (Vr) and event flow-weighted mean concentration (C) given by:

$$Lr = Vr*C \qquad (7.7)$$

thus:

$$E[Lr] = E[C]*E[Vr] + COV[Vr,C] \qquad (7.8)$$

in which E[C] is the expected value of C; E[Vr] is the expected value of Vr; and COV[Vr,C] is the covariance of Vr and C. If Vr and C are independent, their covariance is zero. According to the U.S. Nationwide Urban Runoff Program (U.S. EPA, 1983), C was found to be generally uncorrelated with Vr for a number of urban sites. Further analyses by Wallace (1980) also indicate that the correlation between pollutant concentration and runoff volume is weak. Therefore, the expected event runoff load (E[Lr]) is given by:

$$E[Lr] \approx E[C]*E[Vr] \qquad (7.9)$$

and the average annual runoff pollution load (ALr) can be estimated by:

$$ALr = NR*E[Lr] \qquad (7.10)$$

in which NR is the average annual number of rainfall events.

The constant concentration approach described above requires the determination of the expected value of event concentration, E[C]. If no runoff quality data are available for a catchment, E[C] may be selected from the literature (e.g., U.S. EPA, 1983). However, it is important that the expected event concentration be selected on the basis of land use, location, hydrology and drainage system characteristics. If field data on runoff quality are available, E[C] may be found from deterministic simulation (e.g., STORM) or statistical models (e.g., Driver and Tasker, 1988).

7.6 Control System Performance

The effectiveness of a runoff control system is usually measured by its long term average performance and its extreme event performance (e.g., average annual overflow volume or 5 year overflow volume). Two different approaches have been proposed for specifying runoff control system performance. The first approach is based on input control which requires the capture of a certain volume of rainfall or runoff. The second approach requires that the output from a runoff control system be limited to a certain fraction of the input. Input control performance such as the control of the 2 year rainfall volume or a half an inch of runoff volume cannot really represent the effectiveness of a runoff control system since different watershed and drainage systems react differently to the same input rainfall and runoff. As a result, it is the output control performance measures such as the average annual percent of runoff volume controlled (Cr) that are useful for describing the performance of a runoff control facility. The proposed methodology emphasizes the output control performance in the analysis of runoff control systems.

Long-term quantity and quality control performance of a runoff control measure can be specified by: i) quantity performance measures such as the average annual percent of runoff volume controlled (Cr) and the average annual number of overflows (Ns); and ii) quality performance measures such as the average annual percent of runoff pollution load controlled (Cp) and average annual number of overflows (Ns). Both Cr and Cp can be used, for example, to represent the situation in which only a certain fraction of runoff control can provide the required benefit. On the other hand, Ns can be used, for example, to represent the situation in which every overflow event, no matter how large the overflow volume, causes damage and damage reduction is measured by the number of overflows abated.

Extreme event performance of a runoff control measure is important for quantity problems such as flooding and quality problems such as bacterial contamination because the time scale of the effect is usually short. Extreme event quantity and quality performance of control measures can be specified by: i) quantity performance measures such as the N-year event overflow volume (Pn); and ii) quality performance measures such as the N-year event pollution mass load (Ln). For instance, Pn may be used to measure the acceptable volume of overflow which the receiving water can accommodate while Ln may be used to measure the maximum allowable event pollution load in the receiving water.

The output performance of urban runoff control systems has been estimated by three different approaches: design storm event analysis, continuous simulation, and statistical models. Each approach has its own advantages and disadvantages.

Most drainage engineers are familiar with the design storm event analysis which is easy to apply and understand. However, there are numerous drawbacks of applying the design storm concept in designing urban runoff control systems (Adams and Howard, 1986).

Continuous simulation has significant advantages over the design storm approach because the entire recorded rainfall history is employed for the prediction of system performance.

Additionally, it allows detailed modelling of urban runoff processes such as the buildup and washoff of pollutants. On the other hand, continuous simulation is time-consuming for preliminary planning purposes.

Statistical models, such as the analytical probabilistic models employed in the proposed methodology, have the advantage of predicting the system performance from statistics of complete rainfall records rather than from the complete records themselves. Thus, system performance of various combinations of control measures can be investigated efficiently by the statistical models. However, statistical models lack flexibility in the detailed modelling of the control system performance because simplifying assumptions have to be made for the derivation of the relationship between input rainfall characteristics and system output performance characteristics. If statistical models predict system performance in close agreement with that obtained from calibrated continuous simulation models, they would be very useful for preliminary planning of urban runoff control systems. The proposed methodology employs analytical probabilistic models to analyze output performance of runoff control systems for this reason.

Following the pioneering work of Howard (1976), a series of developments on analytical probabilistic models was undertaken at the University of Toronto. Adams and Bontje (1983) developed a software package of analytical probabilistic models called Statistical Urban Drainage Simulator (SUDS) for the prediction of quantity performance. Details of the derivation can be found in Adams and Bontje (1983) and Adams and Zukovs (1987). Li (1991) derived analytical probabilistic models for runoff quality transformations through a storage-treatment system. The concept of the quantity and quality models is discussed in the following paragraphs. The reader is recommended to consult the aforementioned literature for details on the modelling procedure.

The layouts of the urban runoff control systems under consideration are presented graphically in Figure 7.1. The quantity control models, coded in SUDS, (Adams and Bontje, 1983) consider runoff from a catchment which is collected at a potential downstream storage site of capacity S and released at a controlled outflow rate OMEGA. Runoff which exceeds the storage capacity and controlled release rate is spilled to the receiving water. The overflow volume is determined by the amount of runoff and the storage contents within the storage reservoir.

Volume balance relationships of rainfall-runoff-overflow are developed among the rainfall characteristics v, b, t and the catchment and drainage system characteristics PHI, Sd, S, OMEGA. These relationships are then transformed onto the joint probability space of v, b, t. The probability of overflow per rainfall event and the expected magnitude of overflow per rainfall event can then be derived. The control system performance measures such as Cr, Ns, and Pn are then derived in terms of rainfall, catchment, and drainage system characteristics (z, l, k, Sd, PHI, S, OMEGA). The depression storage Sd is used to model upstream storage and the runoff coefficient PHI is used to model upstream runoff reduction such as an infiltration facility. The storage capacity S is used to model the downstream storage reservoir volume and the controlled outflow rate OMEGA is used to model the outflow release rate from storage.

The approach used by Adams and Bontje (1983) is also used to model the transformation of runoff quality through a down-stream storage-treatment system (Figure 7.1). If the runoff event volume is less than the combined capacity of available storage volume and constant controlled outflow rate to a treatment facility, no overflow occurs at the storage reservoir and the only source of pollution comes from the treated effluent. For larger runoff event volumes, part of the runoff is overflowed from the reservoir into the receiving water in addition to the treated effluent. For storm sewer systems, outfall treatment units may utilize physical or physical-chemical treatment operations of moderate efficiency. As a result, the treated effluent becomes an important long term source of pollution in addition to overflow from storage.

These quality models (Li, 1991), which have been coded in a software package called Extended SUDS, account for the pollution contribution from treated effluent and storage overflow. Mass balance relationships have been developed among the rainfall characteristics v, b, t, the catchment and drainage system characteristics PHI, Sd, S, OMEGA, and the treatment efficiency of storage and treatment facilities N for both overflow and non-overflow conditions. These relationships are then transformed onto the joint probability space of v, b, t. The probability of exceeding a certain amount of pollution load per rainfall event and the expected magnitude of pollution load per rainfall event can then be determined. The control system performance measures such as Cp and Ln are then derived in terms of rainfall, catchment, and drainage system characteristics, as well as treatment efficiency (z, l, k, Sd, PHI, S, OMEGA, N). Upstream and downstream control systems are modelled by Sd, PHI, S, OMEGA in a manner similar to that in SUDS.

Both the SUDS and Extended SUDS models have been compared with the continuous STORM model (HEC, 1974). Kauffman (1987) found that the analytical model SUDS compared reasonably well with STORM simulation results for catchments with an average of less than 120 runoff events per year. Li (1991) compared SUDS and Extended SUDS with STORM using the data from Barrington catchment in East York, Ontario and found that Cr, Ns, Cp, Pn predicted by SUDS and Extended SUDS were in good agreement with those simulated by STORM. Figure 7.3 illustrates the comparison between the analytical models and STORM for the Barrington catchment. However, Ln predicted by SUDS and STORM was found not in good agreement and the discrepancy might be attributed to the different approach in modelling runoff pollutant generation.

Since the analytical probabilistic models are generally able to predict control system performance in close agreement with the simulation model STORM, they have been suggested for use in the preliminary analysis of existing runoff control systems and the preliminary planning of runoff control system design and rehabilitation.

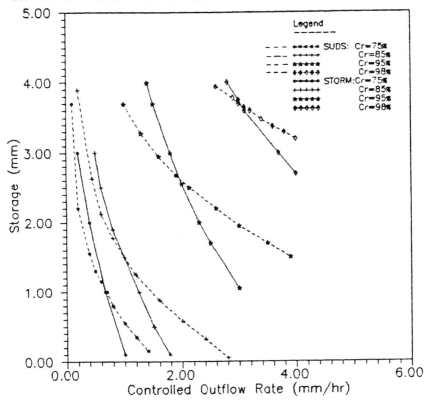

Figure 7.3: Comparison of Cr isoquants (S and OMEGA mixes) from STORM and SUDS.

7.7 Least-cost Analysis

If the existing runoff condition or the existing runoff control system performance is undesirable or unable to achieve the required level of runoff control, additional runoff control measures must be provided. Preliminary planning of runoff control systems requires the determination of the most promising combinations of control measures that can achieve the required level of control so as to focus the design level analysis on those combinations of control measures. To determine the most promising combinations of control measures, cost-effectiveness relationships for different combinations of control measures should be established. By comparing those relationships, the most promising combinations can be identified.

Cost-effectiveness relationships of different combinations of control measures can be established by determining the least-cost mixes of control measures that achieve different levels of quantity and/or quality control. A constrained cost minimization technique is employed to determine the least-cost mix of control measures that can achieve a certain level of control. The formulation of the least-cost analysis is as follows:

$$minimize \quad CT = fc \ [Xi,...,Xn] \quad ; \quad i=1,...n \qquad (7.11)$$

$$subject \quad to \quad Yk[Xi,...,Xn] \leq Yko \quad ; \quad k=1,...,m \qquad (7.12)$$

in which CT is the total cost of providing the required control measures; fc[Xi,...,Xn] is the total cost of providing n types of control measures (e.g., S, OMEGA); n is the total number of feasible control measures; Yk[Xi,...,Xn] is the control performance measure k as a function of control measures Xi to Xn; Yko is the required control performance measure k; Xi is the control measure i; m is the total number of required control performance measures (e.g., Cr and Ns).

A graphical illustration of the optimization procedure is shown in Figure 7.4. The performance isoquants (isoquants 1 and 2) represent the combinations of control measures which achieve the same level of performance while the isocost curves (c1, c2, c3) indicate the combinations of control measures with the same total cost.

The top figure depicts the situation in which the required performance isoquants intersect. The tangency point A is the least-cost mix of control measures 1 and 2 for isoquant 1 while tangency point B is the least-cost mix of control measures for isoquant 2. The intersection point C between the isoquants is the control measures mix which achieves both isoquants exactly. However, the least-cost mix of control measures which can achieve both isoquants is point B because this combination of control measures can achieve the control of isoquant 2 and a

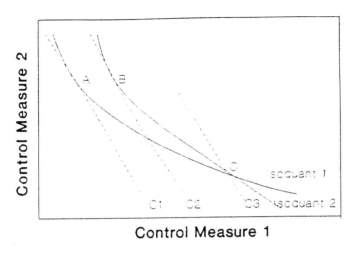

Control Measure 1

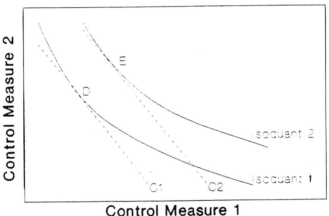

Control Measure 1

Figure 7.4:   Illustration of the determination of least-cost mix of control measures for integrated quantity and quality control.

greater control than isoquant 1.

The lower figure indicates the situation in which the performance isoquants do not intersect. For this particular situation, one of the isoquants is dominant. As a result, the least-cost mix of control measures is point E. This two-dimensional analysis can be extended to multiple mixes of control measures such as S, OMEGA, Sd and PHI. A microcomputer program has been written for this purpose to facilitate preliminary quantity and quality management planning of urban runoff control measures.

7.8 Sensitivity Analysis

The evaluation of engineering alternatives is subject to uncertainty both in model selection and in parameter estimation.   The proposed planning methodology requires analytical models to predict control system performance and optimization models to determine the least-cost mixes of control measures. Sensitivity analysis should be conducted for each application of the methodology so that the impacts of the model assumptions and the parameter values on the analysis results can be evaluated.

Various assumptions have been made to derive the analytical probabilistic models. As a result, the sensitivity of the analytical models should be investigated in each application of the models. Input rainfall characteristics and treatment efficiency should also be varied to test the sensitivity of control performance. From these analyses, the upper and lower bounds of control system performance can be determined.

7.9 Application of Methodology

A catchment in the Barrington area of East York, Toronto is used to demonstrate the methodology. The study area is single family residential served by a storm sewer system. Its area is 17.4 ha and the estimated percent impervious is 60%. Rainfall data was collected by the East York Engineering Department. The statistics of the data were found to be comparable to the closest permanent rain gauge located at the Toronto Bloor Street. Quantity and quality data on the storm overflows were recorded by the Ministry of Environment (Mill, 1977; Kronis, 1982). The purpose of this study is to determine the most promising combinations of control measures that can achieve long term quantity and quality control performance.

Forty-three years of continuous rainfall record at the Toronto Bloor Street gauging station were first analyzed to determine the probability density functions and statistics of rainfall characteristics such as event volume, duration, interevent time, and average intensity. It was found that exponential distributions could be used to describe these characteristics.

The analytical probabilistic model results from SUDS were then compared with the continuous simulation model STORM in terms of the performance isoquants of Cr, Cp, Ns. The performance isoquants predicted by SUDS were generally in good agreement with those simulated by STORM. Figure 7.3 presents the comparison of Cr isoquants between SUDS and STORM. Other comparisons of performance isoquants indicate similar results.

Four different types of control measures, namely upstream and downstream storage (Sd and S), runoff reduction facilities (PHI) such as infiltration facilities, and downstream treatment rate (OMEGA), are modelled by the proposed methodology. There are many combinations of these four types of control measures that can achieve the required performance. The combinations of upstream runoff reduction (PHI), downstream storage volume (S) and treatment rate (OMEGA) are examined in this study. Other combinations of control measures have been investigated with the proposed methodology for the Barrington catchment (Li, 1991).

The cost functions for the downstream detention basin (S) and microscreening facility (OMEGA) are taken from literature (Pavoni, 1977; Wiegand, 1986) while the cost function of upstream runoff reduction facilities (PHI) is assumed.

Quantity performance is modelled by Cr which ranges from 10% to 98%, while quality performance is modelled by Ns which ranges from 1 to 50 spills per year. The total and marginal costs of providing various combinations of Cr and Ns are indicated in Figure 7.5. It is noted that both the total and marginal costs increase rapidly when Cr is higher than about 70% and Ns is below about 5 spills per year. For instance, the total cost to

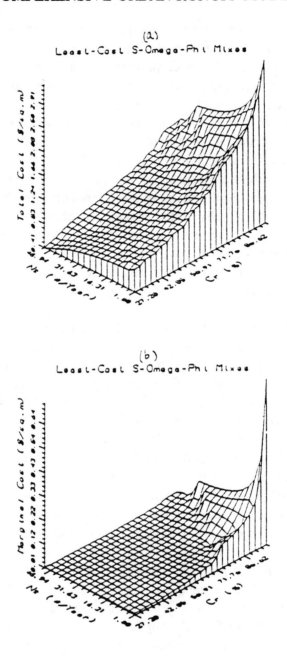

Figure 7.5:   Total and marginal cost surfaces for integrated Cr and Ns control.

provide 98.5% runoff control and to limit spills to 1 per year is about $3.40 per square meter of catchment while the marginal cost is about $0.80 per square meter. The least-cost combination of S and OMEGA to provide this performance is 6.62 mm and 1.2 mm/hr, respectively.

Other least-cost combinations of control measures (e.g., S and PHI, OMEGA and PHI) are also determined and compared with those of S, OMEGA, and PHI. The most promising combination of control measures for achieving the Cr and Ns control requirement is the S and OMEGA combination (the downstream storage and treatment systems).

The sensitivities of performance measures Cr, Ns and Cp are investigated with respect to the rainfall parameters (z,l,k) and the treatment efficiency N of the downstream storage/treatment system. Each interevent time definition (IETD) results in a set of values for the rainfall characteristics. It is found that the performance measures are generally not sensitive to the variation of IETD. However, the performance measure Cp is sensitive to the variation of N as would be expected.

The sensitivities of the total and marginal costs of achieving Cr, Ns, and Cp are investigated with respect to the rainfall parameters (z,l,k) and the parameters of the cost functions for control measures. The total and marginal costs of the least-cost mix of control measures are found to be generally insensitive to the variation of IETD and sensitive to the variation of cost parameters of the downstream storage/treatment system. The uncertainty in cost parameters can cause up to 50% change in the total costs and 70% change in the marginal cost of achieving high levels of runoff quantity and quality control. These changes in cost are attributed to the changes in the design (i.e., magnitudes of S and OMEGA). Therefore, it is imperative to use accurate local cost data to conduct the runoff control planning.

7.10 Conclusions

As the deterioration of urban water quality continues and the need to control runoff quantity problems such as flooding and erosion still prevails in many areas, an integrated approach to control both urban runoff quantity and quality problems becomes eminent. The techniques discussed above provide a comprehensive and systematic methodology for preliminary planning and screening of urban runoff quantity and quality control systems.

With the development of analytical probabilistic models such as SUDS and Extended SUDS, preliminary prediction of urban runoff control system performance can be efficient and reliable. Although the assumptions of the analytical models may not be perfectly satisfied in every application, the models may still be useful for preliminary evaluation of urban runoff control system alternatives.

Least-cost analysis of urban runoff control systems provide important information such as cost-effectiveness relationships and expansion paths for achieving various levels of runoff control. This information is useful for specification of the target design performance and subsequent design level analyses of control systems. Since the cost and performance of control systems are explicitly taken into consideration, the proposed runoff control planning methodology extends conventional engineering analysis which emphasizes an approximate performance analysis of runoff control with rather arbitrary design levels.

Future improvements to the methodology are encouraged to incorporate receiving water quality analysis in urban runoff control system planning. The probabilistic model of stream quality proposed by Di Toro (1984) offers some insight to modelling the impact of urban runoff on receiving water quality. The analytical probabilistic models SUDS and Extended SUDS should also be further verified with field data as well as other continuous simulation models.

7.11 References

Adams, B.J., and Bontje, J.B., (1983). Microcomputer applications of analytical models for urban drainage design, in *Emerging Computer Techniques for Stormwater and Flood Management*, ASCE, 138-162.

Adams, B.J., Fraser, H.G., Howard, C.D.D., and Hanafy, M.S., (1986). Meteorological data analysis for drainage system design, *Journal of Environmental Engineering*, ASCE, 112(5):827-848.

Adams, B.J., and Howard, C.D.D., (1986). Design Storm Pathology, *Canadian Water Resources Journal*, 11(3):49-55.

Adams, B.J., and Zukovs, G., (1987). Probabilistic models for combined sewer system rehabilitation analysis, in *Systems Analysis in Water Quality Management*, ed. by M.B. Beck, Pergamon Press, 97-108.

Benjamin, J.R., and Cornell, C.A., (1970), *Probability, Statistics, and Decision for Civil Engineers*, McGraw- Hill Book Company.

Di Toro, D.M., (1984). Probability model of stream quality due to runoff, *Journal of Environmental Engineering*, ASCE, 110(3):607-628.

Driver, N.E., and Tasker, G.D., (1988). Techniques for estimation of storm-runoff loads, volumes, and selected constituent concentrations in urban watersheds in United States, U.S.G.S. Report No. 88-191, U.S.G.S., Colorado.

Flatt, P.E., and Howard, C.D.D., (1978). Preliminary screening procedure for economic storage treatment tradeoffs in storm water control, in *Proceedings, International Symposium on Urban Storm Water Management*, University of Kentucky, 219-227.

Howard, C.D.D., (1976). Theory of storage and treatment plant overflows, *Journal of Environmental Engineering*, ASCE, 102(4):709-722.

Huber, W.C., Heaney, J.P., Nix, S.J., Dickinson, R.E., and Polman, D.J., (1983). Stormwater Management Model Users Manual, U.S. EPA, Cincinnati, Ohio.

Hydrologic Engineering Center, (1974). Manual of the storage, treatment, overflow and runoff model (STORM), U.S Army Corps of Engineers, Davis, California.

Kauffman, G., (1987). A comparison of analytical and simulation models for drainage system design: SUDS versus STORM, MASc Thesis, University of Toronto, Toronto, Ontario, Canada.

Kronis, H., (1982). Physical-chemical treatment and disinfection of stormwater, Research Report No. 88, Ministry of Environment, Ontario.

Lager, J.A., Didriksson, T., and Otte, G.B., (1976). Development and application of a simplified stormwater management model, Report No. EPA-600/2-76-218, U.S. EPA, Cincinnati, Ohio.

Li, J., (1991). Comprehensive urban runoff control planning, PhD Thesis, University of Toronto, Toronto, Ontario, Canada.

Mills, G.W., (1977). Water quality of urban storm water runoff in the Borough of East York, Research Report No. 66, Ministry of Environment, Ontario.

Nix, S.J., Heaney, J.P., and Huber, W.C., (1983). Analysis of storage release systems in urban stormwater quality management: A methodology,in *Proceedings, International Symposium on Urban Hydrology, Hydraulics and Sediment Control*, University of Kentucky, 19-29.

Nix, S.J., and Heaney, J.P., (1988). Optimization of storm water storage release strategies, *Water Resources Research*, 24(11):1831-1838.

Pavoni, T., (1978). *Handbook of water quality Management planning*, Van Nostrand Reinhold Co., 205-206.

Sullivan, R.H., Hurst, W.D., Kipp, T.M., Heaney, J.P., Huber, W.C., and Nix, S.J., (1978). Evaluation of the magnitude and significance of pollution from urban storm water runoff in Ontario, Research Report No. 81, Ministry of Environment, Ontario.

U.S. EPA, (1983). Results of the nationwide urban runoff program, NTIS PB84-185545.

Wiegand, C., (1986). Cost of urban runoff quality controls, in *Proceedings, Urban Runoff Quality Impact and Quality Enhancement Technology*, ASCE, Cincinnati, Ohio, 367-380.

Wallace, R.C., (1980). Statistical modelling of water quality parameters in urban runoff, Master's Technical Report, University of Florida, Gainesville, Florida.

7.12 Nomenclature

AL_r	:	Average annual runoff pollution load, kg
b	:	Interevent time, hr
c_i	:	Isocost cost function i, $
C	:	Event flow-weighted mean concentration, mg/L
C_r	:	Average annual percent of runoff volume controlled, %
C_p	:	Average annual percent of runoff pollution controlled, %

CT	:	Total cost of control measures, $
E(b)	:	Mean value of b, hr
E(i)	:	Mean value of i, mm/hr
E(t)	:	Mean value of t, hr
E(v)	:	Mean value of v, mm
fc	:	Cost function of control measures, $
i	:	Event average intensity, mm/hr
IETD	:	Interevent time definition, hr
k	:	Reciprocal of mean of b, 1/hr
l	:	Reciprocal of mean of t, 1/hr
Lr	:	Event runoff pollution load, kg
Ln	:	N-year event pollution load, kg
m	:	Number of control performance measures
Ns	:	Average annual number of spills
pdf	:	probability density function
PHI	:	Runoff coefficient, dimensionless
Pn	:	N-year event overflow volume, mm
S	:	Downstream storage capacity, mm
Sd	:	Depression storage, mm
t	:	Event rainfall duration, hr
v	:	Event rainfall volume, mm
Vr	:	Event runoff volume, mm
w	:	Reciprocal of mean of i, hr/mm
Xi,	:	Control measure type i
Xn,	:	Control measure type n
Yk	:	Control performance type k
Yko	:	Required control performance type k
z	:	Reciprocal of mean of v, 1/mm
OMEGA	:	Downstream outflow control rate, mm/hr

Chapter 8

Remote Sensing Inputs for Flash Flood Forecasting in Urban Areas

N. Kouwen and E.D. Soulis
Department of Civil Engineering
University of Waterloo
Waterloo, Ontario, Canada. N2L 3GI

8.1 Introduction

Urban rainfall-runoff models were the first hydrologic simulation models that computed runoff separately for different land cover classes, namely pervious and impervious areas.   The two contributions were added and routed together through sewers and surface drainage systems. This method of simulating storm runoff has been successfully applied to a larger number of land use classes, namely impervious, barren or low vegetation, intermediate height vegetation, forests, wetlands and water covered areas.

The problem encountered was the definition of parameter values, the most important being the effective saturated conductivity of the soil and the overland flow resistance.   For urban areas, these values are well known for impervious areas. Thus only the values for the pervious areas needed to be optimised. However, for non urban areas, or for urban areas with

0-87371-898-4/93/$0.00 + $.50

177

a greater diversity of land cover classes, these parameters (and others) needed to be determined for each land cover class. This can only be accomplished on watersheds where there are at least as many streamflow gauges as there are land cover classes and where there is a diversity of the mix of classes.

Remotely sensed data offers in each gauged area the detailed real-time information on watershed conditions and rainfall amounts necessary to solve this problem. This information is of particular importance in flood forecasting because the hydrologic processes that most directly affect flash floods occur at the land surface and are greatly influenced by the terrain, or land cover characteristics of the watershed. Weather radar now provides detailed real-time rainfall data at frequent intervals. The timing and areal distribution of rainfall also directly affect the flood hydrographs. It thus appears that data management systems and hydrologic models designed to make maximum use of this remotely sensed data would offer improved flash flood forecasting.

This paper describes the application of the WATFLOOD data management system and the SIMPLE rainfall-runoff model to a heavily urbanised watershed that exceeds a size that can be readily modelled by a conventional urban runoff model. The problems associated with parameter optimisation and the estimation of rainfall distributions is discussed.

8.2 Watflood Data Management System

WATFLOOD is an integrated set of computer programs to forecast flood flows for watersheds having response times ranging from one hour to several weeks. A shell program organises all the menus used for data input and correction functions and acts as the manager for the hydrologic model SIMPLE and utility programs. The emphasis of the WATFLOOD system is on making optimal use of remotely sensed data. Radar rainfall data, LANDSAT or SPOT land cover data can thus be directly incorporated in the hydrologic modelling.

WATFLOOD is unique in its ability to preserve the distributed nature of a watershed's hydrologic and meteorologic characteristics without sacrificing computational efficiency. This has been accomplished through the use of **Grouped Response Units** (GRU's), in which process parameters are tied to land cover and land cover mixes vary from basin element to basin element.

The system is completely modular but has a consistent data structure throughout. It has been under continuous development since 1972. Several MSc and PhD research programmes have provided the rationale incorporated in the software.

8.2.1 WATFLOOD components

The WATFLOOD system consists of a user interface program that presents a set of menus that allow the user to carry out a number of tasks that are required to produce a flood forecast. Separate computer programs create basin files, convert radar Constant Altitude Precipitation Index maps (CAPPI's) to rainfall maps, calibrate radar weather data with ground level rain gage data in a real-time environment, perform a simulation of the hydrology of a watershed, and plot flow forecast at various scales. While the WATFLOOD system has been designed to optimally use remotely sensed data, conventional data can just as readily be used to set up and operate the system.

The system's most important component is a hydrologic simulation model that is self-calibrating in terms of watershed parameters and initial conditions. Since 1972, SIMPLE has served as a numerical laboratory to experiment with various rainfall-runoff modelling techniques and data utilisation. The model and data management system are unique in the following ways:

1. It is a geo-referenced system that was designed from the outset to incorporate remotely sensed data such as weather radar, LANDSAT or SPOT land cover data, GOES snow

cover data, or any other hydrologically significant remotely sensed data such as soil moisture or percent of snow covered area.

2. The model was used to develop the "Grouped Response Unit" (GRU) approach to hydrologic modelling. This is a method of weighting the watershed response to land cover distribution instead of the usual parameter weighting schemes. The GRU method allows the transfer of parameters to other watersheds without concern about the relative amounts of the various land cover types. The grouped response unit is an extension of urban storm water modelling where the contributions from pervious and impervious areas are calculated separately and added prior to sewer and streamflow routing.

3. The model has been calibrated and validated in time on one watershed and validated in time and space on several others. This is not often possible with other models, which have weighted parameters instead of weighted responses. SIMPLE incorporates a pattern search optimisation algorithm. The pattern search is said not to be affected by the inter-dependency of a model's parameters.

4. The flood forecasting system is completely self-contained with respect to data input and can be run on a laptop computer, thus making it a stand alone system that can continue during power failures or data interruptions. The entire system is menu-driven with all the complicated activities hidden from the final user, who may not have detailed knowledge of hydrologic modelling, remotely sensed data or computing systems.

5. The flood forecasting system is designed to operate automatically, using real-time weather radar precipitation data as well as conventional rainfall and streamflow data. It can be run as a background task on a computer to activate an alert if a potential for flooding occurs. In addition, it can give

specific, detailed local forecasts for the purpose of evacuation, road closures, emergency vehicle routing and the like.

8.3 Hydrologic Model

SIMPLE is a physically based simulation model of the hydrologic budget of a watershed programmed in FORTRAN 77. Because the model is aimed at flood forecasting, only those processes which dominate flood flows are included. These processes are: interception, infiltration, interflow, baseflow, overland and channel routing. The model is limited to these only because to introduce others serves only to introduce more degrees of freedom, which would make calibration more difficult. At the same time, the exclusion of other processes (such as evaporation, direct interaction of the river with the groundwater reservoir) will necessarily limit the use of the model to those areas where these additional factors are not of concern.

Furthermore, the model has been kept simple to allow its use on micro-computers within a time frame suitable for forecasting. Typically, the program executes in 4 - 6 seconds for a 120 hour simulation on a 486/33 computer. The size of the watershed does not affect running time, since the size of the grid is scaled to provide just enough resolution for the drainage pattern. Normally about 30 to 60 elements will suffice. On the IBM-AT or compatible computer a maximum of approximately 200 elements can be accommodated.

This following sections describe the watershed model in detail. The values of many parameters need to be determined and while some may be assigned standard well known values, others may be subject to great variations and uncertainty. Where possible, standard values are used, but those parameters which cannot be predicted are calibrated using a pattern search optimization technique. In the following sections, those parameters which are optimized are identified.

The modelling process begins with the addition of rainfall to the watershed. The various processes are described below.

8.3.1 Interception

Interception of precipitation by vegetation is calculated in SIMPLE using the equation given by Linsley et al. (1949)

$$V = (S_i + C_p E_a t_R)(1 - e^{-kP}) \qquad (8.1)$$

where:

V   is depth of interception from the beginning of the storm,
S_i is storage capacity per unit of projected area,
C_P is ratio of vegetal surface area to its projected area,
E_a is evaporation rate per unit per unit of surface area,
t_R is the duration of the rainfall,
k   is a constant,
P   is the precipitation since the beginning of the storm.

In SIMPLE, C_p is assumed to be 100 and E_a at.00025 mm/hr. The values for S_i are set for each month and each land cover class and t_R is taken as the time from the beginning of rainfall. The product of C_P and E_a is used as a single parameter A7.

8.3.2 Surface Storage

The ASCE Manual of Engineering Practice No. 37 for the design of sanitary and storm sewers (ASCE, 1969) gives typical values of retention for various surface types. Table 8.1 is a listing of depression storage for various conditions and values are seen to vary greatly. Because of the uncertainty associated with depression storage, this is one of the parameters included for optimization, but it is ranked 5th out of 5 in priority.

As with interception, it is assumed that the limiting value of depression storage S_d is reached exponentially (Linsley et al., 1949):

$$D_s = S_d \, (1 - e^{-kP_e}) \qquad\qquad (8.2)$$

where D_s is the depression storage, P_e is the accumulated rainfall excess and k is a constant.

Table 8.1. Surface retention values (ASCE, 1968).

Type of Surface:	Retention (mm) S_d
Impervious Urban Areas	1.25
Pervious Urban Areas	3.0
Smooth Cultivated land	1.3 - 3.0
Good pasture	5.0
Forest litter	8.0

8.3.3 Infiltration

Due to the importance of the infiltration process in runoff calculations, but also because infiltration capacity is such a highly variable quantity, this process requires a great deal of attention in any hydrologic model. Many formulae are used (see for instance Viessmann et al., 1977) and any choice is open to criticism.

In keeping with the underlying philosophy of keeping the model based on identifiable physical processes, the Philip formula (Philip, 1954) was chosen as representing the important physical aspects of infiltration process. It also readily incorporates the notion of surface detention. The Philip formula is identical to the Green-Ampt equation (Green & Ampt, 1911) except that it includes the head due to surface ponding as well as the capillary potential. The Green-Ampt approach assumes the ponding head is insignificant when compared to the potential head. The Philip formula (Philip, 1954) expresses the rate of infiltration as:

$$\frac{dF}{dt} = k \left(1 + \frac{(m - m_o)(P + H)}{F}\right)$$ (8.3)

where:

F   is the total depth of infiltrated water in mm, t is time in seconds,

k   is permeability in mm/sec,

m   is the average moisture content of the soil to the depth of the wetting front,

m_o is initial soil moisture content,

P   is the capillary potential at the wetting front in mm,

H   is detention storage.

Equation 8.3 represents the physical process of infiltration since the pressure gradient acting on the infiltrating water is used to determine the flow using Darcy's Law.   Because of the uncertainty of the effective value of k over the basin, it is an optimized parameter.

Initially, the infiltration capacity is very high because of the shallow depth of the wetting front.   This causes a very large pressure gradient inducing high infiltration.   However, as the wetting front descends, the pressure gradient is quickly reduced, thus reducing the potential infiltration rate. Using the information in Philip (1954) relating permeability to capillary potential, the following relationship provides the capillary potential:

$$CP = 250 \log(k) + 100$$ (8.4)

where:

CP is the capillary potential in mm and
k   is the permeability in nmvs.

Water depth on the soil surface is continually modified to reflect the net precipitation input, infiltration, and overland flow discharge.

8.3.4 Interflow

Infiltrated water initially is what is commonly referred to as the Upper Zone Storage (UZS).   Water within this layer percolates downward or is exfiltrated to nearby water courses and called interflow. In the model, percolation downward is ignored because in most cases when dealing with single rainfall events, the path from the UZS to the stream via the groundwater reservoir is too long in duration to contribute appreciably to streamflow during the event. Of course exceptions do occur and must be recognized.

Interflow is represented by a simple storage-discharge relation:

$$QINT = REC*WAC \qquad (8.5)$$

where:

QINT is interflow in m3/sec,
REC is a coefficient (optimized),
WAC is water accumulation in the UZS region.

REC is a coefficient which cannot be predicted and is therefore determined through optimization.

8.3.5 Overland flow

When the infiltration capacity is exceeded by the water supply, and the depression storage has been satisfied, water is discharged to the channel drainage system. The relationship employed is based on the Manning formula and takes the form:

$$Q1 = (D1 - Ds)^{1.67} Sl \ A/R3 \qquad (8.6)$$

where:

Q1 is the channel inflow from each land cover class to the element's major drainage system,

D1 is average depth of water stored on the element,

Ds is depression storage capacity (optimized),

Sl is average overland slope, determined by entering the number of contours within a basin element and the contour interval,

A is the area of the basin element,

R3 is a roughness parameter (optimized).

In SIMPLE, Eqs. 8.1 through 8.66 are used separately for each land class in each computational element.

8.3.6 Base flow

The base flow in SIMPLE is determined from a measured hydrograph at the basin outlet. The baseflow contributed by each basin sub-element is found by simply prorating it to the total basin area. A recession constant is used to gradually diminish the pre-rainfall flow over the duration of the event being modelled.

8.3.7 Total runoff

The total inflow to the river system is found by adding the surface runoff, the interflow and the baseflow for the land cover classes. These flows are then added to flows entering the channel from upstream and routed to the downstream basin element.

8.4 Routing Model

The routing of water through the channel system is accomplished using a storage routing technique. This is an adequate approach for upstream routing and is suitable because it is based on river cross-section and profile data. The method involves a straightforward application of the continuity equation

$$\frac{I_1 + I_2}{2} - \frac{O_1 + O_2}{2} = \frac{S_2 - S_1}{\Delta t} \qquad (8.7)$$

where:

$I_{1,2}$ is inflow to the reach consisting of overland flow, interflow, baseflow, and channel flow from all contributing upstream basin elements,

$O_{1,2}$ is the outflow from the reach,

$S_{1,2}$ is storage in the reach,

Δt is the time step of the routing.

The subscripts 1 and 2 indicate the quantities at the beginning and the end of the time step. The flow is related to the storage through the Manning formula:

$$O = (1/R2)(AX)^{1.33} S_o^{1/2} \qquad (8.8)$$

where:

O is flow in m3/s,

$R2$ is the channel roughness parameter (optimized),

AX is channel cross-section area which is related to storage by dividing the storage by the channel length,

S_o is channel slope.

A change in this relationship occurs when the flow exceeds the channel capacity and the flow spills into the flood plain. One requirement for running SIMPLE is a relationship which gives the channel capacity at any point in the basin. This is accomplished by measuring the channel cross-section area at various points in the watershed and fitting a relationship such that the channel cross-section area is given as a function of drainage area. This relationship is used to determine if the flow exceeds the channel's capacity at any point at any time.

8.4.1 Data Requirements

The data required for the model can be put into three categories: watershed data, hydrologic parameters, and hydrometric data. The following watershed data is required for each element: the elevation of the streambed at the halfway point; element area; drainage direction to receiving element; surface slope as indicated by contour density; stream density; and percentage of land cover classes. The parameter list includes: depression storage; saturated conductivity; interflow storage-discharge coefficient; overland flow roughness coefficient; and river roughness coefficients. Finally, hourly rainfall amounts and measured hydrographs are required to calibrate the model.

8.4.2 Parameter Optimisation

The model features the Hooke and Jeeves (1961) automatic pattern search optimization algorithm taken from Monro (1971). The program can be run to automatically determine which combination of parameters best fit measured conditions. The optimised parameters are: saturated conductivity; the interflow storage-discharge coefficient; surface roughness; and stream roughness. In this study, the only parameter changed for Black Creek was the stream roughness. Obviously, this parameter is unique for each watershed.

8.5 Study Area

The WATFLOOD system was first calibrated on the Grand River watershed above Cambridge (Galt), Ontario, Canada. Three rainfall - runoff events were used to estimate the model's parameter values. Most parameters were based on accepted values but some, for instance, permeability, surface roughness, and an interflow depletion factor were optimized separately for each of four major land cover class. A fourth parameter was optimized for river roughness, using five regional values to allow for differences in river conveyance characteristics in various parts of the watershed. In total, seventeen parameters were determined using the Hooke and Jeeves (1961) pattern search algorithm.

Since the advent of micro computers, the application of optimisation to models such as SIMPLE has become affordable. During the past eight years, at least 500,000 iterations of the model have been conducted to arrive at an acceptable parameter set. Usually, the parameters have to be recalibrated when the model is changed. This often results in 2000 to 5000 iterations. First, a sensitivity analysis of the model parameters is carried out. The objectives of this part of the exercise is to find the valid range of the parameters. All processes have to contribute properly to the total runoff. Limits are set on the parameters and the automatic pattern search routine is applied to the model. When the error cannot be further reduced, a check is made to ensure that the relative value of the parameters is approximately correct. For instance, the saturated conductivity should be larger for forested land than for pasture or lawns. If discrepancies occur, new values with the proper relationships are assigned as starting values and the optimisation repeated. This process continues until the best fit is obtained for a number of events and the parameter values are physically correct, although not always directly comparable to laboratory values for say permeability or channel roughness. These discrepancies are caused by for instance, pooling of surface runoff, flow concentrations and other effects.

The model was then validated on eight more events on the

Grand River as well as other events on the Saugeen River, the Rouge River, Duffins Creek, Lynde Creek, Oshawa Creek and the Humber River. The results of these calibration studies have been reported elsewhere (Kouwen et al, 1992; Kouwen et al., 1990). These watersheds are primarily rural watersheds and when SIMPLE was being written, its application to urban watersheds was not contemplated, although a provision was made at the beginning to include impervious areas as a separate land cover class. The model was developed for the Grand River basin and the urban area being a relatively small fraction, was not considered in detail. It was treated differently from the pervious areas by setting interception, depression storage and infiltration equal to zero and the surface roughness was set to 0. 1 times the surface roughness for lightly vegetated areas.

When the model was applied to the Humber River basin in the Western part of Metropolitan Toronto, it became apparent that the model performed poorly for most of the Humber River basin but very well in the totally urbanized Black Creek subwatershed. This led to a more detailed investigation of the model's performance on the Black Creek watershed.

Figure 8.1 is a map of the Humber River watershed with the Black Creek sub-watershed shown on its East side. Also shown are the streamflow and rainfall gauge locations and the 4 km by 4 km UTM coordinate grid used to subdivide the watershed. The rainfall-runoff process is modelled separately for each watershed element. Black Creek has a drainage area of 58 km^2 and is modelled with six elements. Table 8.2 shows the characteristics of each of the six contributing elements as determined from a Landsat MSS composite image taken on May 5, 1987.

The Landsat image was classified by using the reflectance in three bands. The pixel size was 79 m by 79 m and as a result, many pixels contain mixed land cover classes. This results in many pixels that cannot be classified as belonging to a specific land use class or it can result in pixels being wrongly classified. While the percentages listed in Table 8.2 seem reasonable, there is room for improvement, especially if remote sensing imagery with a greater resolution is used.

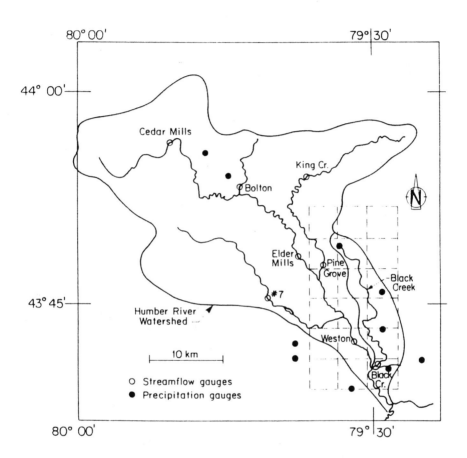

Figure 8.1: Map of the Humber River watershed.

In spite of these difficulties, the application of SIMPLE to the Humber River watershed resulted in much better agreement between the measured and computed hydrographs for Black Creek gauge than for the other locations. Most of the error is probably due to the poor distribution of the raingage locations. However, as shown on Figure 8.1, the Black Creek watershed has very good raingage coverage. In fact, there is almost one gauge in each computational element, which is about ideal for the purpose at hand.

Table 8.2: Basin land cover classifications.

Element			Percent land cover				
No.	Area	Imp.	Grass	Forest	Brush	Wetl.	Water
48	3	32	05	05	56	01	01
56	15	46	08	01	43	00	01
59	2	51	13	01	34	00	00
63	16	55	07	01	36	02	00
65	19	34	02	01	60	06	00
74	16	29	01	02	61	06	00

Figure 8.2 is an example of the output produced by SIMPLE. The figure shows the measured and computed hydrograph for eight locations on the Humber River and its tributaries. The eight one is for Black Creek. The heavier line followed by the dots is the measured hydrograph while the thin line is the computed hydrograph. The vertical line denotes the end of the rainfall and the time at which a flow forecast might be made.

8.6 Modelling results

The computed hydrographs in the upper, rural parts of the

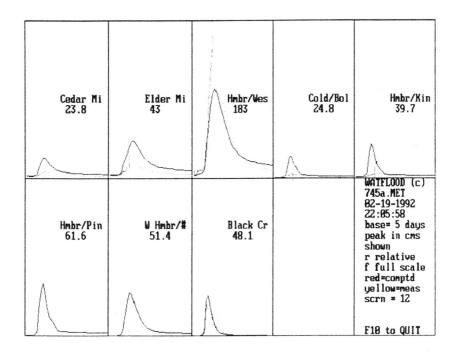

Figure 8.2: Humber River hydrographs.

watershed are all too high while the two hydrographs in the lower Humber show much better agreement, the hydrograph for Black Creek being the better of the two. Figure 8.3 is an enlargement of the Black Creek hydrograph for the May 1974 event. The computed hydrograph shape is in close agreement with the measured hydrograph.

Considering that the model was developed for rural watersheds and calibrated on the Grand River Watershed, this good result on an urban watershed was not expected. Many of the computed hydrographs on the other watersheds (Saugeen River, Eastern Metropolitan Toronto rivers, and the Grand River) are in good also in good agreement with the measured hydrographs. The most probable cause for the poor results on the Humber River is the very poor distribution of the raingages

(Figure 8.1).

The inability to obtain a better fit between the computed and measured hydrographs for the upper reaches of the Humber River points to the obvious need for a denser raingage network if rainfall events that exhibit areal non-uniformity are to be properly modelled. From the foregoing results, it might be concluded that one gauge for each computational elements yields satisfactory results. Obviously, this is not a practical solution for larger watersheds. For Black Creek, there happens to be one raingage for each element.

The good modelling results can be attributed in part to having a dense raingage network. However, the model also appears to be represent the urban rainfall-runoff-routing process very well. This is an important finding because it shows that urban runoff modelling does not have to be carried out at as detailed a scale as might previously have been thought.

While specialised equipment is required to process Landsat imagery, the effort involved in processing the data and setting up the necessary data files for SIMPLE is minimal for an experienced operator. Also, the remaining watershed data required to run SIMPLE can be obtained from a 1:50,000 scale topographic map. In fact, all the data requirements can be

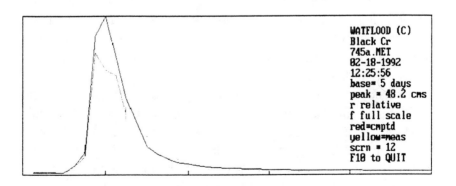

Figure 8.3: Black Creek - May 1974.

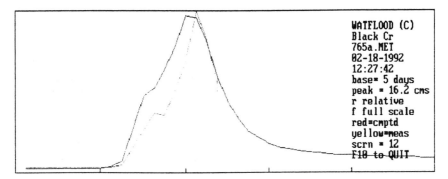

Figure 8.4: May 1976 hydrograph.

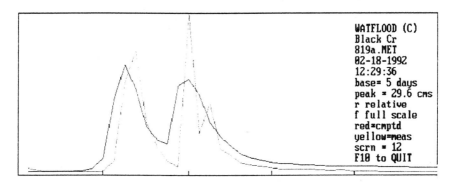

Figure 8.5: September 1981 hydrograph.

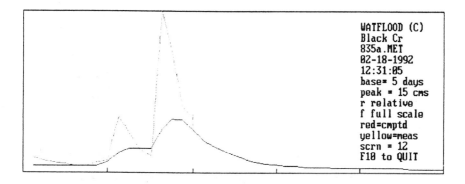

Figure 8.6: May 1983 hydrograph.

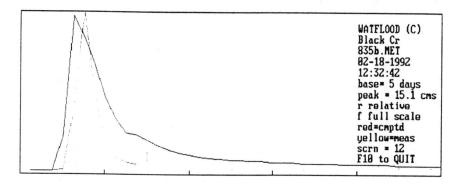

Figure 8.7: May 1983 hydrograph.

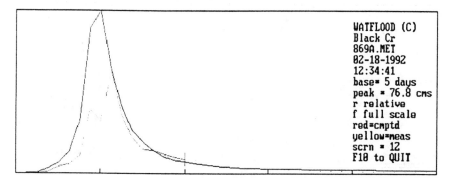

Figure 8.8: September 1986 hydrograph.

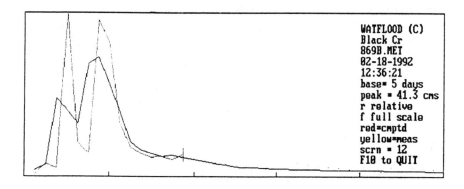

Figure 8.9: September 1986 hydrograph.

obtained from conventional sources but would involve more effort. The only data that requires a visit to the watershed is a table of channel cross-sectional area versus drainage area. In summary, the data that was used to obtain the hydrographs shown in Figures 8.3 through 8.8, would not normally be considered as "detailed data". It is certainly less detailed than the normal data requirements for urban runoff models. It is not intended that SIMPLE be used as a design tool for storm water management facilities. It is intended as an operational model, that is for flood forecasting or reservoir operation. For this purpose, it appears that remotely sensed land cover data adequately delineates the land cover characteristics, and hence the hydrologic characteristics of an urban watershed.

Radar is sometimes quoted as being the ideal rainfall measuring device but many problems with weather radar remain (Wilson, 1976; Browning and Collier, 1989). Still, it is generally agreed that radar improves rainfall estimates when rainfall gauges are spaced at a lower densities than approximately one gauge per 300 to 400 km^2. Similarly, radar rainfall measurements adjusted with raingage measured rainfall amounts improved flow forecasts when the raingage density was lower than the same range (Cooper, 1988). Based on these figures, it would appear that in general, the flow estimates on the Humber River and its tributaries could be improved with the use of radar rainfall data. However, the raingage density for the Black Creek watershed is approximately one gauge per 15 km. As a result, radar rainfall data is unlikely to improve flow forecasts on Black Creek.

8.7 Summary and Conclusions

The Grouped Response Unit is a familiar concept to urban runoff modellers. The runoff contributions from different land cover or land use classes are computed separately and then added prior to streamflow routing. Instead of using just two land cover classes, pervious and impervious, SIMPLE allows runoff contributions to be calculated for up to six different classes: impervious, barren ground or light vegetation, forests, crops and low vegetation, wetlands, and water covered areas. The model was designed for rural applications but its application to the Humber River watershed near Metropolitan Toronto revealed it simulated the hydrology for a heavily urbanised watershed very well. Computed hydrographs are compared to measured hydrographs for seven events.

The modelling results shown in Figures 8.3 through 8.8 show that a fairly coarse representation of an urban watershed can still yield good flow predictions provided that the proper percentages of land cover and good rainfall estimates are incorporated in the simulation.

Finally, the use of GRU's is based on the assumption that model parameters are associated with land cover. Each class has its own set of parameters and these should be transferable from one watershed to another without further calibration. The permeability for impervious area is obviously zero, but, other than river roughness, all other parameters used in the simulation were those transferred from the Grand River watershed. These parameters were also successfully applied to the Saugeen and Eastern Metropolitan watersheds (Kouwen et al., 1990)

8.8 Acknowledgements

The image analysis was carried out by Alain Pietroniro, Diane Freeman, Michelle Tittley and Diane Grady. Their efforts are gratefully acknowledged.

8.9 References

American Society of Civil Engineers, (1969), "Design and Construction of Sanitary and Storm Sewers", *Manuals and Reports of Engineering Practice,* No. 37, New York.

Browning, K.A. and C.G. Collier, 1989, "Nowcasting of Precipitation Systems", *Rev. of Geophysics, Vol.* 27, No.3, Aug., pp. 345-370

Cooper, TE., 1988, *Measuring the Enhancement Weather Radar Provides a Rain Gauge Network for Streamflow Forecasting,* M.A.Sc. Thesis, Department of Civil Engineering, University of Waterloo, Ontario.

Green, W.H., and G.A. Ampt, (191 1), "Studies in Soil Physics. 1. Flow of Air and Water Through Soils", *J. Agricultural Science,* Vol. 4, pp. 1-24.

Hooke, R. and T.A. Jeeves, (1961), 'Direct Search' Solution of Numerical and Statistical Problems. *J. Assoc. Comp. Mach.* Vol 8, No 2, pp. 212-229

Kouwen, N., E.D. Soulis, A. Pietroniro and J. Donald, 1990, "Flash Flood Forecasting with a Rainfall-Runoff Model Designed for Remote Sensing Inputs and Geographic Information Systems", *Proceedings, International Symposium on Remote Sensing in Water Resources,* Enschede, Netherlands, August, 805-814.

Linsley, R.K., M.A. Kohler and J.L.H. Paulhus, (1949) *"Applied Hydrology",* McGraw - Hill Book Company, New York, N.Y.

Monro, J.C. (1971), Direct Search Optimization in Mathematical Modelling and a Watershed Application. *NOAA Technical Memorandum.* NWS HYDRO- 12. April.

Philip, J.R., (1954), "An Infiltration Equation with Physical Significance", *Soil Science,* Vol. 77, Jan.-Jun., pp. 153-157.

Viessman, W, J.W. Knapp, G.L. Lewis and T.E. Harbaugh. *(1977), Introduction to Hydrology, Harper &* Row. N.Y.

Wilson, J.W., 1971, "Integration of Radar and Raingage Data for Improved Rainfall Measurement", *J. Applied Meteorology,* Vol. 9, June, pp. 489-497.

Chapter 9

Use of a Multiple Linear Regression Model to Estimate Stormwater Pollutant Loading

Thomas A. Hodge and Louis J. Armstrong
Woodward-Clyde Consultants
500 12th St, Oakland, California 94607-4014

It is widely hypothesized that the quality of stormwater runoff varies by land use. Therefore, pollution loading from a particular watershed can be formulated as a function of the area of the various land use types that form it. In order to develop an estimate of regional pollutant loads, it is desirable to derive a best estimate of the quality of storm runoff from different land uses in the region. A multiple linear regression model was developed from water quality and land use data collected during the 1989-90 wet weather season in Alameda County, California. This model was used to estimate runoff pollutant concentrations by land use for this region. The resultant pollution concentrations were used to estimate the pollutant loads for an average year. Both single storm event data and averaged data from eleven storm events occurring during one wet weather season were used. The F-test was applied to statistically test the validity of the model.

0-87371-898-4/93/$0.00 + $.50
© 1993 by Lewis Publishers

9.1 Introduction

As cities and counties start to implement the United States federal regulations regarding storm water pollution, they are collecting a substantial amount of water quality data. This water quality data is being used to estimate impacts due to pollutant loading to receiving waters. Since runoff and pollutant concentrations can vary by land use, many of the loading calculations are done for individual land use types. Small homogeneous drainage areas are typically sampled to estimate storm water concentrations by land use. However, it is difficult to collect sufficient data from drainage areas with homogeneous land use. There are also questions concerning the representativeness of the selected sampling sites. By focusing on small sampling stations with homogeneous land uses, large catchments which drain areas of heterogeneous land use are ignored. Sampling sites may therefore represent a much smaller portion of a study area than if larger drainage areas had been included.

We present a method, using a multiple linear regression model (MLR), which avoids some of these problems. Data from ten relatively homogeneous land use stations and five heterogeneous stream stations in Alameda County, California (Figure 9.1) were used. A total of eleven storms were sampled but no one station was sampled during more than eight storms.

9.2 Review of Prior Studies

The Clean Water Act has spurred a number of studies to investigate the magnitude and nature of pollution from urban storm water runoff. The most comprehensive of these is the Nationwide Urban Runoff Program (NURP) of the U.S. Environmental Protection Agency (EPA, 1983). This study analyzed runoff for chemical constituents at 85 stations nationwide. Other more recent studies have generally involved a specific region of the country and fewer constituents.

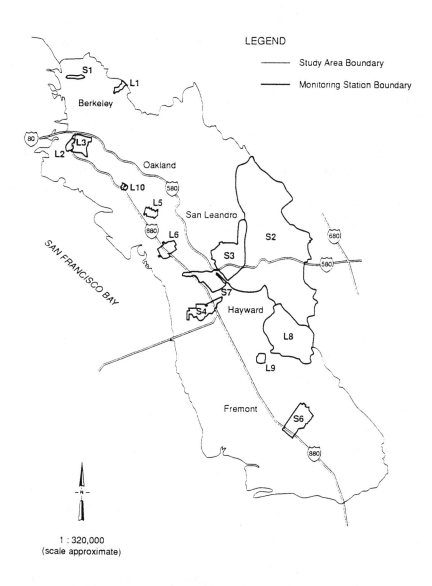

Figure 9.1: Monitoring stations in Alameda County.

In all studies reviewed, investigators have attempted to identify the relationship between land use and stormwater quality. It seems apparent that such a relationship should exist. It is expected that industrial areas are generally more polluted than residential areas. Except for differences between urban and non-urban runoff, the NURP study was unable to verify that a relationship with land use category holds. NURP found that there was no statistically significant variations among urban sites, and asserted that the land use category is virtually useless in predicting station variation (EPA, 1983). Despite this, NURP computed median storm event mean concentrations (EMCs) for residential, mixed, commercial and open land use categories. An EMC is a flow weighted mean concentration for a storm event. These EMCs were presented in the NURP Final Report accompanied by suggestions of caution on their use.

Several studies have used land use specific EMCs, either using NURP data or locally collected data, to compute pollutant load from a region. Woodward-Clyde Consultants (WCC) estimated pollutant loads from Santa Clara Valley, California using data from local sampling stations (WCC 199 la). Sampling stations were chosen based on uniformity of land use. A representative concentration of runoff by land use was computed by averaging the site mean concentrations from stations with similar land use. Runoff volumes were computed using the EPA's Storm Water Management Model (SWMM). Volumes of runoff from each land use type were multiplied by the representative concentration to compute loads. Other stations, with mixed land uses, were used for verification of load estimates for large watersheds. A bias correction factor was applied to total load estimates based on the verification error. The need for this factor indicates difficulties with applying data from small watersheds to large ones.

The Aquatic Habitat Institute (AHI) used data from WCC's Santa Clara Valley study and several other studies to compute pollutant load to the entire San Francisco Bay (Gunther, 1991). AHI used runoff coefficients to compute runoff volume. Runoff coefficients are simply the average proportion of rainfall which appears as runoff, and can be estimated from the percent impervious surface for a watershed. Constituent concentrations were computed similarly to the Santa Clara study, but with an expanded data set.

Silverman, Stenstrom and Fam (1988), developed an approach that is the most similar to what we present. They computed the load of oil and grease to San Francisco Bay using data collected at fifteen stations within the region. Silverman, et al. also used runoff coefficients to compute flow volumes, though they were developed from measured flow data at five of the stations. Land uses from census tract data were used to determine the percent of land tributary to each station identified as residential, commercial/industrial or undeveloped. A total of 34 samples were taken at these stations. With land use as independent variables and oil and grease concentration as the dependent variable, a regression model was solved using 34 equations and three unknowns. A negative coefficient for undeveloped area was set to zero, as negative values were thought to be meaningless.

9.3 Model Description

Using storm runoff and concentration data for total lead and total copper, a system of equations was developed to predict the mean concentration for each land use. For this model the runoffs were estimated using a calibrated Storm Water Management Model (SWMM). The equations were formulated such that the sum of the products of the runoff from each land use type times the concentrations associated with each land use type would equal the product of the total runoff times the total concentration, or the total load. The generic equation is of the form:

$$\sum M_a X_a = Y \qquad (9.1)$$

where:

>X$_a$ is the runoff generated from land use "a", as estimated
> from the SWMM model,
>M$_a$ is the unknown concentration for land use "a",
>Y is the total measured load at the monitoring station.

The equations were normalized by dividing by the flow volume. This provided a unit flow matrix which has the added advantage that the independent variable (Y) and the unknown coefficient (M) are in the same units of concentration.

Since there were more sampling data than land use categories, i.e., more equations than unknowns, the system of equations is considered to be over-determined. If a system of equations is over determined, then there is no uniquely valid solution to the system. To resolve this problem, a multiple linear regression was used to find the best fit to the system of equations. Multiple linear regression (MLR) is a method of finding the line that best fits the collected data when more than one variable exists.

The model was tested for three different systems of equations. For all three systems, the land use categories used were: open, commercial, residential, and industrial. Transportation was combined with commercial since the proportion of flow to the sampling stations contributed by transportation areas was low, yielding poor estimates.

For the first system, annual site mean concentrations were used for the fifteen sampling stations. A site mean concentration is the mean of the EMCs for a sampling station. The system then consisted of a 15 by 4 matrix. The advantage of using the annual site mean concentrations is that variations in concentrations between storm events have been incorporated into the mean. This model was also easy to formulate since site mean concentrations are regularly reported and limit the amount of data handling needed. However, one problem using site mean concentrations is that the stations vary by which storms were sampled and averaged during the year. Due to the difficulty in sampling many stations at one time, different storms were sampled among stations. Thus the inputs to the model (the site mean concentrations) are not truly comparable between stations.

For the second system, EMCs for individual storms were used. This method produces near uniformity in storm characteristics among samples. The problem with this method was that the amount of data available was not consistent for all storms. This system of equations appeared encouraging for one storm where thirteen stations were sampled, but was unreliable when there was less data. There were no storms where all stations were sampled simultaneously.

The last system of equations used event mean concentrations for all storms. This system provides the most data to run the MLR and resulted in a 93 by 4 matrix. This system loses the uniformity of storm conditions, but provides a much more robust model due to the large data set and therefore is the preferred model.

Statistical parameters from the regression models are presented in Table 9.1. Results from both the preferred model and the simplified model using site mean concentrations are provided for comparison. As we see, the R^2 is lower for the final version of the model using all storms. This is deceiving since the degree of freedom is different.

To statistically test the validity of the MLR models, the F-test was applied (Draper and Smith, 1966). The F value is substantially higher for the preferred model which indicates better confidence in the validity of the model. From comparisons with published F distribution tables (Gibra, 1973), our confidence that the coefficients are significantly different from zero (i.e. that there is a relationship with land use) is better than 99 percent for copper and 95 percent for lead. However, the R^2 does tell us that less than 20 percent of the variability in the sample data is due to land use.

One additional advantage of using all the sample data rather than the station mean values is that the error estimates generated from the regression are the total error (Draper and Smith, 1966). This makes the computation of confidence intervals much easier.

Using mean values introduces an additional error about the mean for each station.

The model assumes that concentrations are uncorrelated with runoff volume. This assumption is supported by the results of the NURP study (EPA, 1983), which found no statistically significant relationship between storm volume and EMC.

Table 9.1: Summary of statistical parameters.

	All EMCs		Site Means	
Constituents	Copper	Lead	Copper	Lead
Variables	4	4	4	4
Degrees of Freedom	89	89	11	11
R Squared	0.17	0.12	0.43	0.29
F-Value	4.67	3.15	2.06	1.13
Confidence Level	>99%	>95%	>75%	<75%

9.4 Results

As mentioned previously, the model formulation which was most successful was that using all sample data in a single regression analysis. This section describes the results of that model.

A box plot of the computed concentration of the two metals in storm water is displayed in Figure 9.2. The bars represent the mean estimate and the 90 percent confidence interval of the mean concentration by land use. The confidence interval assumes that the errors from the regression line follow a log-t distribution with 89 degrees of freedom. A log transform is used since measurements of concentration are bounded by zero on the low end but unbounded at the high end. A t-distribution is assumed since the true variance about the mean is unknown and can only be estimated by the variance of the sample data. The t-distribution approaches the normal distribution as the degrees of freedom approaches infinity.

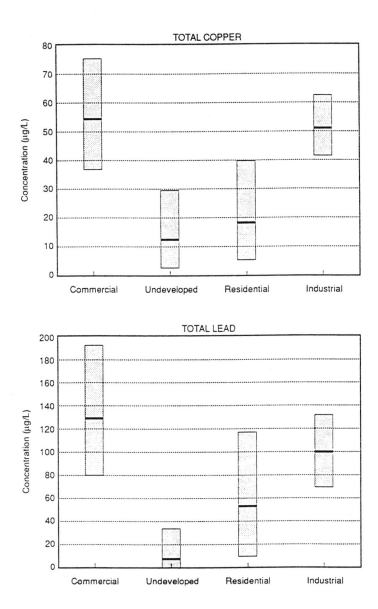

Figure 9.2:   Mean concentration and 90% confidence bounds
               by land use.

From the box plots we can see that there is considerable overlap between the confidence intervals for commercial and industrial land uses. This indicates that storm water concentration estimates for these two land uses are not significantly different. Statistically, there is more confidence in the differences between the mean values of open or residential areas versus the commercial and industrial areas.

Figure 9.3 provides a comparison of the results of the Alameda County model with NURP estimates and the results a study in Santa Clara Valley (WCC 1991a). The Alameda County model shows greater variation among land uses than the NURP or Santa Clara results. This is to be expected as the other studies were forced to assume that all sampling stations were homogeneous. The regression model needs no such assumption but rather can incorporate heterogeneous stations. Additionally, the other studies were unable to distinguish as many land use types. NURP did not include an industrial category and the Santa Clara report could not separate residential and commercial due to the intermingling of land use zones. All studies do mark a significant difference between urban and undeveloped land.

Table 9.2 provides a summary of the important parameters and the results of the model. We see that the proportion of runoff from different land use categories is quite different from their relative areas. Transportation was modeled in SWMM to compute runoff volumes, but combined with the commercial category to compute the water quality, as explained in the previous section. Figure 9.4 provides a graphical presentation of the load fractions. Although open areas are the largest of any land use, they produce less than 2 percent of the total pollutant load due to low runoff volume and pollutant concentration. Residential areas produce 63 percent of total runoff, but only 40 percent of total load for copper and 48 percent for lead. Transportation, commercial, and industrial areas combined represent 50 to 60 percent of the pollutant load due to high metals concentrations and high runoff volume, but represent only 20 percent of the urban land area. This indicates that targeting those areas for pollution control measures could present substantial cost savings over region wide measures, while still addressing the bulk of the problem.

Table 9.2: Summary of parameters and results of the model.

	Open	Residential	Commercial	Transportation	Industrial	Total
Area (square miles)	190	112	13	7	24	345
% Area	55	32	4	2	7	100
Median Year Flow (af/year)	3000	48000	8100	5900	10200	76000
% Flow	4	63	11	8	13	100
Concentration (ug/l)						
Copper	12	18	54	54	51	
Lead	8	54	129	129	99	
Estimated Load (kg/year)						
Copper	44	1070	540	390	640	2685
Lead	30	3200	1290	940	1250	6710

9.5 Conclusions

From the results of this study we can conclude that there is a statistically significant relation between storm water concentrations of certain metals and land use. While the relationship is verifiable, it explains less than 20 percent of the variability in sampling data. Pollutant concentration is related to many other factors which were not analyzed in this study, for example, storm characteristics. Additionally, the relationship is stronger when comparing open or residential areas versus commercial or industrial areas. Differences between commercial and industrial areas are less apparent.

The MLR method presented for formulating regional water quality parameters should be explored further. This method allows selection of a wide range of sampling locations and therefore allows a greater portion of the study area to be represented in the load model. The method also provides a more complete error estimate. The analysis can easily be performed using inexpensive commercial software. While runoff volumes for this study were computed using a hydrologic computer model, this is not necessary for the method.

Until the other parameters governing storm water pollution are

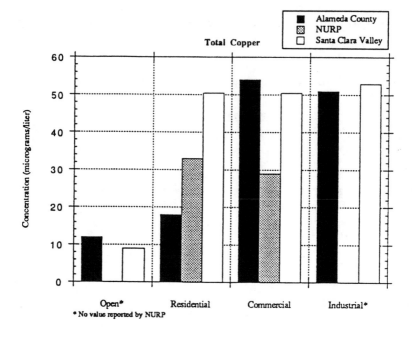

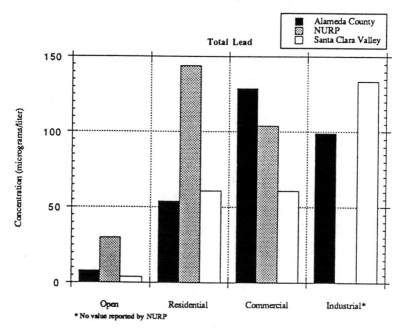

Figure 9.3: Comparison of concentration with other studies.

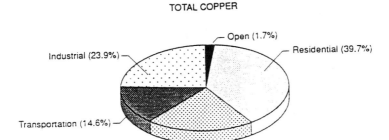

Figure 9.4: Pollutant load from study area by land use type.

better understood, the MLR approach provides a relatively cost-effective way of computing loads based on water quality data collected from homogeneous and heterogeneous drainage areas.

9.6 References

Draper, N.R., and Smith, H. (1966). Applied Regression Analysis. John Wiley and Sons, Inc., New York, New York.

Gibra, I.N. (1973). Probability and Statistical Inference for Scientists and Engineers. Prentice-Hall, Inc., Englewood Cliffs, New Jersey.

Gunther, Andrew (1991), The Loading of Toxic Contaminants to the San Francisco Bay-Delta in Urban Runoff. The Aquatic Habitat Institute, Richmond, California.

Mendenhall, W. (1987). Introduction to Probability and Statistics. Duxbury Press, Boston, Massachusetts.

Silverman, G.S., Stenstrom, M.K., and Fam, S. (1988). Land use Considerations in Reducing Oil and Grease in Urban Stormwater Runoff. J. Environmental Systems, Vol. 18(1):31 - 47.

United States Geological Survey, (1990). Techniques for Estimation of Storm-Runoff Loads, Volumes, and Selected Constituent Concentrations in Urban Watersheds in the United States. Water Supply Paper 2363, by Nancy E. Driver and Gary D. Tasker.

Woodward-Clyde Consultants, (Feb. 1991). Santa Clara Valley Nonpoint Source Study Volume 1: Loads Assessment Report. Submitted to Santa Clara Valley Water District.

Woodward-Clyde Consultants, (May 1991). Eugene Storm Drainage Water Quality Program, Loadings Report. Prepared for the City of Eugene, Oregon, Department of Public Works, Engineering Division.

Woodward-Clyde Consultants, (Oct. 1991). Alameda County Urban Runoff Clean Water Program Loads Assessment. Submitted to Alameda County Flood Control and Water Conservation District.

Chapter 10

Calibration of PCSWMM to Estimate Metals, PCBs and HCB in CSOs from an Industrial Sewershed

K.N. Irvine, B.G. Loganathan, E.J. Pratt, H.C. Sikka
Center for Environmental Research and Education
State University College at Buffalo
1300 Elmwood Ave., Buffalo, NY 14222

10.1 Introduction

The Buffalo River, New York, is one of forty-three "Areas of Concern" identified by the International Joint Commission as exhibiting environmental impairment. The impairments identified for the Buffalo River in a level I Remedial Action Plan (New York State Department of Environmental Conservation (NYSDEC), 1989) included: restrictions on fish and wildlife consumption; fish tumors and other deformities; degradation of benthos; restrictions on disposal of dredged sediment; and loss of fish and wildlife habitat. In addition, it was noted that degradation of fish and wildlife populations and bird or animal deformities and/or reproduction problems were likely. The NYSDEC (1989) identified multiple possible pollutant sources to the Buffalo River, including: combined sewer overflows; direct industrial discharges; leaching from inactive hazardous waste sites; water column interaction with historically contaminated bed

0-87371-898-4/93/$0.00 + $.50
© 1993 by Lewis Publishers

sediment; and upstream point and nonpoint sources such as municipal wastewater treatment plants and agricultural runoff.

A mass balance evaluation of pollutant level dynamics and loadings within the Buffalo River Area of Concern (AOC) has been initiated under the U.S. Environmental Protection Agency's ARCS (Assessment and Remediation of Contaminated Sediment) Program. An objective of the mass balance evaluation is to apply a combination of hydrodynamic, sediment transport and chemical fate models to estimate pollutant movement through the lower Buffalo River (Wang and Martin, 1991). The modelling results will be used to guide the selection of remediation strategies for the river. The ability of the chemical fate model to accurately reflect pollutant dynamics within the river is predicated, in part, on reliable estimates of pollutant loads to the river from the various sources.

Combined sewer overflows (CSOs) are a possible source of organic compounds, metals and bacteria to the Buffalo River and this chapter communicates initial results of an ongoing CSO evaluation. A personal computer (PC) version of the Stormwater Management Model (PCSWMM) is being used for planning level estimates of overflow quantity to the river. Model calibration for overflow quantity estimates and characterization of overflow quality from a major outfall servicing a large, industrialized sewershed are discussed in this chapter. Model sensitivity, the range of calibrated parametric values, model inaccuracies and pollutant level characterization also are discussed in terms of using modelled results for planning level evaluations of CSO impacts. Ultimately, the CSO pollutant loading estimates will be incorporated into the mass balance evaluation of pollutant level dynamics for the Buffalo River.

Levels of total PCBs, HCB, Mn, V and Cu are examined in this chapter. Organochlorine compounds such as PCBs and HCB are of concern as widespread, persistent and harmful contaminants in view of environmental quality and health (Safe, 1987; Tanabe, 1988; Loganathan et al., 1990; Loganathan and Kannan, 1991). The metals, Mn, V and Cu were chosen for evaluation because elevated levels may have a negative impact on animal, fish and human health (e.g. Irvine et al., in press; Forstner and Wittmann, 1983), data are available from past and ongoing Buffalo River studies (e.g. NYSDEC, 1989) and because of limitations on the types of elements that can be determined by short-lived instrumental neutron activation analysis, the analytical methodology available for this study.

10.2 Study Area

10.2.1 The Buffalo River

The Buffalo River is located in western New York State and drains an area of 1244 km^2, primarily to the south and east of the city of Buffalo. Landuse within the watershed varies. The upper portion of the watershed is characterized primarily by woods and farmland. However, tributaries to the river also pass through several small communities, receiving both industrial and municipal discharges.

The Buffalo River AOC spatially extends from the mouth of the Buffalo River to the point upstream at which backwater effects during Lake Erie's highest monthly average level does not impact river flow (Figure 10.1). Historically, the AOC was heavily industrialized, with activities such as: steel production; coking operations; oil refining; chemical and dye production; and flour milling. Industrial activity has declined along the river in the last decade and steel production, coking operations and oil refining have ceased. The AOC below highway 62 (Figure 10.1) is a navigable channel and is maintained at a depth of approximately 7 meters by the Buffalo District Army Corps of Engineers.

Historical (1940-1985) monthly mean inflows into the upper end of the AOC range between 45 m^{3}s^{-1} in March and 3.3 m^{3}s^{-1} in July (Meredith and Rumer, 1987). The AOC can exhibit esturine-like characteristics during lowflow periods when levels at the eastern end of Lake Erie increase due to higher velocity, south-westerly winds. Flow reversals and thermal stratification

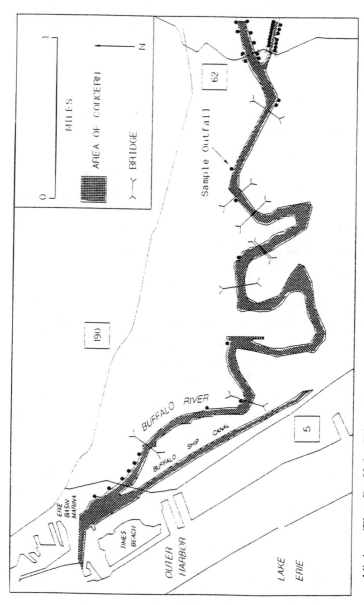

Figure 10.1:   The Buffalo River Area of Concern and location of combined sewer outfalls (after NYSDEC, 1989). "Sample Outfall" indicates the location of the Babcock St. sewershed outfall.

between lake and river water have been observed for several kilometers upstream during these times of wind setup.

The Buffalo River Improvement Corporation (BRIC) augments flows within the AOC through pumping operations that transfer water from Lake Erie to industries along the river. The industries discharge the lake water to the river after using it for cooling and other processing purposes. The BRIC flow augmentation has diluted pollutant concentrations within the river and decreased residence times (Sauer, 1979). Prior to the industrial decline, the BRIC often contributed 90% of the total river flow during the drier summer months (Sauer, 1979). Pumping rates from the lake have declined along with industrial activity, averaging around 0.7 m^3s^{-1} in recent times (J. Dietz, BRIC, pers. comm.). A more detailed discussion of the physical, industrial and cultural characteristics within the Buffalo River AOC is provided by Irvine et al. (1991).

10.2.2 Sewer System and Sample Sewershed Characteristics

The city of Buffalo is serviced primarily by a combined sewer system and there are a total of 39 combined sewer outfalls to the Buffalo River AOC (Figure 10.1). The outfalls drain various landuses, including: industrial; commercial; residential; and railway yards. Industries are granted permits to discharge waste and processing effluents to the sewer system. In the absence of overflows, these effluents are treated at the Bird Island Sewage Treatment Plant.

This chapter focuses on the detailed assessment of overflow quantity and quality associated with one of the outfalls in Figure 10.1. Sampling and detailed modelling are planned for additional outfall points to the river, but the Babcock St. sewershed was chosen for initial study because of several factors. First, the sewershed is large (contributing area of approximately 256 ha) and the outfall represents one of the major overflow points from the entire city (Calocerinos and Spina, 1989). The sewershed is industrialized and represents a high potential source of various types of pollutants. The sewer system for the sewershed was easily defined and the overflow was easily instrumented in a secure location. Finally, rainfall data for this area were available from the Buffalo Sewer Authority (BSA).

Landuse within the Babcock St. sewershed primarily is industrial (20% heavy and 67% light industry) with smaller percentages of residential (11%) and commercial (2%) zones. There are a variety of industries located in the sewershed, including those related to food processing, automobiles and automobile recycling, metal workings, china and chemical production. The industrial properties often include large, open pervious areas. The Erie-Lackawana rail lines occupy the central portion of the sewershed. The BSA estimates that the average industrial waste flow in the Babcock St. sewershed is 2,419,152 liters per day. Residential structures are a mix of primarily double and multi-family dwellings of a relatively high density. High-rise apartment structures are not present. The BSA estimates that the average residential waste flow in the Babcock St. sewershed is 1,355,682 liters per day.

As shown in the sewershed schematic (Figure 10.2) a 1.8 meter diameter brick conduit carries flow from the sewershed into the primary overflow chamber. The overflow weir is 4.9 meters long and 1.7 meters in height. Interevent sanitary flow enters the chamber and is deflected to the right by the weir (Figure 10.2). The flow leaves the chamber by a 0.61 meter diameter pipe. Excess combined flow during storms is routed over the weir and through a 1.8 meter diameter pipe to the Buffalo River.

10.3 Methodology

10.3.1 Field Sampling

Overflow rates were recorded at 5-minute time steps using a Montedoro-Whitney System Q flow device. The System Q consists of a datalogger connected to a probe that measures average (directional) flow velocity and flow depth. The flow

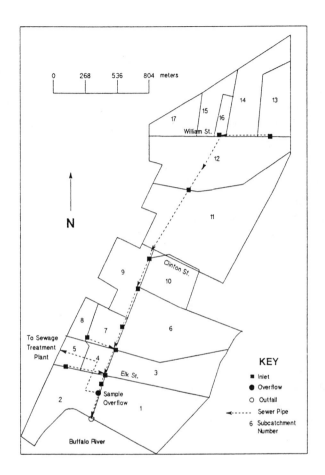

Figure 10.2: Schematic of the Babcock St. sewershed. Subcatchment areas and the sewer system used in the model are shown.

probe was fixed to the top of the overflow weir in the primary overflow chamber. All data was uploaded from the datalogger to a laptop computer and was used to calculate overflow rates. The System Q was connected to a Sigma Streamline Model 701 Discrete Pump Sampler that was used to obtain samples for overflow quality analyses. The System Q communicates with the pump sampler to provide a flow-proportioned, composite sample for each overflow event. The sampling system was programmed to take 250 ml samples at 250 m^3 overflow intervals. The intake for the pump sampler was fixed at a level several centimeters in front of, and below the crest of the weir. The system was installed in the primary overflow chamber for two field seasons. The first field season began June 25, 1990 and ended December 4, 1990. The second field season began April 12, 1991 and ended November 4, 1991.

Rainfall intensity data at 5-minute intervals was obtained from the BSA raingauge located on the roof of the South Buffalo Pump Station, approximately 900 meters east of the outfall. The raingauge is a tipping-bucket type with a sensitivity of 0.25 mm of rain.

10.3.2 Analytical

Levels of Mn, Cu and V were determined using short-lived instrumental neutron activation analysis (INAA) at the McMaster Nuclear Reactor, Hamilton, Ontario. The overflow samples were wet-filtered using Millipore filters (minimum pore size of 0.45 microns) to separately determine metals levels associated with the particulate and dissolved phases. The mass of filtered sediment used in the analyses ranged between 0.001 and 0.155 gm, while 5 ml of filtrate was analyzed. No pre or post irradiation chemistry is required. Irradiation, delay and count times, details of the reactor flux and characteristics of the detector used for the analyses are provided in Vermette et al. (1987) and Irvine (1989). The efficacy of INAA for the analysis of these types of samples is discussed by Irvine et al. (in press). In general, INAA results for water and sediment samples are comparable to more frequently used analytical methods such as atomic absorption and inductively coupled plasma.

Total PCBs and HCB also were analysed in the dissolved and particulate phase of the CSO samples following the procedures described in the Federal Register (1984) and Loganathan et al. (1990), with some modifications. Approximately 1 litre of sample was filtered through 0.45 micron Whatmann Glass Fibre Filters using Millipore filter systems to separate the dissolved and particulate phases. The dissolved phase was extracted thrice (60 ml each time) using methylene chloride in a separatory funnel. The extract was passed through anhydrous sodium sulfate and collected in an Erhlenmeyer flask. The volume of the extract was brought down to 10 ml using K-D concentrator and transferred to hexane. The sample extract was treated with 5% fuming H_2SO_4 in concentrated sulfuric acid and washed with hexane-washed water. The extract was concentrated to 1 ml and an aliquot of this extract was injected into a gas chromatograph.

The particulate phase was soxhlet extracted for 16 hours using methylene chloride. The methylene chloride extract was K-D concentrated and transferred to hexane. The extract was subjected to silica gel column chromatography for separation of PCBs from pesticides. The fraction containing PCBs and HCB were K-D concentrated and the extract was treated with 5% fuming H_2SO_4 in concentrated sulfuric acid and washed with hexane-washed water. The extract was concentrated to 1 ml using nitrogen gas and injected into a gas chromatograph.

A gas chromatograph (Varian model GC-3400) equipped with a ^{63}Ni electron capture detector and automatic sampler (Varian model 8100) was used for quantification of PCBs and HCB. A capillary column, DB-5 (J&W Scientific) having dimensions of 30m x 0.25 mm i.d. and 0.25 micron film thickness was used. The temperature was programmed from 160° to 230°C at a rate of 2°C min^{-1} with initial and final hold times of 10 minutes, respectively. The injector and detector temperatures were kept at 250°C and 300°C, respectively. Nitrogen was used both as carrier and makeup gas. HCB and PCBs were quantified using individually resolved peaks with corresponding standard peaks. PCBs standard containing 1:1:1:1 mixture of 1242,1248,1254 and 1260 was used for PCBs quantification.

The dissolved, particulate phases and blanks were spiked with surrogate standards (2,4,5,6-tetrachloro-*m*-xylene and 2,2',3,4,4',5-,6,6'-octachlorobiphenyl) in order to check the efficiency of the analytical procedure. Hexane-washed distilled water was used as blank. Recoveries of the surrogate standard were 100 +/- 20%. Analytical results were not corrected for the surrogate standard recoveries.

10.3.3 Modelling

The RUNOFF and TRANSPORT blocks of PCSWMM, (version 3.2c, Computational Hydraulics Inc.) were used in all model runs. It was decided to use TRANSPORT in this study because surcharging does not frequently occur in the Babcock St. sewershed and the block also provides the ability to route water quality parameters, which cannot be done with EXTRAN. Future research will consider pollutant scour and deposition within the sewer network.

It was determined through preliminary modelling of the inflows to and outflows from the overflow chamber that flows exceeding 0.32 m^3s^{-1} would discharge to the Buffalo River. Therefore, rather than providing a detailed representation of the overflow weir hydraulics, any modelled flows greater than 0.32 m^3s^{-1} were considered input to the Buffalo River. The sewershed discretization and general system configuration used in the model are shown in Figure 10.2.

10.4 Results

10.4.1 Model Calibration

A total of 8 overflow events were observed during the 1990 field season and 9 overflow events were observed during the 1991 field season. An overflow "event" arbitrarily was defined as a period of overflow separated from other overflows by a minimum of 2 hours. Overflow event characteristics are summarized in Table 10.1 and corresponding rainfall characteristics are summarized in Table 10.2. The BSA had removed all raingauges prior to the December 3, 1990 event and rainfall data therefore are not included for this event in Table 10.2. It was observed that this event was generated by a rain-on-snow process. The rainfall records for the May 29, July 8 and September 26, 1991 events have not been processed to date.

The model has been calibrated for 8 events, as denoted in Table 10.1. These events were selected for calibration because a full rainfall record was available; a range of event characteristics (overflow volume, peak overflow rate, overflow duration) was represented; and in several cases samples of overflow quality had been obtained.

The calibration events exhibiting the best and worst fit to the observed data are shown in Figure 10.3. The accuracy of fit for the peak overflow rate and event volume for all calibrated events is summarized in Figure 10.4. It was decided that for planning level estimates of pollutant loadings, an accurate representation of event volume was most important. Calibration therefore focused on minimizing the error of estimate for event volume and this is reflected in the nearly linear 1:1 plot in Figure 10.4a. The average (absolute) prediction error of overflow volume for the 8 events was 4% and the average (absolute) prediction error of peak overflow rate was 24%. The calibration results for the Babcock St. sewershed compare favorably with results reported in the literature (e.g. Warwick and Tadepalli, 1991; Nix et al., 1991; Zaghloul and Al-Shurbaji, 1990).

Three parameters provided the greatest flexibility in model calibration for the sewershed: subcatchment surface slope; subcatchment width; and per cent imperviousness. The range of values for these three parameters from the calibration runs is presented in Table 10.3. Subcatchment width, W, is defined as:

$$W = A/l \qquad (10.1)$$

where A is the total subcatchment area and l is the subcatchment length. Subcatchment width is a conceptual parameter that is difficult to measure precisely and therefore can be used to calibrate hydrograph shape (James and Robinson, 1986; Nix et al.,1991). Warwick and Tadepalli (1991) noted that although

Table 10.1: Overflow Characteristics

Event Date	Event Volume m³	Peak Overflow Rate m³s⁻¹	Event Duration hr
Jul 9, 1990*	3021	0.458	2.6
Aug 13, 1990*	5245	0.471	4.8
Aug 28, 1990	933	0.298	1.6
Sep 5, 1990* event "a"	1020	0.393	1.3
Sep 5, 1990* event "b"	1716	0.429	1.75
Sep 7, 1990*	1306	0.349	1.6
Oct 11, 1990*	815	0.327	1.2
Dec 3, 1990	11817	0.671	10.5
Apr 19, 1991	1667	0.490	1.4
Apr 20, 1991	3218	0.507	2.4
Apr 21, 1991	3417	0.456	3.2
Apr 22, 1991	10554	0.472	9.2
May 29, 1991	0.018	0.00005	0.25
Jul 4, 1991*	4060	1.11	2.8
Jul 7, 1991*	597	0.224	1
Jul 8, 1991	32	0.066	0.3
Sep 26, 1991	578	0.174	1.5

* model calibration event

Table 10.2: Rainfall Characteristics

Event Date	Event Depth mm	Peak 5 min Intensity mm hr^{-1}	Peak 15 min Intensity mm hr^{-1}	Event Duration hr
Jul 9, 1990	24.6	48.8	38.6	3
Aug 13, 1990	35.3	51.8	21.3	11.6
Aug 28, 1990	23.1	61.0	46.8	2.6
Sep 5, 1990 Event "a"	23.9	94.5	44.7	1
Sep 5, 1990 Event "b"	8.9	61.0	28.5	0.75
Sep 7, 1990	16.0	24.4	19.3	3.2
Oct 11, 1990	17.5	6.1	5.1	9.4
Dec 3, 1990	NA[*]	NA	NA	NA
Apr 19, 1991	15.2	9.1	9.1	3.25
Apr 20, 1991	10.0	6.1	5.1	4.1
Apr 21, 1991	10.1	3.0	3.0	4.1
Apr 22, 1991	16.7	3.0	3.0	10.0
May 29, 1991	NA	NA	NA	NA
Jul 4, 1991	46.8	61.0	45.7	2.2
Jul 7, 1991	14.1	42.7	26.4	2.2
Jul 8, 1991	NA	NA	NA	NA
Sep 26, 1991	NA	NA	NA	NA

[*] data not available

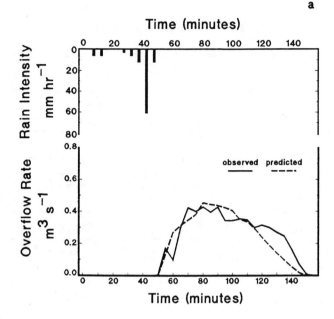

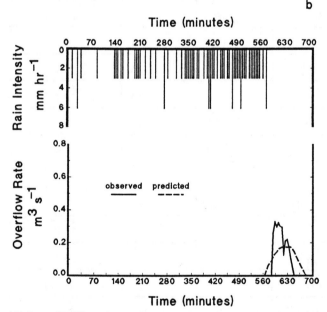

Figure 10.3:   Calibration results for the best fit event (a)
               occurring on Sept. 5, 1990 (event "b"); and the
               worst fit event (b) occurring Oct. 11, 1990.

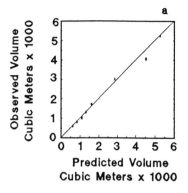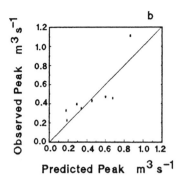

Figure 10.4: Calibration results for 8 events: event volume
(a); and event peak overflow rate (b).

from aerial photographs, not all impervious areas are directly
absolute per cent imperviousness can be accurately measured
connected in a hydraulic sense. Therefore, the effective per cent
imperviousness could be used as a calibration parameter. It might
be expected that the effective per cent imperviousness would
vary, depending on storm characteristics and relative contributions
from different parts of the sewershed. A similar conclusion might
be drawn for the "lumped" approach to representing subcatchment
surface slope.

The range of calibrated values for subcatchment width, slope
and per cent imperviousness suggests that there will be some
uncertainty when using the model to estimate long term (e.g.
annual) pollutant loadings. A quantitative analysis of the relation
ships between parametric values, antecedent conditions and storm
characteristics has not yet been accomplished. Qualitatively, it
appears that parameter values should be selected to maximize
runoff for long duration, low intensity rainfall events (e.g. August
11, 1990; Figure 10.3) and for short duration events that
experience wet antecedent conditions (e.g. September 5, 1990,
event "b"; Figure 10.3). Parameter values should be selected to
minimize runoff for rainfall events having extended periods
(e.g.15 minutes or more) of high intensity (greater than

Table 10.3:   Parametric values, subcatchment width, slope and
              percent imperviousness.

Subcatchment	Area (ha)	Width (m)	Slope (m/m)	Per Cent Impervious
1	14.1	115-450	.0005-.0017	10-40
2	9.1	75-320	.0005-.0017	10-40
3	13.5	75-340	.0005-.0017	10-30
4	8.8	75-315	.0004-.0016	10-30
5	5.4	75-350	.0005-.0017	10-30
6	24.5	130-550	.0005-.0017	10-30
7	8.6	75-345	.0005-.0019	10-30
8	4.4	75-300	.0005-.0019	10-30
9	24.8	100-450	.0004-.0016	10-23
10	23.9	100-475	.0004-.0016	10-30
11	35.0	105-575	.0004-.0018	5-12
12	23.9	120-590	.0004-.0018	10-50
13	16.2	105-515	.0005-.0019	10-40
14	19.2	105-525	.0005-.0021	5-12
15	6.0	75-300	.0005-.0021	5-12
16	3.4	75-275	.0005-.0021	10-45
17	15.0	120-560	.0004-.0021	10-50

approximately 40 mm hr^{-1}). A continuous modelling approach
also could be used to eliminate the need to specify antecedent
conditions. This type of approach is discussed, for example, by

Robinson and James (1981). Calibration procedures for continuous modelling may be different than event-based calibration.

Typical values for other parameters used in the model calibration are presented in Table 10.4. Values for these parameters initially were selected using the PCSWMM RUNOFF Manual (James and Robinson, 1986) and Calocerinos and Spina (1989) as reference. The Manning's n values appear towards the high end of those typically used, but this is reasonable for the Babcock St. sewershed. The pervious surfaces in the sewershed can be rugged, abandoned industrial land and railway yards. Many of the impervious surfaces are in poor repair, thereby increasing surface roughness. These same surface characteristics also would produce greater depression storage.

Although model results, in general, were reasonable, it appears that some model inaccuracy may result from the simplifying assumptions used to estimate runoff from pervious surfaces. The practice of identifying single minimum and maximum infiltration rates for each subcatchment should be re-examined in future modelling efforts. Ando et al., (1986) found large variations in the infiltration characteristics of urban areas, depending on landuse. The large proportion of pervious land in the south Buffalo area suggests that a closer examination of infiltration characteristics would be appropriate.

10.4.2 Model Sensitivity

The sensitivity of model results to the three parameters that provided the greatest flexibility in calibration was examined using the rainfall input of the September 7, 1990 storm event. The actual calibrated model results are shown in Figure 10.5. This event was chosen for a sensitivity analysis because it had a well defined rainfall and runoff period and antecedent dry period; the calibration results were reasonable; and the event was of moderate size. Subcatchment width, surface slope and per cent imperviousness were varied within the observed range of calibration values that had been determined from the eight calibration runs. The parametric values (Table 10.3) that would produce the minimum and maximum flow from each subcatchment were taken as the extremes for the sensitivity analysis. The value for each parameter subsequently was increased at intervals of 25% of the range between the minimum and maximum values. The other parametric values were held constant at the previously calibrated level.

A total of fifteen sensitivity runs were done (five runs for each of the three parameters) and the results are presented in Figure 10.6. For this example, the model appears least sensitive to changes in slope. Sensitivity to changes in subcatchment width and per cent imperviousness varies over the range of the sample runs. The difference between the maximum and minimum parametric values is greater for the subcatchment width than for per cent imperviousness and in general it appears that the model is more sensitive to changes in per cent imperviousness.

10.4.3 Analytical Results

Analytical results currently are available for three overflow "periods". The first overflow period represents the overflow events that occurred on August 28; September 5 and September 7, 1990. The four events were composited in one sample since the birth of Gordon William Irvine (September 3, 1990; 7 lbs. 4 oz.) precluded a field visit to collect samples and reset the system between events. The second overflow period represents four overflow events that occurred between April 20 and April 22, 1991. The third overflow period represents the event of July 4, 1991. The analytical results for various metals, HCB and total PCBs are presented in Tables 10.5 and 10.6. Results from other studies are presented in Table 10.7 for comparison purposes. The results available in the literature often do not report levels for the solid and dissolved phases separately.

Table 10.4:   Typical parametric values, Babcock St. Sewershed
              calibration.

Parameter	Value
Surface Manning's n impervious pervious	0.022-0.025 0.32-0.33
Conduit Manning's n concrete brick	.017 .025
Depression Storage (mm) impervious pervious	2.5-4.5; 4.0 was typical 8.0-9.9; 9.9 was typical
Per Cent Zero Detention	5-25; 5 was typical
Horton Infiltration maximum rate (mm hr^{-1}) minimum rate (mm hr^{-1}) decay coefficient (sec^{-1})	76.2 13.2 .01330-.03330

The NYSDEC target limit on total PCBs levels in water is 1 ng l^{-1}, which is based on the standard to protect wildlife from the toxic effects of eating contaminated fish (NYSDEC, 1989). The total PCBs levels in the Babcock St. overflows therefore are of some concern. The BSA does not permit discharge of PCBs to the sewer system and to date the source(s) of the total PCBs have not been determined, although research in this area is planned.

Levels of metals, total PCBs and HCB typically are lower and less variable in the dissolved phase than the particulate phase. Although the analytical results represent a minimum of four events and a maximum of nine events, more overflow samples should be obtained to thoroughly evaluate pollutant level dynamics and more accurately quantify level variability. This information is essential in evaluating the uncertainty related to long-term pollutant loading estimates. For example, long-term

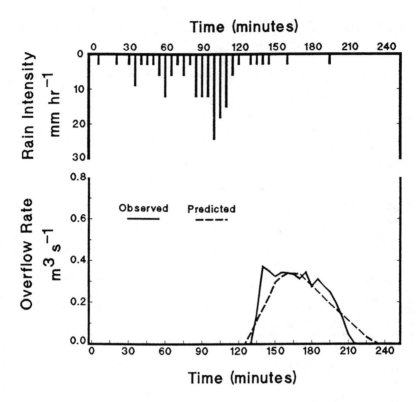

Figure 10.5:   Calibration results for overflow event of
              September 7, 1990.

Table 10.5: Metal levels from sampled overflows.

	Particulate Phase, ppm			Dissolved Phase, ppm		
Period	Mn	V	Cu	Mn	V	Cu
Sep, 1990	448	86	414	.750	.0025	.131
Jul 4, 1991	1229	100	478	ND*	ND	ND

* not determined for the event

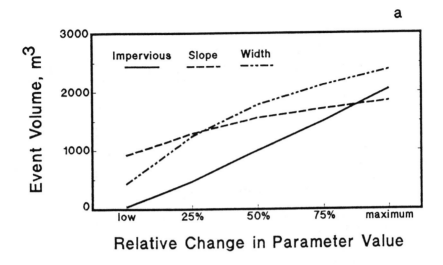

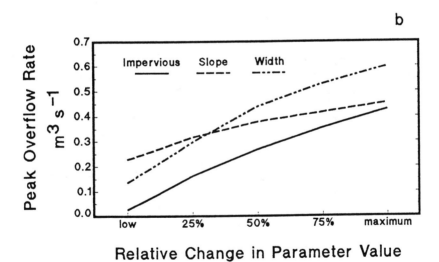

Figure 10.6:   Sensitivity analysis for the Babcock St.
               sewershed model using the September 1990
               rainfall data.

Table 10.6: Total PCBs and HCB levels from sampled overflows.

	Particulate Phase, ng l^{-1}		Dissolved Phase, ng l^{-1}	
Period	Total PCBs	HCB	Total PCBs	HCB
Sep, 1990	400	6	30	BDL[*]
Apr, 1991	ND[**]	ND	20	.045
Jul 4, 1991	970	16	38	.98

[*] below instrument detection limit of 3 ng ml^{-1}
[**] not determined for this event

Table 10.7: Comparison total PCBs, HCB and metals levels.

Parameter	Level	Comments	Reference
total PCBs[*]	463 ng l^{-1}	mean level, 3 events, separate storm system	Granier et al., 1990
total PCBs[*]	27-179 ng l^{-1}	samples from combined systems, 3 Canadian cities	Marsalek and Ng, 1989
HCB[*]	0.46-98 ng l^{-1}	samples from combined systems, 3 Canadian cities	Marsalek and Ng, 1989
V[**]	0.00393 ppm 70 ppm	dissolved phase particulate phase	Droppo, 1987
Mn[**]	0.209 ppm 424 ppm	dissolved phase particulate phase	Droppo, 1987

[*] particulate and dissolved phase not separately analysed
[**] results from one CSO event, Hamilton, Ontario.

loadings to the Buffalo River from the CSOs currently are being estimated as:

$$Lp_j = (sum\ V_i)Cp_j \qquad (10.2)$$

where Lp_j is the loading of pollutant, j; V_i is the overflow volume for each event in the time period under consideration; and Cp_j is a "representative" level for pollutant, j. Equation (10.2) is calculated separately for both the dissolved and particulate phases. A "representative" level could be an average value from several events for each site. One standard deviation around the average could be used as a simple measure of uncertainty for Cp_j. Alternatively, a range for the loading estimate could be calculated using the minimum and maximum observed levels for pollutant, j. To illustrate this latter approach, the 1991 overflow volume results (observed) were used with the available total PCBs data (Table 10.6) and the results are presented in Table 10.8.

Table 10.8: Total PCBs loadings, 1991 field season.

Phase	Loading, gm	
	Minimum Cp_j	Maximum Cp_j
Dissolved	0.5	0.9
Particulate	9.6	23.4

The results in Table 10.8 do not consider uncertainty that would be introduced by modelling overflow volume, since the observed volumes were used in the calculations. However, if loading estimates are to be made for years in which overflow volume was not monitored, model estimate uncertainty will need to be considered.

10.5 Conclusions

The calibration procedure and results were discussed for the application of PCSWMM to an industrialized sewershed that overflows to the Buffalo River, New York. The average (absolute) prediction error of overflow volume for eight calibration events was 4% and the average (absolute) prediction error of peak overflow rate was 24%. These calibration results compare favorably with results reported in the literature. Three parameters had the greatest flexibility in model calibration for the sample sewershed: subcatchment surface slope; subcatchment width and per cent imperviousness. Sensitivity analysis indicated that, in general, the model for the sewershed was least sensitive to changes in surface slope and most sensitive to changes in percent imperviousness.

The levels of total PCBs, HCB, Mn, V and Cu associated with the particulate and dissolved phases of several sampled overflows were reported. The total PCBs levels are of some concern, as they exceed state-targeted guidelines for water quality. Future research in the U.S. should concentrate on determining pollutant levels separately for the particulate and dissolved phases because proposed amendments to the Clean Water Act explicitly consider sediment quality in regulations.

An approach to estimate long term pollutant loadings to the Buffalo River from the sampled sewershed was discussed. Overflow volumes from the 1991 field season and data on total PCBs levels were used to illustrate the estimation approach. A total of eight years of rainfall data are available for the south Buffalo area. The calibrated model will be used with these data to estimate overflow volumes from the sewershed on a longer term than presented in this chapter. Additional overflow samples will be taken in 1992 for quality analysis. These additional samples will be used to evaluate the uncertainty of the "representative" pollutant levels in the calculation of long term pollutant loadings. The loading estimate approach also will be applied to other sewersheds contributing to the Buffalo River as a first step in identifying appropriate CSO remedial measures.

10.6 Acknowledgements

Funding for this research partially was provided by the U.S. EPA, Great Lakes National Program Office (grant X99502401) and from the Research Foundation of SUNY. Several individuals assisted in the collection of field data and provision of background data for this study, including Mr. Dennis Torok, SUNY College at Buffalo; Mr. Greg McCorkhill, P.E., Calocerinos and Spina; Mr. Jim Caruso, Mr. Frank DiMascio, P.E. and other personnel at the Buffalo Sewer Authority.

10.7 References

Ando, Y., Takahasi, Y., Izumi, K. and Kanao, K. (1986). Urban flood runoff modelling considering infiltration of various land uses. Proceedings of the International Symposium on Comparison of Urban Drainage Models with Real Catchment Data, Dubrovnik, Yugoslavia, Pergamon Press, pp. 229-238.

Calocerinos and Spina, (1989). Buffalo Combined Sewer Overflow Phase II Study. State Project C-36-1044-01-3.

Droppo, I.G. (1987). The Effectiveness of a Stormwater Detention Pond in Enhancing Water Quality. unpubl. B.A. Thesis, McMaster University, Hamilton, Ont., 70p.

Federal Register (1984). Method 608. Organochlorine Pesticides and PCBs. 49:89-104.

Forstner, U. and Wittmann, G.T.W. (1983). Metal Pollution in the Aquatic Environment, Second Edition. Springer-Verlag, Berlin, 345p.

Granier, L., Chevreuil, M., Carru, A.M. and Letolle, R. (1990). Urban runoff pollution by organochlorines (polychlorinated biphenyls and lindane) and heavy metals (lead, zinc and chromium). *Chemosphere*, 21(9):1101-1107.

Irvine, K.N. (1989). The Effect of Pervious Land on Pollutant Movement Through an Urban Environment. unpubl. Ph.D. Thesis, McMaster University, Hamilton, Ont., 211p.

Irvine, K.N., Stein, G.P. and Singer, J.K. (1991). An Environmental Guidebook to the Buffalo River. Center for Environmental Research and Education, SUNY College at Buffalo, 30p.

Irvine, K.N., Murray, S.D., Drake, J.J. and Vermette, S.J. (1989). Spatial and temporal variability of dry dustfall and associated trace elements: Hamilton, Canada. *Environmental Technology Letters*, 10:527-540.

Irvine, K.N., Drake, J.J. and James, W. (in press). Particulate trace element concentrations and loadings associated with erosion of pervious urban land. *Environmental Technology Letters*.

James, W. and Robinson, M. (1986). User Manual PCSWMM3 RUNOFF Module, 3rd ed. CHI Report No. R143, 196p.

Loganathan, B.G., Tanabe, S., Tanaka, H., Watanabe, S., Miyazaki, N., Amano, M. and Tatsukawa, R. (1990). Comparison of organochlorine residue levels in the striped dolphin from Western North Pacific, 1978-79 and 1986. *Mar. Pollut. Bull.*, 21:435-439.

Loganathan, B.G. and Kannan, K. (1991). Time perspectives of organochlorine contamination in the global environment. *Mar. Pollut. Bull.*, 22:582-584.

Marsalek, J. and Ng, H.Y.F. (1989). Evaluation of pollution loadings from urban nonpoint sources: Methodology and applications. *J. Great Lakes Res.*, 15(3):444-451.

Meredith, D.D. and Rumer, R.R. (1987). Sediment Dynamics in the Buffalo River. Report, Dept. of Civil Engineering, SUNY at Buffalo, 171p.

NYSDEC, (1989). Buffalo River Remedial Action Plan.

Nix, S.J., Moffa, P.E. and Davis, D.P. (1991). The practice of combined sewer system modeling. *Water Resources Bulletin*, 27(2):189-200.

Robinson, M. and James, W. (1981). Continuous SWMM Quality Modelling for the City of Hamilton Using Atmospheric Environment Service Data. Proceedings Stormwater and Water Quality Management Modeling and SWMM Users Group Meeting, Sept. 28-29, 1981, Niagara Falls, Ont., McMaster University, Hamilton, pp. 469-491.

Safe, S. (1987). PCBs and human health. Environmental Toxin Series, Volume 1, Springer-Verlag, Berlin, pp. 133-145.

Sauer, D.E. (1979). An Environmental History of the Buffalo River. Buffalo Color Corporation, 88p.

Tanabe, S. (1988). PCB problems in the future: Foresight from current knowledge. *Environ. Pollut.*, 50:5-29.

Vermette, S.J., Irvine, K.N. and Drake, J.J. (1987). Elemental and size distribution characteristics of urban sediments: Hamilton, Canada. *Environmental Technology Letters*, 8:619-634.

Wang, P.F. and Martin, J.L. (1991). Temperature and conductivity modeling for the Buffalo River. *J. Great Lakes Res.*, 17(4):495-503.

Warwick, J.J. and Tadepalli, P. (1991). Efficacy of SWMM Application. *Journal of Water Resources Planning and Management*, 117(3):352-366.

Zaghloul, N.A. and Al-Shurbaji, A.M. (1990). A storm water management model for urban areas in Kuwait. *Water Resources Bulletin*, 26(4):563-575.

Chapter 11

Microcomputer-Based RTC of CSOs in an Industrialized City

Mark Stirrup and Mark Robinson
Regional Municipality of Hamilton-Wentworth
71 Main Street West, Hamilton, Ontario L8N 3T4

The City of Hamilton is still largely served by combined sewers. During dry weather, and small storm events, combined sewage is conveyed to the Region's Woodward Avenue STP. During larger rainfall events, specifically high intensity thunderstorms, the combined discharge of stormwater runoff and sanitary wastewater may exceed the capacity of the sanitary interceptors and STP. Automatic regulators installed at various points in the system limit flow to the interceptors resulting in CSOs. The advent of inexpensive, computationally powerful microcomputers has introduced the possibility of economically controlling these regulators in real-time to minimize CSOs. While a small number of applications of this technology exist elsewhere, this approach has not been sufficiently tested in Ontario.

In 1989 the Regional Municipality of Hamilton-Wentworth initiated a demonstration project to design, install and evaluate a microcomputer-based real-time control system for managing CSOs.

0-87371-898-4/93/$0.00 + $.50
© 1993 by Lewis Publishers

243

This paper reviews the following components of the Royal Avenue RTC Demonstration Project:

1. the design and installation of a real-time rainfall and flow monitoring network and telemetry system;

2. the design and installation of a microcomputer controller for operating the automatic sluice gate in the Royal Avenue CSO Regulator;

3. the estimation of CSO frequencies and volumes at Royal Avenue (with and without RTC) via continuous modelling, using the SWMM RUNOFF and TRANSPORT modules; and

4. the development of software for real-time data acquisition, database management and processing, and RTC of the Royal Avenue CSO regulator.

11.1 Introduction

Hamilton is one of the largest cities in Canada which is still predominantly serviced by combined sewers. The lower City, bounded by the Harbour headline in the north and the Niagara Escarpment in the south, is serviced almost entirely by combined sewers, as are the older, northernmost portions of the Hamilton Mountain. Newer suburban areas, mostly on the Mountain, are serviced by separate sanitary and storm sewer systems.

Two major sanitary interceptors service the City of Hamilton. Combined sewage collected from the lower City, and sanitary sewage from the West Mountain is conveyed to the Woodward Avenue Sewage Treatment Plant (STP) via the Western Sanitary Interceptor (WSI) which runs eastward along Burlington and Barton Streets. The Eastern or Redhill Creek Interceptor (RHCSI) runs parallel to the Redhill Creek, transporting sanitary and combined sewage flows generated by the remainder of the

Mountain to the STP. During dry-weather and small storm events, sanitary and storm flow is conveyed to the STP for treatment. Larger rainfall events, specifically high intensity thunderstorms, often generate flows in the combined sewer system in excess of design capacities. These flows may overload the sanitary interceptors and/or STP. In order to avoid flooding the STP or surcharging the sewer system, flow is diverted to the receiving waters as Combined Sewer Overflow (CSO).

The City's combined sewer system contains approximately 165 diversion structures, with 24 CSOs to local receiving waters (Chedoke Creek, Cootes Paradise, Redhill Creek and Hamilton Harbour).

A number of these diversion structures employ motorized gates which can be automatically opened or closed to regulate flow to the sanitary interceptor. These gates are activated either automatically by local water level sensors, or remotely by operators at the STP. The operator's decision to open or close a gate is frequently based upon prior operating experience using system status indicators including wet-well depth and flow readings upstream of a small number of regulator locations throughout the City. The dynamic spatial and temporal properties of rainfall events, especially thunderstorms, can generate runoff responses in which some sections of the sewer system are overloaded while others, which may be in close proximity, are flowing at only a fraction of their capacity. As a result, gate closure may be initiated well before interceptor capacity is fully utilized and deactivated long after flows subside. While such a procedure results in maximum safety for the STP and a lower probability of local basement flooding, CSO frequencies and volumes are likely greater than necessary. The potential for more precise control of these automatic diversion structures is evident.

Schilling (1985) surveyed several cities in North America which have investigated or installed RTC systems. From his survey it appears that RTC of CSO is feasible for various sizes of combined sewer systems. However, Schilling also identified a wide gap between research and applications of systems control concepts for CSO networks.

Recent technological advances have introduced the possibility of using microcomputers for controlling CSO diversion structures and/or storage in real-time to minimize CSO in wet and dry weather, in lieu of building costly supervisory control systems. While a small number of applications of real-time control (RTC) of CSO exist elsewhere, this approach has not been sufficiently tested to date in Ontario.

In 1987 the Region made application to the Ontario Ministry of the Environment (MOE) for funding of a demonstration project to evaluate microcomputer-based RTC of CSOs. Funding was provided equally by the Region through its Capital Budget and the MOE under the Province's Pollution Control Planning (PCP) Studies Program.

The main objectives of this project were to demonstrate microcomputer-based RTC at the Royal Avenue CSO Regulator, located near the intersection of Royal Avenue and Stroud Road, in West Hamilton and determine whether wider application of this new technology is feasible. The design, installation and evaluation of the proposed RTC system constituted the bulk of the project which was begun in 1989 and has been conducted entirely in-house by Regional staff.

This paper reviews the following components of the Royal Avenue RTC Demonstration Project:

1. the design and installation of a real-time rainfall and flow monitoring network, telemetry system, and microcomputer controller for operating the automatic sluice gate in the Royal Avenue CSO Regulator;

2. the development of software for real-time data acquisition, data processing and database management and RTC of the Royal Avenue CSO regulator; and

3. the estimation of CSO frequencies and volumes at Royal Avenue (with and without RTC) via continuous modelling, using the PCSWMM3 RUNOFF and TRANSPORT modules.

11.2 Real-Time Control of CSOs

The objective of a RTC system is to operate CSO regulators to make most efficient use of system storage, and to take full advantage of treatment facilities. RTC can eliminate unnecessary or premature diversions of combined sewage by utilizing in-line storage, and by controlling off-line storage facilities, detaining CSO until the sanitary interceptor and STP can accommodate it. RTC systems may also be able to divert unavoidable overflows to less sensitive receiving areas, and may reduce future storage requirements for CSO control.

Remote monitoring and control techniques are aimed at enhancing the operator's knowledge of the performance of the combined sewer system during a storm event and providing a means for effecting its operation. A computerized RTC system would continuously monitor the combined sewer system and use the data collected in real-time to operate CSO regulators and/or storage/treatment facilities within the system to minimize CSO.

A RTC system gathers and analyzes data acquired from monitoring stations and flow regulators in the combined sewer system. This data may include wind speed and direction, rainfall intensities, sewer depths and discharges, storage levels, regulator settings or positions and pollutant concentrations. Information is generally transmitted to a central computing facility, which controls the affected portion of the sewer system. The data is then immediately available for use by an operator, in a supervisory mode, or as input to an operational computer model, which usually comprises techniques for flow forecasting and optimization, in an automatic mode.

Combined sewer systems are particularly suited to RTC since they are seldom fully utilized, carrying less than 10% of capacity for roug^Xxy 90% of the time (Schilling, 1985). During most storms there will be unused volume within some major conduits. Furthermore, the design of storm and combined sewer systems has traditionally been based upon design storms derived from rainfall intensity-duration-frequency (IDF) curves. The limitations of this design methodology have been summarized by James and

Robinson (1982):

1.   the procedure used to construct the design storm's synthetic hyetograph does not differentiate between storm types, and as a result, the design storm represents an amalgamation of many storm types (thunderstorms, cyclonic rains, etc.);

2.   design storms specify a single rainfall distribution (i.e. hyetograph shape), and thus do not accurately reflect the real spatial and temporal variations of rainfall during storm events; and

3.   current practice specifies a design storm with a particular return period, but does not specify a design antecedent dry period. Antecedent conditions can greatly affect the runoff and pollutant loads generated by a rainfall event. Consequently, the recurrence interval of the design rainfall event cannot be transferred to runoff or pollutant volume.

Without RTC, a combined sewer system designed in this manner can only be expected to perform optimally when it is loaded with its design storm (Schilling, 1990). The likelihood of such an event occurring in nature is minimal. During real storms, spatial and temporal rainfall variations may result in surcharging in some portions of the combined sewer system, while others are significantly underutilized. For example, small localized storms might generate CSOs even though storage is available elsewhere in the system.

The effect of storm speed and direction on the response of Hamilton's storm and combined sewer system to thunderstorms has been found to be significant. In Hamilton, storms coming from the southwest, and thus following the general direction of the drainage network, can result in significantly higher peak flows than storms of similar magnitude travelling in other directions. A critical storm speed has been found to exist for a drainage basin, where storm travel time over the basin is comparable to its characteristic runoff response time (Scheckenberger, 1983). A given storm will yield the highest peak flows when travelling at this speed, in the direction of the drainage network.

A combined sewer system is said to be controlled in real-time if monitored data is used to operate CSO regulators during actual events. Generally a RTC system consists of the following components:

1. a regulator which can be manipulated to affect system performance (eg. a sluice gate);

2. one or more sensors to monitor the state of the system (eg. raingauges and/or flowmeters);

3. a controller to activate (i.e. open or close) the regulator;

4. a telemetry system which retrieves measured data from the sensors and transmits signals to the controller to actuate the gate; and

5. software to analyze the collected data, estimate future conditions and direct control of the regulator in real-time.

Regulator control strategies are generally classified as either reactive or adaptive. Reactive strategies are generally developed off-line, and simply react to the current discharge and storage conditions. Adaptive strategies comprise on-line forecasting and optimization. Control is based on continuous updating of flow forecasts in real-time. Real-time data acquisition is an essential component of an adaptive control scheme. The observed data is used to update forecast model parameters, thereby improving system performance.

Large scale deterministic computer models such as SWMM (Huber et al., 1983) are not presently suited to running in real-time at fine timesteps (5 to 15 minutes), on 286 and 386-based microcomputers. Such models do not provide updates of forecasts in a computationally simple manner (O'Connell, 1980), although some methods have been suggested (O'Connell and Clarke, 1981). On the other hand, simple input-output methods, such as transfer function (TF) models, can provide accurate forecasts with very little computational effort. Patry (1983) provided a comprehensive analysis of water quantity and quality forecasting for urban catchments and developed both statistical and simple conceptual models based on deterministic and stochastic concepts.

In developing an optimization model, the objective function, whatever it may be, should penalize overflows and credit direct (and stored) throughflows to the STP (Bradford, 1976). By assigning different weights to overflows in the optimization model, based on their pollutant concentrations or their expected impacts on receiving waters, overflow sites could be selectively chosen so as to minimize their overall impact on the environment.

11.3 Real-Time Rainfall and Flow Monitoring

The design and operation of an RTC system relies heavily on telemetred rainfall data. Appropriate knowledge of rainfall intensity and storm dynamics is essential for hydrologic modelling and system forecasting.

Accurate measurement of depth and/or discharge in sewers and at control structures is also essential for the implementation of RTC of CSO. Real-time measurements can be used on their own as a decision variable for control purposes, or used to update or correct for errors in the performance of rainfall-runoff forecasting models.

A monitoring network consisting of ten electronic dataloggers and tipping bucket raingauges (TBRGs), and two ultrasonic flow monitors was installed in 1989 and has been operated continuously.

Independent, multi-channel electronic dataloggers were utilized to receive signals from these sensors, translate them into numerical quantities and temporarily store the values. Permanent storage of this data was felt to be best suited to a central computing facility with high capacity, high speed access storage and system backup. By having the dataloggers independent from the sensors, monitoring system components could be interchanged which provided greater flexibility and reduced down time when taking equipment out of service for repair.

At present, the two flow sensors and three of the raingauges are connected to a Central Microcomputer Controller (CMC) by modem and dial-up telephone lines to permit real-time collection of rainfall and flow data. Data is currently transmitted at 1200 baud. The feasibility of using dial-up lines was evaluated during the project.

Raingauge locations were selected to encircle the Royal Avenue CSO regulator and hence provide advance information pertaining to storms coming from the main cardinal directions in particular the southwest, the prevailing storm direction in the area. The TBRG at Royal Avenue provides the detailed information required to forecast flows at the Royal Avenue CSO regulator.

The monitoring network provides the following information:

1. historical rainfall and flow data for the calibration and validation of the PCSWMM3 model of the combined sewer system;

2. historical rainfall and flow data for the continuous PCSWMM3 simulations; and

3. real-time rainfall and flow data for operating the Royal Avenue RTC Demonstration System.

A 286-based machine currently functions as the CMC, handling real-time data retrieval. To date, this machine has performed adequately as the CMC for the Royal Avenue RTC Demonstration System. Results of the study evaluation indicate that the use of a 286-based machine as the CMC for a large-scale RTC system would not be feasible due to hardware limitations and that 386 or 486-based equipment is more appropriate.

Data from the raingauge stations not equipped with modems is collected during site visits using a laptop microcomputer. This machine can also be used to program the data loggers (eg. select timestep, communications protocols, etc.) and perform some simple data processing on site. This approach facilitates prompt identification of instrument difficulties and allows in-situ bucket calibration. Minor maintenance problems can be addressed immediately and the laptop can be used to verify their resolution before departing the site. Instruments in need of more major repairs can be immediately replaced with working spares. This avoids the situation in which an inoperative sensor remains in the field undetected until data is processed and analyzed; often at a much later date.

In operating the system for the three year period it has been observed that the availability and quality of the dial-up lines are not sufficient for real-time data acquisition. Transmission errors would limit the performance of the RTC system. In future work only dedicated telephone lines will be utilized. It should be noted however, that for bulk retrieval of data, dial-up lines were adequate.

11.4 Data Acquisition and Management

Specialized software was required to provide a means of communicating with the remote electronic dataloggers for programming and reconfiguring the dataloggers, retrieving the recorded rainfall and flow data, processing the recorded raw data and displaying it in a more understandable format. A centralized database management system (DBMS) was required to archive the large volumes of collected data and to provide facilities for online querying and retrieval of archived data.

The HWRTC (Hamilton-Wentworth Real-Time Control) software package was developed by the project team for operating the Royal Avenue RTC Demonstration System. The Data Acquisition and Management component of the HWRTC provides access to the following utilities:

1. HWCOMM
2. HWDPRO
3. HWDBMS

HWCOMM (Hamilton-Wentworth Datalogger **Comm-unications**) is a collection of utilities developed to provide a means of communicating with the project's electronic dataloggers. HWCOMM supports RS232 serial communications with IBM-PC compatibles through a 4800 baud direct connection or via a 1200 baud modem on dedicated or dial-up telephone lines. These utilities are used to program or reconfigure certain characteristics of the dataloggers, and to retrieve recorded data from the logger's memory. Data from the three telemetered monitoring stations is retrieved by the CMC. Data from the remaining remote sites is retrieved on-site with a small laptop PC.

The data contained in the datafiles retrieved by HWCOMM is stored in a compressed format. HWDPRO (**Hamilton-Wentworth Data Pro**cessor) performs two main tasks: processing the data contained in the raw datafiles collected by HWCOMM, prior to appending to the central database; and converting the raw datafiles into the format needed to append the data to the central database. HWDPRO facilitates viewing acquired data in the field to check for station problems and generates a summary of the recorded rainfall and/or flow events contained in the raw datafiles.

HWDBMS (Hamilton-Wentworth DataBase Management System) is a permanent, centralized database management system, created by the project team to archive, retrieve, and process the time-series data acquired by the rainfall and flow monitoring network. Rainfall and flow data can be retrieved and displayed in a number of formats, at any user-specified timestep, for any user-specified time period. Data may be displayed in either metric or imperial units, as selected by the user. Available data formats include rainfall event summaries, flow event summaries, daily and monthly rainfall volumes, and daily and monthly flow volumes. Rainfall intensities may also be written in SWMM RUNOFF module format (E1 and E2 data groups), and subsequently imported into a RUNOFF module datafile. This option is especially useful for creating datafiles for continuous modelling, and was used to prepare datafiles for the continuous simulations conducted during the study.

11.5 Royal Avenue CSO Regulator Controller

The 2325 mm x 2850 mm Royal Avenue Box Sewer collects combined sewage from a catchment approximately 444 ha in area, located above and below the Niagara Escarpment, and conveys it to the Royal Avenue CSO regulator. The area below the Escarpment is serviced predominantly by combined sewers, while the drainage network above the Escarpment is separated. The Royal Avenue sewer receives separated stormwater at four inlet points from a total drainage area of 142 ha on the West Mountain. Stormwater entering at the most downstream inlet and combined sewage being conveyed in the Royal Avenue sewer from the west are prevented from mixing under low flow conditions by a 670 mm high wall which extends about 40 m west from the outlet (i.e. to a point which is upstream of the stormwater inlet).

Located just upstream of the outfall are two pipes, connecting the combined sewer to the Royal Avenue CSO Regulator Chamber. The larger 525 mm sewer conveys flow to the regulator chamber. The smaller 300 mm 'tell-tale' pipe, connected to a stilling well in the regulator chamber, was installed to allow measurement of the water level in the combined sewer, via a float recorder housed in the stilling well.

A motorized gate is located over the end of the 525 mm pipe leading from the combined sewer. This gate controls the flow of sewage into the 900 mm sanitary interceptor, which in turn conveys combined sewage to the STP via the WSI.

During dry-weather and smaller runoff events, sewage collected in the northern portion of the Royal Avenue catchment, and stormwater entering at the three upstream inlets is conveyed through the open CSO regulator to the 900 mm sanitary interceptor, to be treated at the STP. Separated stormwater runoff from the southern portion of the basin, above the Escarpment and entering at the most downstream inlet, is discharged to the Chedoke Creek.

When the capacity of the regulator (i.e. the 525 mm sewer) is exceeded, excess flow will be diverted to the Chedoke Creek. If the gate is left open, approximately 0.6 cms of combined sewage will be intercepted and conveyed to the 900 mm sanitary interceptor. However, if the gate is closed, all flow will be diverted to the Chedoke Creek.

The motorized gate at Royal Avenue is an example of an automatic CSO regulator, and was felt to be well-suited for use in development of RTC technology. Originally constructed around 1969, the regulator was equipped with a remote measurement and control system, supplied by Bristol Babcock Inc, consisting of a float-type liquid level gauge, liquid level transmitter and receiver, and gate control panel. The installed system allowed the gate to be operated remotely by staff at the STP, or automatically actuated by the float recorder when it reached predetermined open or close setpoints. By the time this project began, the automatic gate control had been abandoned in favour of remote operation from the STP. The float gauge and transmitter were located inside the Royal Avenue regulator chamber. The recording flow receiver, gate control panel and power supplies for the transmitter and receiver were installed at the STP. The transmitter and receiver are connected by two dedicated Bell Canada telephone lines. A manually operated switch on the gate control panel determines how the regulator at Royal Avenue is controlled. The gate switch has three positions; AUTO, OPEN and CLOSE.

In the AUTO position, the operation of the gate was originally driven by the float gauge. The gate was set to close as soon as CSO to the Chedoke Creek began, and to reopen five minutes after overflow ceased. This level is approximately equivalent to the capacity of the inlet to the regulator. As long as the switch was set to AUTO, the gate would be operated in this manner. The gate's position was monitored with the aid of indicator lights on the control panel. Automatic operation of the gate can be overridden by an operator at any time by moving the switch to either the OPEN or CLOSE position.

Under automatic control the level at which the gate was opened and closed remained constant from event to event. This system lacked the capability to adapt to changing conditions, because it relied on a single piece of information which did not accurately reflect conditions elsewhere in the combined sewer system.

In the RTC system developed, the CMC decides to open or close the gate after analyzing real-time rainfall and flow data from selected sites in and around the Royal Avenue CSO catchment. The status of other regulators in the system and the inflows to the STP should also be considered for future modifications of the system. The conditions which warrant opening or closing the gate will vary from event to event.

The CMC must generate a signal which will actuate the motorized gate at Royal Avenue. This interface is provided by the electronic datalogger, through its Auxiliary (AUX) Output Channel, a software selectable 12 volt DC (switched) signal. The datalogger is connected directly to the CMC through its serial port.

The gate is actuated by the RX relay in the transmitter in the Royal Avenue Regulator Chamber. To operate this relay, a signal voltage of 120 volts is required. The 120 volt signal is derived from the datalogger's 12 volt DC auxiliary output (AUX) with the assistance of a Gate Control Interface Module designed by the project team. The AUX signal is fed through this interface, which is connected to one of the spare card slots offered by the logger, and is used to operate any 120 volt AC device placed across the interface module, including the RX relay which operates the gate at Royal Avenue. The AUX signal is activated by the CMC via software. When AUX is turned on (and left on), the gate motor's RX relay is energized and the gate is closed. Conversely, if AUX is turned off, the RX relay is deactivated and the gate reopens.

11.6 Analysis of Royal Avenue CSO

Analysis of a combined sewer system, to determine CSO frequencies and volumes, should be carried out using a continuous hydrologic simulation package, for as long a term as practical given the available rainfall record.   Continuous simulation provides more accurate estimates of antecedent moisture conditions and surface pollutant buildup. If the combined sewer system being modeled includes off-line storage tanks, initial levels of storage in these facilities are also known. Accurate accounting of antecedent conditions is required if reliable estimates of CSO frequencies and volume are to be obtained.

A suite of water quantity models was assembled with the following objectives:

1. estimation of current (i.e. no RTC) CSO frequencies and volumes;

2. generation of continuous rainfall-runoff time-series to provide a basis for development of real-time flow forecasting software; and

3. development of a RTC simulation model for estimation of future (i.e. with RTC) CSO frequencies and volumes.

These models are capable of simulating rainfall-runoff processes, flow routing, in-line and off-line storage, and the hydraulic operation of diversion structures, in continuous mode. The RUNOFF and TRANSPORT modules of PCSWMM3, a microcomputer version of the United States Environmental Protection Agency's (USEPA) Stormwater Management Model (SWMM), Version III, form the basis for the model simulations. The estimation of current CSO frequencies and volumes at the Royal Avenue Regulator, using PCSWMM3, indicates how the combined sewer system currently functions, without RTC, and provides a basis for the subsequent theoretical evaluation of proposed RTC strategies.

In addition, the rainfall-runoff time-series generated by the continuous SWMM simulations could be used to develop simplified flow forecasting models to predict the response of the Royal Avenue combined sewer system to recorded rainfall.

The RUNOFF model for the Royal Avenue catchment was calibrated using observed rainfall and flow data collected at the Royal Avenue CSO in 1990. Each calibration run consisted of a continuous simulation for May to July inclusive. Eight events during this period were calibrated. Original parameters were taken from an earlier RUNOFF model of the Chedoke Creek Drainage Basin. The Royal Avenue CSO Regulator is situated within this larger basin. The original Chedoke Creek model was calibrated using flow records collected downstream of Royal Avenue. The rainfall and flow data collected at Royal Avenue in 1990 allowed recalibration of the RUNOFF model for this drainage basin. The results of the Royal Avenue RUNOFF model calibration are presented in Figure 11.1.

Total volumes and peak flows for all eight calibration events fall within 20% of observed values. The adequacy of the calibrated Royal Avenue RUNOFF model was then checked using an independent data set of eight events, observed during July to October, 1990. Simulated volumes for all eight events fell within 20% of observed volumes and peak flows for seven of the eight events were within 20% of observed peaks.

The calibrated, validated SWMM model was run to estimate CSO frequencies and volumes at the Royal Avenue Regulator for 1989, 1990 and 1991. These estimates provide a basis for the evaluation of proposed strategies for controlling the gate in the Royal Avenue CSO Regulator.

The large volumes of rainfall data required for these analyses were handled by HWDBMS, the centralized database management system described previously. With HWDBMS, data from any of these gauges can be quickly retrieved and converted to the SWMM RUNOFF rainfall input format (E1 and E2 data groups). This greatly reduced the time and effort required to create datafiles for continuous simulation.

The TRANSPORT model of the box sewer in the vicinity of

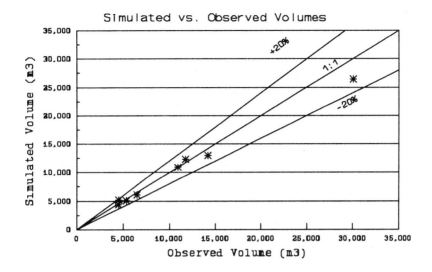

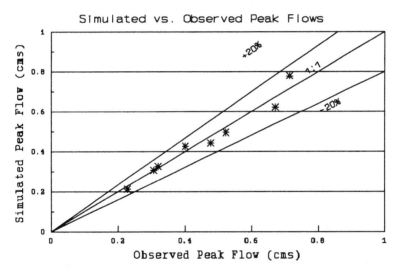

Figure 11.1: Runoff calibration model results.

the outfall and regulator was constructed assuming the gate in the Royal Avenue CSO Regulator was always open and intercepting the maximum volume of combined sewage. Intercepted flows would be conveyed to the STP, unless diverted to local receiving waters by some other downstream CSO regulator. The model provides the following information:

1. the total inflow of combined sewage to the Royal Avenue Box Sewer;

2. the CSO which is diverted over the central dividing wall, from the combined sewer to the separate storm sewer, and subsequently discharged to the Chedoke Creek;

3. the total CSO discharged to Chedoke Creek; and

4. the combined sewage which is intercepted by the Royal Avenue CSO Regulator (when gate is open), for conveyance to the STP.

On average, there were 41 runoff events per year for the period 1989-91 (May through October), generating an average annual flow volume of approximately 454,000 m³ in the Royal Avenue Combined Sewer. Assuming the gate in the regulator is open, this results in 25 CSOs per year, on average, and approximately 83,000 m³ of CSO annually. About 82% of the combined sewage reaching the regulator is intercepted. The other 18% is discharged to the Chedoke Creek. The CSO frequencies and volumes generated by these SWMM simulations are in close agreement with those produced by a longer term analysis performed using 20 years of hourly rainfall recorded at the Mount Hope Airport and RBG AES stations. On average the simulations generated 22 CSOs per year for the 1970-89 period (May through October).

As stated earlier, these analyses assumed the gate in the Royal Avenue Regulator Chamber was open at all times. Records of gate operation kept at the STP indicate that the gate was closed a number of times by operators during this three year period. These closures are generally initiated to decrease the hydraulic load on the STP during severe rainfall events, and prevent pumphouse flooding. While the reasons for the aforementioned gate closures is understood, it is also apparent that they have a significant impact on the total duration and volume of CSO at Royal Avenue, and that there are likely significant benefits to be gained from more precise operation of the gates.

In order to examine the impacts of a control strategy at Royal Avenue, the May to October, 1990 period was simulated assuming the gate closed when the flow reached the following levels: the threshold capacity of the regulator (0.6 cms), 1.25, 1.5 and 2.0 times the threshold capacity. Table 11.1 presents the intercepted flow, CSO volume and duration of gate closure for the above noted conditions as well as for the actual operation of the gate.

Table 11.1: Impact of control strategies on CSO volume.

Control Strategy	Intercept Flow (1000 m³)	(%)	Total CSO (1000 m³)	(%)	Gate Closure (hours)
1	409	82	88	18	0
2	285	57	211	43	142
3	229	46	268	54	83
4	320	65	176	35	41
5	361	73	136	27	22
6	388	78	109	22	10

Note: Strategy 1: Gate open all the time
      Strategy 2: STP operation of gate in 1990

Strategy 3: Gate closed when flow = 0.6 cms
Strategy 4: Gate closed when flow = 0.75 cms
Strategy 5: Gate closed when flow = 0.9 cms
Strategy 6: Gate closed when flow = 1.2 cms

As can be seen from the table, the minimum amount of CSO is generated by leaving the gate open under all conditions. However, this also generates the maximum amount of flow to the STP. It is evident that this volume of flow is unacceptable for operation of the plant since the gate is frequently closed by operators. It is interesting to note that closing the gate at the threshold level of the regulator actually produces more total overflow for the full year than the rather imprecise method of operation currently practised. This indicates that there is likely a better setpoint, in excess of the 0.6 cms, which increases the intercepted flow while not adversely affecting the STP. In the absence of information about levels at other regulators in the system, such a setpoint could form a viable operating strategy for the Royal Avenue regulator.

11.7 RTC Demonstration System

The RTCDEMO (Real-Time Control Demonstration) component of the HWRTC is the operational software utility for the Royal Avenue RTC Demonstration System. RTCDEMO retrieves real-time rainfall and flow data from monitoring stations in and around the Royal Avenue catchment, analyzes and displays this information, and directs control of the gate in the regulator chamber. The proposed RTC system is shown schematically in Figure 11.2.

For the purposes of testing and demonstration of the RTC, the CMC was installed at the STP. It is likely that any permanent RTC system installation will be centered there. A datalogger containing the Gate Control Interface Module described previously was connected directly to the CMC via a 4800 baud serial interface. The dataloggers at the three remote stations

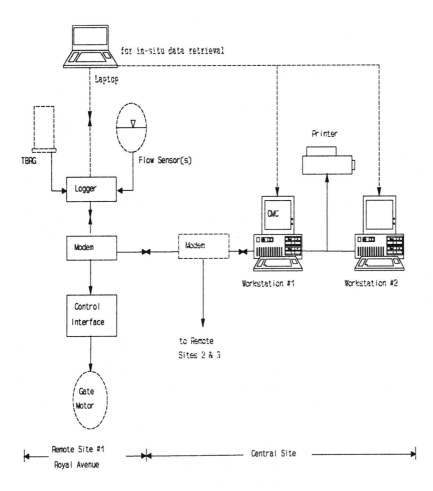

Figure 11.2: Hardware for proposed RTC system.

outfitted with 1200 baud modems, were linked to the CMC via dial-up telephone lines. RTCDEMO accesses the dataloggers at these three stations, retrieves and displays the latest recorded rainfall and flow information. Rainfall recorded by the TBRGs during the previous four time intervals (20 minutes), is presented as five-minute average intensities in mm/hr. Flow-related inform-ation for the Royal Avenue Box Sewer which may be displayed includes measured flow depths (m), measured flowrates (m^3/sec), forecasted flow depths (m), and/or forecasted flowrates (m^3/sec).

The current status of the gate in the Royal Avenue Regulator Chamber is displayed; OPEN or CLOSED, or in the process of OPENING or CLOSING.

The information retrieval cycle requires approximately 2 minutes and is repeated every 5 minutes. The remaining 3 minutes of each data acquisition cycle is available to determine gate setting. Field measurements indicate that the gate takes approximately 100 seconds to open (or close) after the appropriate signal is sent.

The current version of RTCDEMO offers two distinct operating modes; supervisory and automatic control. Supervisory control is the default mode of operation for the gate, and is enabled when RTCDEMO commences. In this mode, RTCDEMO acts as a decision support tool providing information on system conditions to an operator. The operator has responsibility for analyzing available information, deciding on the appropriate gate setting and transmitting the signal which opens or closes the gate. In automatic control mode RTCDEMO provides several algorithms for local control of the Royal Avenue gate. An operator must initially change the RTCDEMO's control mode to automatic from supervisory. At this point, the operator is required to select the algorithm which is to be used for gate control. RTCDEMO will use this algorithm to automatically determine gate setting until supervisory mode is reestablished or until another automatic algorithm is selected.

The structure of RTCDEMO is such that control algorithms can be added, deleted or modified off-line and RTCDEMO reinstalled. The control algorithms can be mathematical equations for rainfall or flow forecasting evaluated using real-time data, actuation setpoints dependent on real-time rainfall or flow data or knowledge-based rules. These options were included to demonstrate some of the parameters and techniques which might be used to direct automatic RTC of Royal Avenue and similar regulators. The design of the software provides a general tool which can be adapted to other regulators within the combined sewer system.

Investigation has shown that control points in the combined

sewer system must be operated as a complete system accounting for treatment plant conditions. It was beyond the scope of this phase of the study to monitor and model the entire city-wide system. The current status of RTCDEMO limits its use to improvement of local control of a regulator. Until a City-wide RTC system is developed, the overall benefits of such local control strategies will be unknown as will requirements for modification to these strategies.

11.8 Conclusions

The Regional Municipality of Hamilton-Wentworth has recently completed a demonstration project to develop and evaluate technology for microcomputer-based real-time control of CSOs. The project was conducted at the Royal Avenue CSO regulator in West Hamilton. The study has found that it is feasible to implement low-cost microcomputer-based control at regulators and control points in a combined sewer system in order to reduce the frequency and volume of combined sewer overflow through more precise operation. 286-based machines were found to be satisfactory for single site operation but it is expected that 386 or 486-based machines will be necessary for a city-wide system.

1. A real-time rainfall and flow monitoring network and telemetry system was installed and operated for three years. Appropriate hardware was evaluated and selected and software was developed for real-time data acquisition, data processing and database management. Dial-up telephone lines were found to be satisfactory for operating a data collection system off-line using periodic bulk retrieval of data but were found to lack the reliability necessary for real-time data acquisition.

2. The PCSWMM3 RUNOFF and TRANSPORT modules were used for continuous modelling of the Royal Avenue catchment to investigate the impact of proposed RTC strategies on CSO frequencies and volumes at the Royal Avenue regulator.

3. A microcomputer-based controller for actuating an automatic sluice gate in the Royal Avenue CSO Regulator in real-time was built, installed and successfully tested.

4. Software for real-time control of the Royal Avenue CSO Regulator in either a supervisory (operator-controlled) or automatic (algorithmic-controlled) mode was developed and tested. The software has been structured for flexibility in adding or deleting algorithms to allow adaptation to other regulators in Hamilton's combined sewer system.

5. A second phase of the study to develop a real-time control system for the complete combined sewer system has begun. This project will require the interaction of all regulators and control points together with the operation of the treatment plant be accounted for. The toolkit of techniques, hardware and software for local control developed in the first phase of the work will be used.

11.9 References

Bradford, B.H. (1976). Real Time Control of Storage in a Combined Sewer System. Proceedings of the International Symposium on Urban Hydrology, Hydraulics, and Sediment Control, University of Kentucky, Lexington, Kentucky: 287-296.

Huber, W.C., Heaney, J.P., Nix, S.J., Dickinson, R.E. and Polmann, D.J. (1981). Storm Water Management Model User's Manual Version III. Office of Research and Development, Municipal Environmental Research Laboratory, USEPA.

James, W. and Robinson, M.A., (1982). Continuous Models Essential for Detention Design. Conference on Stormwater Detention Facilities Planning Design Operation and Maintenance: 163-175.

O'Connell, P.E. (1980). Real-time Hydrological Forecasting and Control. Proceedings of First International Workshop, Institute of Hydrology, Wallingford, Oxon, U.K.

O'Connell, P.E. and Clarke, R.T. (1981). Adaptive Hydrological Forecasting - A Review. Hydrological Sciences Bulletin, 26(2): 179-205.

Patry, G.G. (1983). Real-Time Forecasting of Water Quantity and Quality in Urban Catchments. PhD. Thesis, University of California, Davis.

Scheckenberger, R. (1983). Dynamic Spatially Variable Rain Models for Stormwater Management, M.Eng. Thesis, McMaster University.

Schilling, W. (1985). Real Time Control of Combined Sewers. Proceedings of the Specialty Conference on Computer Applications in Water Resources, Water Resources Planning and Management Division, ASCE, Buffalo, New York: 499-510.

Schilling, W., (1990). Introduction to Simulation and Real Time Control. WPCF Pre-Conference Workshop on Simulation and Real Time Control of Urban Drainage and Wastewater Systems.

Chapter 12

Application of the SWMM4 Model for the Real Time Control of a Storm Trunk Sewer

A. Kwan, D. Yue and J. Hodgson
Stanley Associates Engineering Ltd.
Stanley Technology Centre
10160 - 112 Street, Edmonton, Alberta, T5K 2L6

Recent plans for new subdivision development in North Edmonton will significantly increase the burden on the currently stressed Kennedale storm drainage system. In July 1991, the City retained Stanley Associates Engineering Ltd. to re-evaluate the hydraulic and structural performance of the Kennedale storm trunk. The study was to develop a calibrated stormwater model, to identify the critical storm events, to identify the flood prone areas within the Kennedale storm basin and to define servicing alternatives for both the present and future development areas. The assessment resulted in the recommendation of utilizing real time control to maximize the use of the existing trunk system.

12.1 Introduction

The Kennedale storm trunk contains a north trunk and a south

0-87371-898-4/93/$0.00 + $.50
© 1993 by Lewis Publishers

trunk which join to form the main trunk downstream before discharging into the North Saskatchewan River (Figure 12.1). The south trunk was constructed during the late 1950's and comprises 11 km of pipe ranging in size from 1200 mm to 2400 mm. The north trunk was constructed during the mid 1970's and comprises 10 km of pipe ranging in size from 1650 mm to 2250 mm. The main trunk was constructed in 1980 and comprises 4 km of piping ranging in size from a 3000 mm circular pipe to a 4900 mm x 3050 mm concrete box.

The Kennedale storm trunk presently services 3213 ha of directly

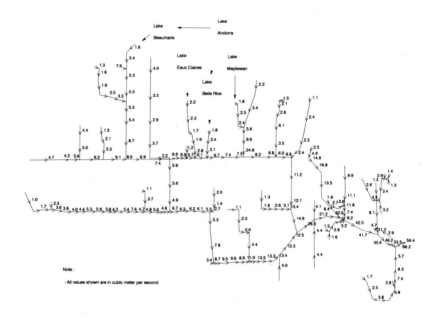

Figure 12.1: Kennedale storm trunk system configuration and capacities.

connected area with an additional 774 ha of lake-regulated area. The lake-regulated designation is for the portion of the service area where the storm runoff will be drained to storage lakes. The discharge from these lakes will be regulated. The future development of North Edmonton will all be lake-regulated and will increase the lake-regulated area to 4249 ha for the Kennedale storm system (Figure 12.2).

Included in the directly connected areas, there is 170 ha of combined sewer area which discharges to the south trunk. Normal dry weather flows are intercepted and conveyed to the City's sewage treatment facilities by means of other combined and sanitary sewer trunks. However, during rainfall events, a portion of the higher wet weather flow overflows into the south trunk.

12.2 Model Development

The Stormwater Management Model (SWMM) is a comprehensive computer model which can simulate rainfall/runoff and dynamic flow conditions (including surcharging) in pipe systems. Version 4 (i.e. SWMM4) was chosen for the analysis of the Kennedale system. Since the off-the-shelf version of SWMM4 can only handle a network of 150 conduits in a personal computer environment, SWMM4 was re-compiled on an HP-UNIX 400 series computer for increased computing efficiency and network capacity (1000 catchments in RUNOFF and 1500 pipe in EXTRAN). The basic system data came from an existing SWMM2 input data file developed by the City of Edmonton in a 1979 study. It was necessary to convert the data to the SWMM4 format while additional data for new areas were added as required. The resulting stormwater model is a lumped hydraulic model for the Kennedale system with approximately 210 conduits representing all piping 1050 mm in diameter and larger. Run times for a six day storm event (with a 5 minute time-step in RUNOFF and a 30 second time-step in EXTRAN) were in the order of two to three hours.

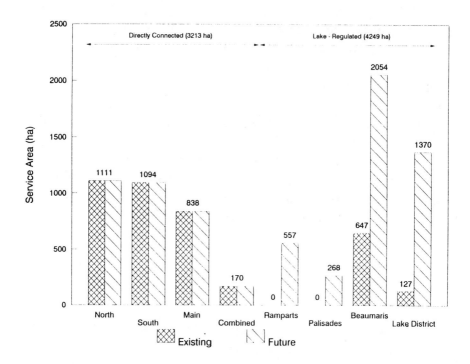

Figure 12.2: Kennedale storm trunk service areas.

The assessment of the Kennedale system included short duration, high intensity design storms as well as long duration, high volume historical storms. Short duration, high intensity storms have put the Kennedale storm trunk into highly surcharged conditions and has resulted in some localized flooding. The long duration, high volume storms have shown that there is a long drawdown period for lake storage. As a result, the modelling concepts for the Kennedale system were divided into short and long storms. The short storms utilized the EXTRAN BLOCK's dynamic routing procedure while the long storms were simulated using the TRANSPORT BLOCK. Both the EXTRAN and TRANSPORT blocks use similar input data and interface with the results from RUNOFF BLOCK (Figure 12.3).

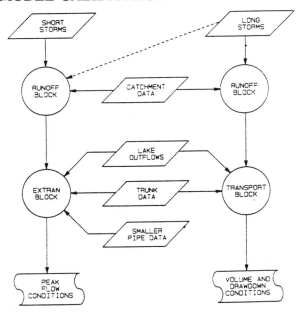

Figure 12.3: Kennedale storm trunk model structure.

Another consideration was the selection of rainfall events. The City of Edmonton has adopted the most severe one and two day rainfall events of 1937, 1978 and 1988 in their City Standards for analysis. However, these events may not be the critical in multiple day rainfall events for lake systems with long drawdown times. Analysis of multiple day rainfall events indicated events which occurred over thirty day periods during 1953, 1983 and 1990 should also be considered in multiple day rainfall event simulations (Figure 12.4).

12.3 Model Calibration

The City of Edmonton operated an extensive monitoring network of rainfall gauges (five), flow monitors (six) and surcharge level gauges (fifteen) in the Kennedale basin during 1991. Some of the stations are part of the City's permanent network. Based on the magnitude of the events and the completeness of the data collected, the rainfall event of 20 to 23 August 1991 was chosen as the calibration event. During this storm event, the surcharge

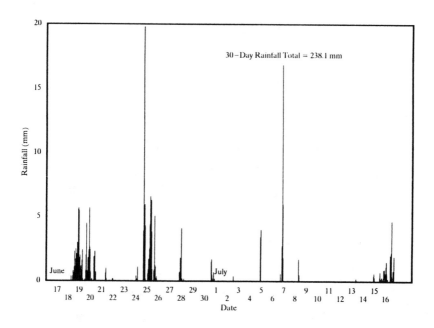

Figure 12.4: Hourly rainfall for 17 June to 16 July 1983.

gauges did not record surcharge at the tributary locations. However, the main trunk and both the north and south trunks did record surcharge. Of the six flow monitors in the Kennedale system, the flow monitor at the outfall was inoperational during the storm event. Therefore, the calibration effort was concentrated on matching the runoff volume and peak flow rates to the five available flow monitors as well as the surcharge level of the surcharge gauges.

Initially, the volume and peak flow parameters were modelled quite accurately with the monitored results. However, the resulting surcharge levels would be significantly lower in the north trunk than indicated by the surcharge gauges. In order to match the surcharge levels, a significantly higher Manning's n would have to be placed at the downstream end of the north

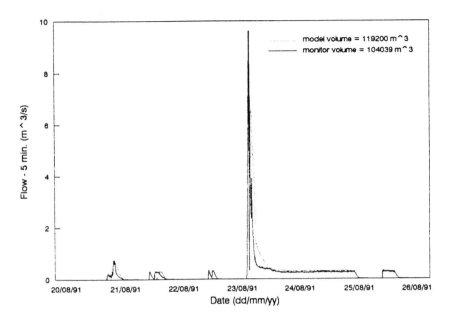

Figure 12.5: North trunk calibration for 20 to 23 August 1991.

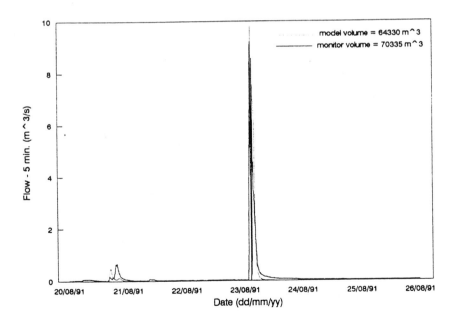

Figure 12.6: South trunk calibration for 20 to 23 August 1991.

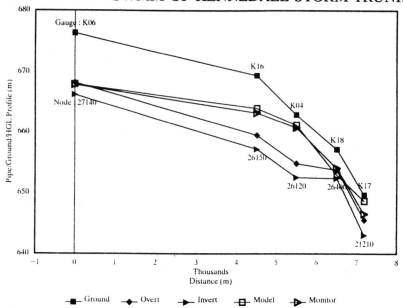

Figure 12.7: North trunk surcharge levels for 20 to 23 Aug. 1991.

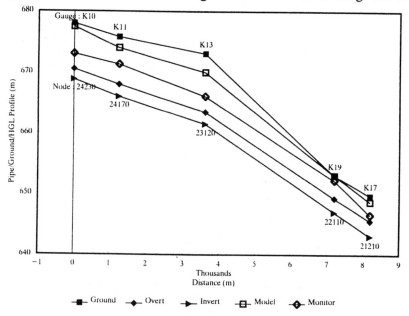

Figure 12.8: South trunk surcharge levels for 20 to 23 Aug. 1991.

trunk. In doing so, the modelled peak flow was less than the monitored flow (Figure 12.5, 12.6, 12.7 and 12.8). A temporary "choke" at the downstream end of the north trunk was believed to be the reason for the inaccuracy of the calibration. Field inspection of the trunk was recommended to confirm if a constriction existed. The subsequent analysis of the Kennedale storm trunk then continued without the 'choke' situation (assuming it would be found and remedied).

A single rainfall event is not sufficient to accurately calibrate a system as big as Kennedale. More calibration work was recommended in order to improve the accuracy of the Kennedale storm system model.

12.4 Structural Assessment

Structural assessment of the Kennedale system has been carried out by inspecting fourteen various locations on the north trunk and the main trunk. The structural condition (i.e. physical characteristics such as cracks, joint separation, etc.) of the trunk was found to be very good. The service condition (i.e. operational characteristics such as debris deposits, seepage, etc.) of the trunk was also good except in some areas at the upstream end of the north trunk.

12.5 Storm Assessment

The hydraulic assessment of the Kennedale system was performed using the calibrated model without the 'choke' situation. Two factors are important in the hydraulic assessment. They are the loading factor rating and the hydraulic gradeline ratings. The loading factor ratings are a measure of the trunk capacity while the hydraulic gradeline ratings are a measure of the surcharge level. Both ratings go from zero to five as the condition is rated from excellent to poor (Table 12.1).

The Kennedale system was first examined with the synthetic

design hyetographs for the two and five year events with the existing system without any lake-regulated inflows. The system performed reasonably well during the two year event (only 4% of the pipe rated poor). However, during the five year event, 60% of the system was found to have poor hydraulic gradeline ratings.

Table 12.1: Hydraulic rating criteria.

Rating	Rating Range	TLF[1]	(m below street level)
1	0 to 1.5	0 to 075	> 5.25
2	1.5 to 2.5	0.75 to 1.25	3.75 to 5.25
3	2.5 to 3.5	1.25 to 1.75	2.25 to 3.75
4	3.5 to 4.5	1.75 to 2.25	0.75 to 2.25
5	4.5 and up	> 2.25	0 to 0.75

Notes: 1.   TLF is the theoretical load factor: TLF $\alpha\ (Qp/Q)^2$, where Qp is the peak flow rate and Q is the pipe non-surcharged full flow capacity.
       2.   HGF is the hydraulic gradeline factor.
       3.   From the City of Edmonton Standard Sewer Condition Rating System report.

These findings are consistent with the earlier investigation by the City of Edmonton.

   Two relief alternatives were investigated for the Kennedale system. They were the construction of a large relief trunk from the North Saskatchewan River and, alternatively, the construction of storage ponds to control the flow at strategic points within the Kennedale system.

12.6 Relief Trunk Alternative

The relief trunk alternative would involve building a relief tunnel along 132 Avenue. This relief trunk would take flows from both the north and south trunk at strategic locations. An overflow outfall system would be constructed parallel to the existing main trunk beginning at the top of the Kennedale ravine. This alternative would relieve most of the system during the five year event with only 11% of the system piping having poor hydraulic gradeline ratings. The total cost for the relief trunk is, however, in the order of $60 million.

12.7 Storage Pond Alternative

The storage pond alternative involves the combination of a connecting tunnel and a system of eleven different storage units. For this alternative, a short (0.65 km) length of tunnel is proposed to connect the existing north trunk at 137 Avenue and 58 Street to the 2400 mm diameter tunnel at Fort Road. In addition, three dry ponds and one buried storage unit would be used in the north trunk service area, while six dry ponds would be located in the south trunk service area. Another dry pond would be utilized in the main trunk service area as well. This alternative would also relieve most of the system during the five year event with 17% of the system piping having poor hydraulic gradeline ratings. The total cost for this alternative is in the order of $15 million.

12.8 Real Time Control for the Trunk

Real time control of all the proposed lake-regulated inflows was found to be necessary for the Kennedale system. As the Kennedale system is already overloaded under the existing conditions, adding lake-regulated inflows would only increase the burden on the existing system. Therefore, a real time control system was proposed which would require no lake outflow to the Kennedale system during any significant storm event. Thus, each lake-regulated system had to store all of the runoff for the design event. The allowable discharge rate for each lake system after the storm event would then be pro-rated according to the respective service areas. The magnitude of flows was determined subject to the constraint that no more than 50% of the Kennedale trunk's capacity was utilized at any point. This provided room to control the system with set points far enough apart not to be affected by fluctuations in flow.

The most difficult aspect of real time control is to accurately forecast the runoff event within the critical time period. Initial calculations showed that the critical time period for the Kennedale system from one end to the other is about 46 minutes. However, a more detailed analysis evaluating the critical reach (draining the flow from the Lake Beaumaris outlet about 10 km along 137 Avenue) indicated a travel time of about 25 minutes. This would be in line with the amount of advanced warning available from rain gauges within the Kennedale System.

At this stage, only a simple response-to-rainfall system is being proposed. Gates will close when a significant rainfall occurs or when downstream water level monitors rise above a set-point. Gates will open when downstream water level monitors drop below their set-points. The benefits of a storm forecast system using gauges outside the Kennedale system will be investigated during the course of system implementation.

12.9 Real Time Control for the Lake Outlets

Lake outlets were proposed to be configured using a system of two weirs and one outlet valve (Figure 12.9). A weir placed at normal water level would maintain lake elevation at a pre-determined level prior to the storm events. During a significant storm event, the outlet valve would be closed in response to either the monitored rainfall or the surcharge levels downstream in the Kennedale system. Within a multiple lake system, a lake's control valve will close when a high water level (HWL) is indicated in downstream lakes. This would detain all subsequent flow, and the lake level would rise as the stored volume of water increases. If the stored storm volume exceeded the available storage and the lake level rose above HWL, the emergency overflow weir would allow water to spill and flow into the pipe leading to the downstream lake or to the trunk. Closure of the outlet valve would not affect overflow from the emergency overflow weir, therefore, a fail-safe outlet would be provided for each lake. Any flow capacity control devices would be located in the second weir wall in order to ensure unrestricted emergency overflow. Under normal operating conditions, the valve would remain closed until the receiving trunk below the downstream lakes had cleared. Once the downstream trunk was clear, all valves would be opened and the lakes would drain to the trunk.

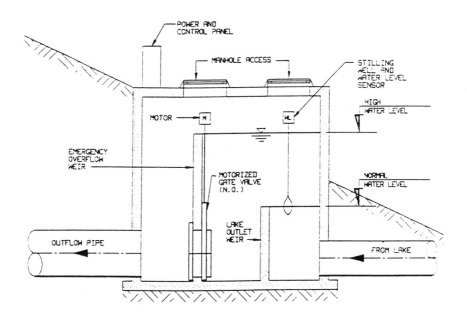

Figure 12.9: Lake outlet configuration.

12.10 Conclusions

The Kennedale storm trunk is a complex system with many unique features. A lumped hydraulic model was developed for the Kennedale storm system using the SWMM4 model. Although the calibration effort was far from comprehensive, some degree of calibration was achieved using the only significant rainfall event available. The structural condition of the trunk was found to be satisfactory while the hydraulic assessment clearly showed the over-stressed condition of the Kennedale system.

Initial investigation showed that a relief trunk would significantly improve the hydraulic condition of the system. Another alternative investigated involved the construction of eleven storage units and a short tunnel. In addition, real time control of all lake-regulated inflows was found to be necessary for the Kennedale storm system. The use of storage and real time control allowed for the addition of runoff from a considerable amount of new development north of the existing service area.

12.11 Acknowledgements

The authors of this work wish to acknowledge the city of Edmonton for funding this project. In particular, Sid Lodewyk and Konrad Siu, who work for the City and provided valuable input throughout this investigation. The pipe rating criteria used for the system assessment were developed by the City's Drainage Branch and is the property of the city of Edmonton. The City of Edmonton continues to support real-time control system investigations as a cost effective approach for maximizing the use of existing infrastructure. Phase II of this investigation, also funded by the City, is a focused real-time control implementation study to be conducted during 1992.

Chapter 13

Computer Integrated On-Line Weather Station and Water Management For Typical Crops

Luis M.J. Carvalho and William James
School of Engineering
University of Guelph

13.1 Questionnaire

A computer tool for farm managers with on-line weather station will help to ameliorate two present day water management concerns: nonpoint source nitrate pollution and the availability of spatially appropriate weather data. Development and successful implementation of a widely acceptable and relevant computer system for farm management necessitates input from the farmers/users. This information includes basic physical farm data (such as farm area) and crop/water cultivation practices. It is also important to gauge the current level of computer technology on the farm and its acceptance. Farm manager's perceptions of the benefits of a computer system, and what components such a program should have, must also be identified. To collect this data a farm/computer survey was drafted by Jim Law and Brett Young. The results of this survey were used as a basis for the design of a widely-applicable computer-based management

0-87371-898-4/93/$0.00 + $.50
© 1993 by Lewis Publishers

system.

A total of one hundred and forty eight surveys were mailed to farms in Southern Ontario. Forty eight were mailed back to the School of Engineering, representing a return rate of 32%. The average farm area was six hundred and thirty acres (Table 13.1), a significant size especially in terms of impact, on local streams, from agricultural runoff. The average number of years the farm operated under the same management was nineteen years. Managers or potential system users are therefore unlikely to change very rapidly so that the project can expect a measure of continuity. Also constant is the wide use of corn and its inclusion in most crop rotations. For these reasons, and the fact that most farm managers have grown the same crops for more than five years, corn is a focus of the computer management system.

Table 13.1: General

630 acres	Average farm area
19 years	Average time under same management
Corn	Most widely cultivated crop
83%	Farmers growing similar crops for over 5 years
92%	Farmers who practice crop rotation
83%	Farmers who will include corn in their crop rotation

Most farm managers have had computer experience (Table 13.2). A slightly smaller percent have access to computers. Most of these computers are IBMs or IBM compatible. The same number that have access to computers are also willing to upgrade their present system. Clearly any computer application geared towards these farm managers must be IBM compatible.

The most popular types of software on the farm were word processors, spread sheets and accounting packages (Table 13.3). Most of the present applications are relatively un-sophisticated

Table 13.2: Computers

63%	Farmers who have access to computers
- 1 Apple	
- 18 IBM	
- 11 Others	
63%	Farmers willing to upgrade computer
83%	Farmers who have computer experience

and deal mainly with financial data management and writing. However it was interesting to observe that farm managers used crop management computer programs in conjunction with their personal computers. Increased use of this software is expected as crop management software was the most popular program farm managers felt should be included in a farm management computer system. Weather was also mentioned in this section, as the second most popular feature. In general there was an overall willingness to acquire and use additional and more sophisticated software.

The environmental section in the survey was designed to identify environmentally-related farm concerns and how the farm managers felt a computer system could be used in this respect (Table 13.4). The features that farm managers thought should be included in a computer system are application rates and dates of farm chemicals and weather. The group most often responsible for the application of chemicals on the farm are the farmers and managers themselves. This strongly suggests that chemical-application software developed specifically for farm managers could be viable and perhaps necessary.

27% of farms have ponds, all of which, save one, have algae problems. This may be an indicator demanding more effective control of fertilizer application on the nearby fields. Further, this could be especially significant to potential water quality problems, as most farms have at least one well. Questions are therefore

Table 13.3: Software

Software presently being used	
68%	Word processing
68%	Spreadsheets
59%	Accounting
36%	Database
Desired features in a computer system	
73%	Crop Management
60%	Weather
56%	Accounting/Tax/Inventory
54%	Livestock/Management

raised as to quality of the well water, extent of possible contamination and well-water uses. Interestingly though, most farm managers reported no water quality problem on their property, an obvious contradiction which should be further investigated in order to assess more clearly present problems or simply gauge water quality perceptions in the farm community.

Animal management, although financially significant to most farms was too diversified (Table 13.5 & 13.6). It did not allow the development of a management system that could be used by a large segment of farm managers, as was the case with the corn crop. This section of the survey does serve to further emphasise the concern farm managers have with manure handling and the environment (Table 13.7).

The farm managers were questioned on the instruments they currently use and those they would like to use, as well as the type of weather data they would like to see included in a computer system. Rain data was cited most (Table 13.8). Almost all the farm managers have a rain gauge and hence few are willing to get another. However, these gauges are thought to be non-recording, totalizing gauges. Such an instrument would not be suitable for

Table 13.4: Environmental

Desired features in computer system	
77%	Fertilizer application rates/dates
75%	Pesticide application rates/dates
67%	Weather
Applicators of farm chemicals	
69%	Farm managers
4%	Contractors
Water and farms	
27%	Farms have ponds
25%	Farms have algae problems
83%	Farms have wells
2	Farms reporting water quality problems
36	Farm reporting no water quality problems

Table 13.5: Animal management.

Farms	Income %
6	Less than 25%
6	Between 25%-50%
9	Between 50%-75%
19	Greater than 75%
6	Did not respond

Table 13.6: Livestock distribution.

TYPE OF LIVESTOCK	FARMS	AVERAGE HEAD COUNT	MIN. HEAD COUNT	MAX. HEAD COUNT
HOGS	8	795	9	4000
DAIRY	18	126	55	300
POULTRY	11	11386	8	50000
MINK	1	1200	-	-
BEEF	15	143	2	600
HORSES	4	4.5	1	10
RABBIT	1	40	-	-
SHEEP	1	150	-	-

Table 13.7: Livestock computer system.

Desired features in computer system	
58%	Feed
58%	Cost
56%	Health
23%	Manure
15%	Environment

a computer system. Inexpensive recording rain gauges however are predicted to find acceptance as there is no manual labour involved with collecting the data. This improvement would also be welcomed by the farm managers if they had a better understanding of recording time steps and how these relate to their farm operation. A curious observation is the willingness of the farm mangers to install humidity meters but humidity as a data type has a relatively small desirability in a computer system.

However, there is no doubt that farm managers would like temperature sensors and some measure of sunlight or radiation (heat units are calculated using temperature). The farm managers seem to know what weather parameters they would like to have but showed some doubt about recording time intervals. Almost 70% are willing to install some meteorological equipment and connect it to their computer.

Table 13.8: Meteorological

INSTRUMENTS	PRESENT	WILLING
Rain Gauge	39	3
Snow Depth Gauge	0	18
Air Temperature	14	19
Humidity Meter	6	26
Anemometer	2	20
Water Levels in Shallow Wells (depth < 15m)	0	16
Water Level in Deep Wells (depth > 15m)	0	13
Hours of Sunlight	0	26

Desired weather data for computer system	
38%	Rain
25%	Heat units
6%	Humidity

In general there was a great deal of interest in a computer-integrated management system for the farm.   81% of the respondents would like to be kept informed of progress while 67% would like to participate. A map displaying the locations of the interested farms was prepared to illustrate the dense spatial coverage a network of computerized weather stations could provide.

One of the most critical factors in this project is the cost that farm managers are willing to incur to purchase the system (Table 13.9). Implementation and development of this project is expected to be self-supporting.   Therefore it is vital that the cost to the farm manager be within an acceptable range. Excluding computer upgrading costs, implementation of the computer system on the farm should be kept, if possible, below $600.

Given that initial acceptance is successful, increasing numbers of farm managers are expected to show interest in the project. This prediction is based, among other things, on the increasing use of computers on farms. J.P. Loranger (1990), indicated that there were 42,000 farmers in Quebec and that a surprising 25% use computers, a high percentage in lieu of the September 1988 Canadian Census report entitled "A Profile Of Farmers With Computers." The report documented that in 1986, just under 2% of Quebec farmers used a computer principally in managing their farm business, while for the whole of Canada the average was 2.6% (Bollman, 1988). The Quebec survey indicates the rapid acceptance and application of computers on the farm. Grant G. Murray quotes a Cornell University Study where "by 1989, fully 15% of U.S. farmers were using Computers" (Murray, 1990), stating that "agriculture may be among the fastest-growing sectors for information technology applications"(Murray, 1990).

As it becomes an essential part of farm management, the computer will be expected to perform three main functions: cognition and communication, data and knowledge management, information processing (Gauthier, 1987).   Implied in these functions is the need for sensors, data links to other systems and interfaces with the operators (Gauthier, 1987).

The benefits to farmers at the farm level could come as

Table 13.9: Expenses farmers are willing to incur to participate in project.

Computer upgrade		
19%	21%	23%
< $2,000	$2,000 - $5,000	$5,000 - $8,000
Crop Management Software		
19%	44%	8%
< $50	$50 - $200	> $200
Livestock Management Software		
19%	40%	10%
< $50	$50 - $200	> $200
Weather Instruments		
27%	33%	8%
< $50	$50 - $100	$100 - $200
Workshops		
42%	33%	2%
< $50	$50 - $100	$100 - $200

improved ability to select variety, planting date, crop mix and to evaluate various management strategies (Mishoe, 1988). To meet such aims and better predict the development of a crop, there is a need to integrate, crop, soil, pest and fertility information (Mishoe, 1988). "There is a large demand for combining crop models with data bases on soils and weather to allow users to evaluate policies and strategies" (Jones, undated).

"Critical to the use of these models is the availability of weather data that are representative of the actual weather for the specific location or region for which these models are applied" (Mishoe, 1982). The implementation of an automatic meteorological measuring and recording system is expected to maintain a higher level of integrity than one necessitating a human operator. A weather station connected to a home computer is expected to reduce or eliminate human error, an inherent problem in non-automatic meteorological stations. The integrity of data collection sensors may also be better assured by examining the data in a timely manner (Titlow, 1988). To do this, the inclusion of a modem for regular communication would be necessary. This feature could yield many additional benefits as it introduces the user to yet another very powerful tool.

The components of a computer integrated system should be designed such that it operates simply and in a user friendly manner. "... to facilitate the application of crop models for those who did not develop them, they must be integrated with data bases and utilities that provide the user access to functions needed for the analysis" (Jones, undated). The only operation the farm manager should be expected to perform is some data entry and analysis of modelling results.

The crop management system should then help evaluate corn management practices using yield prediction as an output. Modelling of this system should be locally relevant both in terms of farm field specifics, such as chemical applications, as well as weather. The system should shelter the user where possible from operational details such as program execution management, data acquisition, data management and data presentation. This system should therefore include: a computer, modem, weather instruments, weather data logging device, crop model, data base and a shell to manage the operation of all the components and provide a user interface.

13.2 Crop Model

The crop model must simulate crop growth and development through the various phenological stages and represent these quantitatively with respect to time. A principal concern in the selection of a corn growth model was that it be complete, commercially available and supported. Such models are not suitable for the computer integrated system as the model must simulate the whole plant and life cycle as influenced by management and environmental factors. The desired full crop growth model could be very complex, so it must be accompanied by documentation and sample input and output files.

A search using farm software catalogues was conducted to find the best model. It was discovered that most of the programs available and relating to corn yield and corn management were not growth simulators but financial programs. The only complete program listed which modelled corn growth was Ceres-Maize.

Ceres-Maize uses weather, soil, management and cultivar parameters to predict among other things, local hydrologic cycle, nitrogen cycle, carbon cycle, phenological development and grain yield (Jones, 1986). These calculations are made with respect to time measured in days. Ceres-Maize organizes the input data in two files, the parameter input file and the weather input file. The weather file contains total radiation, maximum temperature, minimum temperature and total precipitation on a daily basis. The parameter input file contains information relating to soil type, soil water characteristics, soil chemistry, cultivar traits, irrigation and management decisions.

Ceres-Maize (Version 2.1 especially) is one of a number of crop models that use standardized input data sets (Hoogenboom, 1990). These crop models include wheat, rice, sorghum, millet, barley, potato, soybean, peanut and dry bean growth simulation (Hoogenboom, 1990). These models have been incorporated into the International Benchmark Sites Network for Agrotechnology Transfer (IBSNAT) project, part of Decision Support System for Agrotechnology Transfer (DSSAT) (Hoogenboom, 1990). This could lead to further system expansion as it facilitates future adoption of other crop models into the CICMS or integration with DSSAT. Selecting Ceres-Maize therefore forces the construction of a system basic to other crop models. This will facilitate future adoption of other crop models into CICMS or integration with

DSSAT.

13.3 Weather Station

The center-piece of an automatic weather station is the data logging system for collecting and storing weather data. The data logger works in conjunction with a PC and must feature easy transferability of weather data to the PC/XT/AT, '386 or '486. Moreover it should permit the farm computer to be used simultaneously for other operations. This means that the datalogger must also be able to operate in background mode, as XT computers are not able to perform multitasking functions.

The data logger must be able to collect the signals from the weather instruments (a) specified in the CIFM Survey and (b) required by the simulation crop model for corn growth. The most constraining factor for selecting the data logger is the cost. The CIFM farm survey indicated that only four farm managers were willing to spend between $250 and $500 on weather instruments. Most respondents would not pay more than $250. Thus an over-riding constraint of this study was that both the weather instrumentation (sensors) and the data logger were to be inexpensive, which severely restricted the choices.

Various data loggers were investigated from companies such as LI-COR, Geneq Inc and Solus Systems Inc. Keithley and Connect Tech Inc. carry data logging computer cards that satisfy the three main technical criteria: (1) support weather instruments, (2) operate in background mode and (3) easily transfer data to the PC. Unfortunately their cards and accompanying software all exceed $1,000 in cost. Only Digitar offered a data logging system for less than $1,000. For $398(US) Digitar offered (in 1991) a weather station system with a data logging computer card that satisfied the criteria and included an anemometer, wind vane, two thermistors, a precipitation gauge, barometric pressure transducer and operating software.

13.4 PC Weather

The "PcWeather" (PcW) station with expanded software is a product of Digitar of Hayward, California. The PcW station has three main components: (a) the data logger, (b) software and (c) instruments. PcW uses a computer card for data collection and storage. The card has five accessible channels, one digital and four analog. As its power source, the card uses an 8 volt adapter that must always be connected. A power failure may result in data loss. Power loss to the card can result in some operating options being reset to factory values. To prevent this an optional power adapter and charger is available to prevent this type of data loss. The computer card downloads weather data to a hard drive or computer diskette every 30 minutes.

PcW software gives the user a large number of operational and data output options. The most important feature of PcW is its ability to run in either foreground or background mode. The PcW program can be executed either by typing PCWPRO at the DOS prompt or by using the "hot keys" when in background mode. The program can be invoked from within other programs such as Word Perfect by pressing the hot key sequence. The hot keys are a combination of keyboard keys that are pressed simultaneously. The PcW program contains a number of menus from which alarms may be set, paths chosen, instruments calibrated, settings initialized and data presented. The PcW data menu options summarize weather data graphically and numerically. Screens with greater degree of detail can also be displayed as well as graphs illustrating the changing weather in 30 minute intervals over a 24 four hour period.

Operation in background mode allows access to the computer to perform regular computer operations while continuing to collect weather data. The operation of PcW in the background has almost no effect on the normal operation of whatever program is being run in the foreground. The only difference that may be observed is a temporary but short pause when PcW downloads data from the data logging card to the hard drive or diskette once every 30 minutes.

The PcW system will also allow the execution of user programs that access the weather card. In the course of temperature data investigations the BASIC program TEMP23.BAS was written by the author. It displays weather data on the screen in one-minute time-steps and at 2400 hours creates a file in which it stores the day's weather data in one minute intervals.

13.5 Weather Instruments

Some instrument alteration and development was attempted in order to better accommodate the desires of the farm managers and to provide weather data input for the crop model. The first modification was carried out on the precipitation gauge supplied with PcW which was not suitable. Second, an affordable radiation meter had to be designed.

13.5.1 Rain Gauge

Although there are a number of variations on the siphon principle, the common basic principle is a siphon working in combination with a float and a float chamber (Doorenbos, 1976). When a known amount of water is collected a siphon drains water, which causes a float to rise in the float chamber, which in turn causes a switch or lever to transmit a signal (Doorenbos, 1976). The signal then corresponds to a known volume of precipitation.

There are a number of possible sources of error associated with these types of gauges: 1. The rate at which the siphon drains is a limiting condition (Doorenbos, 1976) which, depending on rainfall intensity, could result in significant error. 2. The siphon does not completely drain the collecting container. 3. The amount of water that is at the bottom of the chamber changes with evaporation introducing an additional source of error (Doorenbos, 1976). 4. In general this type of gauge is very delicate and easily affected by dust and insects (Doorenbos, 1976). 5. Gauge calibration is very important and its neglect may cause large errors (Doorenbos, 1976).

The precipitation gauge included in the PcW package (1990) was a small siphon gauge with a float. As the float rose it closed a reed switch. Each switch closure indicated 2.54 mm (0.1") rain. The PcW siphon gauge in addition to the measuring problems already discussed is also subject to error arising from its small size. The collecting funnel has a diameter of 7.1 cm and gauges with a diameter less than 10 cm can produce additional error in total rain collection (Doorenbos, 1976).

The tipping bucket on the other hand comprises a collecting funnel, tipping bucket and a switch. The funnel collects and directs the rain water into a wedge shaped bucket. When the bucket is full, the weight of the water causes the bucket to tip, emptying the water and positioning the second bucket under the funnel. Every time the bucket tips, a switch closes a circuit indicating that a known weight of rain has been collected. This type of gauge experiences some error at the beginning and end of a rain event: the bucket may contain an amount of water in the bucket from a previous event that was not sufficient to tip the bucket. This error causes the gauge to under-measure the total amount of rainfall for that period (thus rainfall intensity is also under-measured).

If this same rain water is present at the beginning of a subsequent rainfall event, then measured rainfall error is passed on to that event. The intensity of the event as well as the total rainfall amount at the beginning of the storm is over-estimated; when the bucket tips for the first time at the beginning of an event it is impossible to estimate the time that rainfall commenced and therefore the duration for calculating intensity. Further, as with siphon gauges, evaporation of the water in the bucket between events causes an under-measurement of total rainfall.

The gauge that requires the least amount of maintenance and attention of a skilled person is the tipping bucket. It will therefore be more likely to yield better and more consistent results than the other gauges given the same amount of care. The tipping bucket gauge will also experience less data recording interruptions as maintenance and repairs will be less frequent. Erroneous data recorded while the gauge was uncalibrated or working improperly could be hard to identify and correct. These reasons make the tipping bucket gauge the most appropriate choice for a farm weather station.

A "Rainwise" tipping bucket rain gauge was purchased for under $150 (Cdn) and can be purchased in the United States for well under $100 (Cdn). The low cost of this instrument is also compatible with the low equipment cost criteria for the acceptability of this weather station. The plastic Rainwise gauge may use a 22.15 cm in diameter collecting funnel which directs water into the 10 cm³ tipping bucket. With this arrangement every tip of the bucket corresponds to 0.26 mm or 0.01 inches of rain.

As a preventive measure it was decided that the rain gauge should be equipped with a device to prevent freezing. A plastic tube was used to raise the gauge funnel from the base to provide space for an automotive light bulb to be installed. The funnel was extended internally to slightly above the tipping bucket to prevent errors due to splash. When air temperature drops below freezing the light is turned on, by the farm manager, using a 12 V DC power adapter. This arrangement also measures approximate frozen or crystallized precipitation. The error however is expected to be high because evaporation losses may be significant (the intensity of the light bulb is not adjustable). On the other hand, this technology is not being used at present at the local farm level; in fact, no winter precipitation of any kind is being measured on the farm. Estimates of frozen precipitation, and intensity, may offer other benefits, but further study is recommended to determine the measurement error.

An improvement to the light bulb arrangement was proposed by Greg Noakes (1991), an undergraduate at the School of Engineering, in a final design project. The basic approach was to use a controller to maintain a metal collecting funnel at just above freezing. This device is a definite improvement but incurs a relatively greater cost. It is also recommended that a comparative study of these two systems be conducted to assess their costs and performance.

13.5.2 Solar Radiation Meters

The purchase of commercially available pyranometers for integration into the system was avoided due to their high cost. Therefore, a number of devices were constructed and tested in an effort to measure total daily short wave radiation affordably. An Eppley pyranometer was used as the reference instrument to calibrate and judge the performance of the experimental pyranometers. The experimental pyranometers can be divided in two groups: 1. those dependant on temperature difference between different surfaces, and 2. those with photoconductive properties. All these devices respond as variable resistors, so that their signal may be measured by one or both of the temperature ports on the PcW data logging card. This approach was adopted because of the greater need for radiation and temperature data than for two temperature readings. This approach will also avoid the need to alter the existing hardware configuration. A description and discussion of these instruments follows.

Temperature Sensors

Three devices were constructed on the principle that the temperature difference between two surfaces of different reflectance could yield a measure of solar radiation. A body at the earth's surface is subject to radiation gains and losses from and to various sources. The gains are: a. incident short wave radiation (direct and reflected) and b. absorbed long wave radiation. The losses are: a. reflected and transmitted shortwave radiation and b. emitted long wave radiation. The difference between the gains and losses is referred to as net radiation (Monteith, 1973).

If two identical bodies were subject to identical radiative conditions, the net radiation each would experience would be identical. However, if one of these bodies has a white surface and the other a black surface, their net radiation will not be the same: one body absorbs radiation in the visible range while the other does not. As the two different bodies receive differing amounts of energy, their temperatures will differ, and different amounts of long wave radiation will be emitted by the bodies, as given by Equation 6.2 (De Jong,1973).

$$R_B\uparrow = \varepsilon\sigma T^4 \qquad\qquad (13.1)$$

$R_B\uparrow$ = Long wave radiation emitted by a surface (Cal.cm^{-2}.day^{-1})

σ = Stefan-Boltzman Const. (Cal.cm^{-2}.day^{-1}.K^{-4})

ε = coefficient of deviation from black body characteristics

T = Temperature (K)

The significance of the difference in long wave radiation emittance by the two bodies must be determined. In particular, their relative importance in comparison to short wave radiation must be resolved. However, since the difference in the long wave radiation emittance is a result of the different amount of short wave radiation received, the difference in net radiation should be directly proportional to the amount of visible radiation. In addition, the temperature difference between the two bodies should correlate with the amount of visible radiation for the same reason. Once these relationships are determined, the total short wave radiation can be calculated by multiplying the flux of visible radiation by 2. There will be some error in this procedure, as it assumes that visible radiation accounts for approximately half of the total short wave radiation at the surface of the bodies (Monteith, 1973). This is not always the case, especially on cloudy days, when the visible radiation fraction is greater, due to cloud moisture absorption of infra-red radiation (Monteith, 1973). The total short wave calculation also assumes that ultra-violet radiation does not contribute appreciably to the energy balance.

The first instrument constructed consisted of a glass dome and ventilated shelter as shown in Figure 13.1. A black thermistor was placed under the glass dome shelter and a second in the vertical ventilated and shaded shelter. The initial observations were encouraging, as the temperature difference between the thermistor and the radiation measurements of the Eppley seemed to respond in a similar manner as shown in Figure 13.2. An attempt was made to correlate the area under these curves. If the temperature difference between the two thermistors could be used as a measure of solar radiation, the area relationship would be evident. The accumulated daily temperature differences between these two thermistors were compared to the total radiation values of the Eppley. Unfortunately, the results did not correlate well. On a second evaluation of this design it was concluded that the two thermistors were not equally exposed to long wave radiation. Although air was able to flow horizontally over the two thermistors the same was not true vertically. This fault was undoubtedly exaggerated by the device's placement on the roof of the engineering building which is covered by a black layer of tar and gravel, and itself acts as a heat emitter.

A second device was constructed which exposed the thermistor to long wave radiation equally, as shown in Figure 13.3. Again the resulting accumulated daily differences of temperature were compared to the short wave radiation measured by the Eppley pyranometer. As with the previous instrument, the results did not correlate well. There are two possible design faults responsible for the lack of a clear relationship: 1. The thermistor was influenced by radiation in the visible spectrum. All or most of the visible radiation should have been reflected but this was not achieved; the white coating on the thermistor is believed not to have been as effective as was necessary. 2. The wire leading to the thermistor, protected by a black rubber, may have conducted radiation absorbed from the visible spectrum to the thermistor.

A third radiation meter was designed to address all the above

RADIATION METER 1

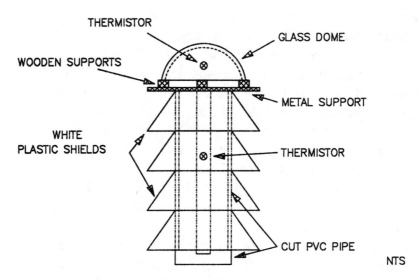

Figure 13.1: Radiation meter 1.

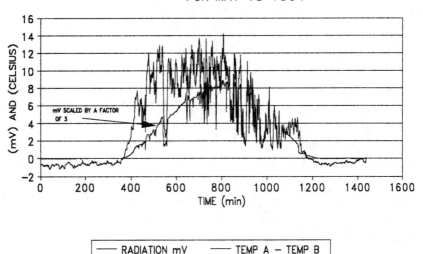

Figure 13.2: Radiation observations using meter 1.

RADIATION METER 2

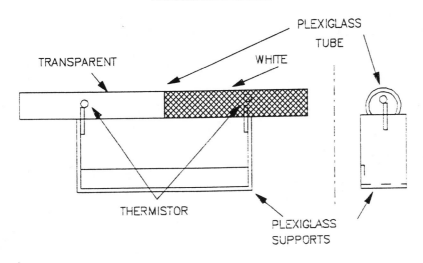

Figure 13.3: Radiation meter 2.

deficiencies in addition to instrument dampening as shown in Figure 13.4. Meter 3 used:

1. two metal (steel) cylinders, one painted black the other white, as shown in Figure 13.4,

2. a white thermistor wire leading to the white disk. This design is expected to behave similarly to the two disks of the Eppley pyranometer, and

3. an increased mass of the sensor which will yield more reliable results as it will better integrate the temperature differences.

This is significant as the time between sampling could be longer, by reducing the number of samples. This design is still to be tested. An important detail in the design of the third meter is the necessity that, in the effort to measure radiation, ambient temperature readings must not be sacrificed.   Therefore, temperature as recorded by the white temperature sensor should be compared to temperature sensors in a thermometer shelter, such as described by Doorenbos (1976) or Hanan (1984).

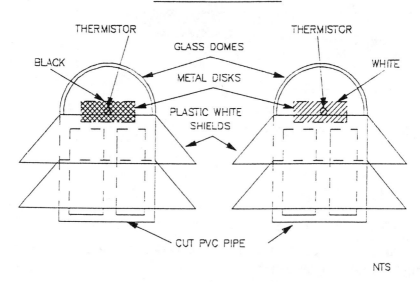

Figure 13.4: Radiation meter 3.

Cadmium Sulphide Photoconductive Cells

The cadmium sulphide photoconductive cell is a light sensitive resistor.   The resistance of this cell decreases with increasing incident illumination, and the spectral response peaks at approximately .6 μm in the visible light range (Norton, 1989). The response of these instruments to visible light may allow them to be used as radiation meters. Their suitability depends on their sensitivity and ease of integration into the CICMS.   Work is currently under way to determine just this. With a unit price of approximately $3.91 (Cdn), if successful, this instrument could easily and affordably be distributed and installed throughout the farming community.

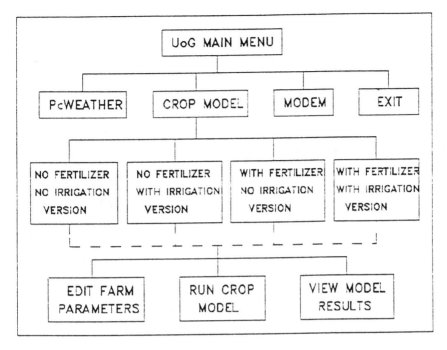

Figure 13.5: Shell's general architecture.

13.6 The Shell

The computer integrated system includes: 1. weather station, 2. data logger, 3. data logger software, 4. a DBMS, and 5. a crop model. To bring together these five components in a compatible and interactive manner, a number of R:Base and FORTRAN programs were written by the authors. It is precisely this integration effort that is the central achievement of this study; the new programs are part of a single overall management system operated by mouse-supported, pull-down menus. The general architecture of the shell and associated applications is illustrated in Figure 13.5.

The system's program manager is executed at the DOS prompt, displaying the main menu; the options are: PcWEATHER, CERES-MAIZE, MODEM and EXIT. Pressing the [F1] key displays a help screen explaining the menu selections.

Selecting PcWEATHER from the pull-down Main Menu displays a second main menu headed CERES-MAIZE. This pop-up menu offers the user the choice of one of four methods of executing Ceres-Maize: 1. Standard-No Irrigation, 2. Standard-With Irrigation, 3. Nitrogen-No Irrigation and 4. Nitrogen-With Irrigation. Ceres-Maize was organized by the writers into four subsections for the computer integrated shell, to help differentiate the four main management choices and the data each version requires. This removes confusion and data entry errors caused by the different data requirements of the four versions. The execution of any of the pop-up menu selections generates nearly identical pull-down menus with the headings EDIT MAIZE, RUN MAIZE, VIEW RESULTS and EXIT. These menus also have help screens that are invoked by pressing the [F1] key, all written by the authors. These menu selection allow easy data input through the use of input forms, value checks, rule checks and perform calculation the user would otherwise have to perform. Weather files are automatically updated, parameter data files generated and the crop model executed without user assistance. The results of the run are subsequently imported into the shell for display.

13.7 References

Bollman, Ray D., September 1988. Social and Economic Studies Division. A Profile of Farmers With Computers - Reasearch Paper Series.

De Jong, B., 1973. Net Radiation Recieved By A Horizantal Surface At The Earth. Rotterdam, Netherlands, Nijgh-Wolters-Noordhoff University Publishers.

Doorenbos, J., 1976. Agro-meteorological Field Stations. Rome: Food and Agriculture Organization of the United Nations

Gauthier, Laurent and Kok, R., 1988. A Microcomputer-based farm management/operating system. Canadian Agricultural Engineering. 30:69-76.

GENEQ INC., 1991. Instrumentation/Systems.

Hanan, Joe J., 1984. Plant Environmental Measurement. Bookmakers Guild, Longmont, Colorado.

Hoogenboom, Gerrit and J.W. Jones and K.J. Boote, 1990. Modeling Growth, Develpment and Yield of Legumes: Current Status of the SOYGRO, PNUTGRO and BEANGRO Models. The American Society of Agricultural Engineers, Columbus, Ohio, June 24-27 1990.

Jones,C.A. and J.R. Kiniry., 1986. CERES-Maize A Simulation Model of Maize Growth and Development. Texas: Texas A&M University Press.

Jones, J.W. undated. Integrating Models with Expert Systems and Data Bases for Decision Making. Climate and Agriculture.

KEITHLEY METRABYTE/ASYST/DAC, 1991. Data Acquisition & Control Hardware and Software for IBM PC/XT/AT, PS/2 and MicroChannel Computers.

Loranger, J. P., July 1990. Agricultural Computing Industry in Quebec. Canadian Association of Agricultural Computing. pg 3

Mishoe, J.W. and Jones, J.W. and Swaney, D.P. and Wilkerson, G.G., 1982. Using Crop Models For Management. I. Integration Of Weather Data. American Society Of Agricultural Engineers, Winter Meeting, Chicago, Illinois, December 14-17, 1982.

Monteith, John Lennox, 1973. Principles of Enironmental Physics. London, England: Edward Arnold Ltd.

Murray, Grant G., 1990. The Enabling Effect - Implications for Agriculture. Presented to the Agricultural Management Systems -Directions and Visions (AMSDAV) Conference, Guelph, Ont., April 30, May 1& 2, 1990.

National Instruments, 1991. IEEE-488 and VXIbus Control, Data Acquisition, and Analysis.

Newark Electronics, 1990. Catelogue 111.

Noakes, Greg, April 15, 1991. Final Report for a Snow Data Collection System. University of Guelph, School of Engineering, unpublished.

Norton, Harry N. 1967. Transducers and Sensors, in: Electronics Engineers' Handbook. Fink, Donald G. and Donald Christiansen. New York: McGraw-Hill Book Company. p. 10-37 to 10-40.

Pc Weather Pro - Installation and Basic Operation Guide. Technology Marketing Inc. Lake Oswego, Oregon, 1987.

Pc Weather Pro - Software Reference. Technology Marketing Inc. Lake Oswego, Oregon, 1987.

Pc Weather Tool Kit - Pc Weather Pro Data File Utilities Software Reference Guide. Technology Marketing Inc. Lake Oswego, Oregon, 1988.

Titlow, J. K. and K.D. Robbins., 1988. Climate Data: Collection, Quality Control and Management. American Society of Agricultural Engineers, International Winter Meeting, Chicago, Illinois, December 13-16, 1988.

Chapter 14

Quality of Stormwater From Residential Areas

Z. John Licsko, Graduate Student
H.R. Whiteley, Associate Professor
R.L. Corsi, Assistant Professor
School of Engineering
University of Guelph
Guelph, Ontario
N1G 2W1

14.1 Introduction

Stormwater from urban residential areas has been found in several studies to contain a variety of pollutants. These include solids, nutrients, chlorides, bacteria, trace organics and heavy metals. The impact of these substances on a receiving environment is controlled by such factors as concentration, duration of discharge, time of discharge and the sensitivity of the receiving environment to a particular pollutant. As point sources of wastewater, residential stormwater can be, for brief durations, substantially in excess of existing water quality standards. Research using biological indicators of environmental impact has also shown that intermittent stormwater discharge can have a measurable impact on the receiving water downstream of storm sewer outfalls

0-87371-898-4/93/$0.00 + $.50
© 1993 by Lewis Publishers

(Garrie 1986, Payne 1990, Willemsen 1990, Gast 1990).

Control of pollutants carried in stormwater from residential areas has not, however, usually been of sufficient concern to warrant specific control measures, although some American municipalities and states have in the last decade implemented substantive stormwater quality programs (Bissonette 1986, Zeno 1986, Livingston 1986, Shaver 1986).

Strategies for improving the quality of residential stormwater can range from elimination of polluting substances to end-of-pipe control. One strategy involves adding quality requirements for structures which are already required for the control of stormwater quantity. These structures include wet and dry detention ponds, wetlands, grassed swales, and groundwater recharge devices. The improvement in stormwater quality produced by these control structures is usually associated with a reduction in the suspended solids concentration in the stormwater. Pollutants such as BOD_5, heavy metals and nutrients are removed to a varying extent by the removal of suspended solids. Nutrients and metals may also be removed from sediment and from water by plant growth in these structures.

Careful design and management of structures used for stormwater quality improvement is needed. Pollutants initially removed may eventually be discharged through re-suspension or re-solubilization unless this is prevented through specific features of the design and operation of the facility.

The research reported here was undertaken to help establish whether low density residential areas and/or commercial development in residential areas warrant stormwater treatment beyond that which currently exists. Specifically, the objectives of this research were to:

1. Characterize the quality of stormwater from two residential catchments and from a commercial catchment typical of those located in residential areas. The focus was on obtaining data on the maximum, and the "steady state" or clean concentration of several stormwater quality parameters.

2. Assess the impact of existing water quantity control structures on the quality of stormwater from two residential catchments.

14.2 Literature Review

Numerous sampling programs have been undertaken in the last three decades to characterize the quality of stormwater from urban areas. In Canada, research undertaken by the International Joint Commission under the Canada-US Great Lakes Water Quality Agreement of 1972 represents the largest single source of information on urban runoff quality. A summary of the pollutants measured, their magnitude, their sources and transport, was presented by O'Neill (1979). Pollutant loadings from urban areas on a $kg \cdot ha^{-1}yr^{-1}$ basis and an analysis of stormwater abatement methods can be found in Marsalek (1978). Additionally, several studies undertaken through joint efforts between municipalities, provincial and federal governments have characterized aspects of stormwater quality and stormwater treatment structures (Ontario Ministry of the Environment 1977, 1982, 1983, & 1991, Canadian Mortgage and Housing Corporation 1980a, 1980b & 1981, Marsalek 1986).

In the United States from 1979-83 the U.S. Geological Survey (USGS) and the Nationwide Urban Runoff Program (NURP) jointly and separately conducted stormwater quantity and quality monitoring in 28 urban catchments (EPA, 1983). Thirteen detention basins were also monitored in the NURP sampling program, nine of which were wet basins or ponds (EPA, 1986). Several earlier research projects also characterized and quantified urban street contaminants in the United States (Wanielista, 1979, Sartor, 1974).

Past studies have highlighted the variability and site-specific characteristics of stormwater quality and of stormwater treatment facilities. As noted in the introduction, the pollutants commonly assessed in these studies included: solids, BOD_5, nutrients, chloride, organics, heavy metals and micro-organisms.

14.3 Description of Sampling Sites

The Fieldstone residential catchment is located in the southern portion of Guelph and is 14.2 ha in area with a density of 12 single family detached houses/ha (gross catchment area). Housing was constructed in about two thirds of the catchment in 1986, with the remaining third of the housing constructed in 1987. A grass-surfaced park with an area of about 1.5 ha forms part of the catchment. Roof downpipes drain directly onto lawns in most instances. Streets, backyards and the park area are all drained by catch basins. The entire catchment drains into a stormwater detention wet pond which was designed to control water flow rates in the catchment to pre-development rates.

The Brandt catchment is located in the northeast corner of Guelph. This catchment covers 18 ha in area and has a single family detached housing density of 11 single detached houses/ha (gross catchment area). The area was developed in 1971. A school and park area, mostly grassed, encompass approximately four hectares of the catchment. Roof downpipes are connected to the stormwater sewer drainage system directly. The catchment drains into a dry storm water detention pond. Each residential catchments had total curb lengths of approximately 3600 m.

The commercial catchment is 1.25 ha and has thirteen retail stores and a gasoline station on the site. Drainage from the parking lot area and the roofs are conveyed to a storm sewer directly into a wetland.

14.3.1 Wet and Dry Pond Characteristics

The surface area of the wet pond at Fieldstone is 0.16 ha (1.1 percent of the catchment area) with an average depth of about 30 cm giving the pond a total volume of approximately 500 m^3. Measured pond overflow rates during storms ranged from a maximum of 20 cm/hr (24.2 mm storm with a two hour duration) to a minimum of 1 cm/hr (8.6 mm storm with a four hour duration).

The dry pond at Brandt encompassed 1.1 ha (6.1 percent of the catchment area). The pond makes use of a naturally occurring grassed wetland. A field on the south and east sides of the pond and a wooded area on the pond's west side sometimes contribute runoff that is in addition to the inflow of stormwater from the catchment. Overflow rates were not monitored for this pond.

14.4 Sampling Procedures

Stormwater quality depends on site and weather conditions which can be highly variable. Consequently, a consistent program of field sampling is important (Huber, 1986). Moreover, a field sampling program for stormwater quality is an expensive endeavour both in terms of the actual sampling and subsequent water quality analyses.

Various sampling methods have been employed in the field to collect data on stormwater quality. Procedures range from street vacuuming and flushing, to pollutant measurements in streams or rivers upstream and downstream of stormwater outfalls. Results are expressed as: mean event concentrations, area unit loadings of pollutants on an annual basis ($kg \cdot ha^{-1} yr^{-1}$), or mass loadings of pollutants per curb length.

In some instances the actual concentrations of pollutants are not as important as is the environmental impact of stormwater and as a result, in situ toxicity testing has been advocated. Both the effect of the intermittent nature of stormwater pollution and of the biological availability of stormwater pollutants can then be assessed by in situ tests (Bascombe, 1990).

Stormwater quality impacts are, however, short term for some water quality parameters (e.g., bacteria and temperature) and long term for others (e.g., nutrient loads). EPA sampling procedures under the National Pollutant Discharge Elimination System (NPDES), for instance require stormwater grab sampling within the first 30 minutes of an event as well as the collection of an event-mean concentration sample. The first sample gives an indication of the maximum concentrations, which is important for some aquatic life, and which, in stormwater toxic management programs, form the basis for acute toxicity testing requirements (Collins, 1991). The event-mean concentration provides information on the longer-term impacts of stormwater pollution such as sediment and nutrient loading.

The choice of sampling procedures, including type of sample collection and list of parameters to be tested, involves balancing cost and effort against expected benefits in defining impacts. In this study, the variability associated with seasonal changes was included by conducting stormwater sampling from March, 1991 to November of 1991. Grab samples were collected at least twice during each event from four or five different locations in the two urban catchments and from one location at the commercial site. At each location, one grab sample was collected as close as possible to the start of each rain event investigated while the second was collected as close as possible to the end of the storm event. Sampling was conducted in this manner to obtain a maximum range of water quality values and a "steady state" value associated with the end-of-event sample. Since both residential sites had sampling completed at both the inlet and outlet of stormwater quantity control ponds this method of sampling also provided a measure of each pond's effectiveness for removing several pollutants.

Water quality parameters tested for in this study included temperature and dissolved oxygen at the time of sample collection, conductivity, pH, biochemical oxygen demand (BOD_5), conductivity, turbidity, fecal and total coliform (MPN/100 ml), sodium, chloride, nitrate, ammonia, total phosphorus, total lead, total solids, suspended solids and dissolved solids. The volatile content of total, suspended and dissolved solids tests were also obtained. This provided a wide range of information on pollutants likely to be included in legislation on stormwater quality.

Also, these test results provide for a comparison of relatively inexpensive tests such as water temperature, dissolved oxygen, pH, conductivity, and turbidity with more expensive and complicated water quality tests. Such a comparison might provide justification for the use of inexpensive surrogate indicators for stormwater quality monitoring.

14.4.1 Sample Analysis

Most sample analyses were completed as outlined in Standard Methods (1985). Exceptions to the procedures outlined in Standard Methods occurred for the analysis of sodium, chloride, total lead and total phosphorus. Sodium and chloride ions were analyzed with the use of Orion ion selective electrodes and analyses were performed as outlined in the supporting documentation for these probes. Total phosphorus and total lead analyses were completed at the Land Resources Water Quality Laboratory (University of Guelph). For both the determination of total lead and of total phosphorus a 25 ml sample was first evaporated. The residue was then digested with sulphuric and nitric acid until the nitric acid had dissipated. The sample was then, with the addition of distilled, deionized water, brought back to the original sample volume of 25 ml. An atomic adsorption spectrophotometer (Varian 300) was then used for the total lead determination, while a spectra-photometer (Technocan auto analyzer) was used in the total phosphorus determination.

14.5 Results

A number of water quality parameters were measured, as mentioned earlier in the stormwater samples collected during this study. Space restrictions, however, permit the discussion of only some of the results obtained. An attempt has been to include those parameters commonly reported for urban stormwater as well as some where interesting trends were observed. Included for discussion were: suspended solids, dissolved solids, BOD_5, fecal coliform, total phosphorus, sodium and chloride.

Similarities in the measured water quality parameters in the two residential catchments resulted in the presentation of an

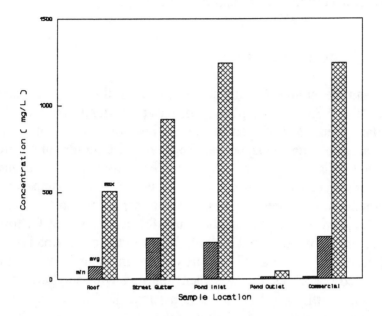

Figure 14.1:   Variations in suspended sediment concentration by
              location.

overall average for data collected in each of these areas. This
includes all four sample locations in each of the two residential
catchments: roof, street gutter, pond inlet and pond outlet.
Exceptions to this method of analysis were noted were applicable.

14.5.1 Suspended Solids

Average concentrations of total suspended solids (tss) from
samples collected were 238 mg/L (street gutter), 212 mg/L
(pond inlets) and (240 mg/L) for the commercial site.   The
concentrations were similar for the various seasons over the
sampling period.   Waller (1980) reported an annual mean
concentration of suspended solids from the surface runoff in

urban areas in Ontario of 170 mg/L.

The concentration of suspended solids at the pond inlets showed a larger range than did the street gutter flow, presumably due to deposition and re-suspension of sediment in the sewer system as shown in Figure 14.1.

Suspended solids concentrations in roof runoff exhibited a high first flush concentration, but this quickly diminished. Based on the time of sampling in terms of an estimated accumulation of rainfall the quality of roof water approached that of rainwater after 2-3 mm of rainfall.

The concentration of tss in the street gutter and pond inlet did not approach that of rainwater until between 10-15 mm of rainfall had accumulated. This is consistent with research by Wanielista (1977) that indicates that up to 95% of pollutants are washed off in the first 12.7 mm of rainfall for watersheds less than 40 ha in size. For watersheds over 40 ha the percentage of pollutants washed off in the first 12.7 mm of rainfall reduces with increased watershed area. At the commercial site, which has an area of 1.25 ha, for two of the four storms for which data were collected, the suspended solids concentration dropped by an average of 95% after 6.5 mm of rainfall.

Suspended solids at both the dry and wet ponds were effectively removed by detention with an average removal proportion for the two ponds of 94 percent. Seasonal effects, however, were apparent in the functioning of both ponds. In late winter and early spring the wet pond can be frozen and flows from the inlet pass over and under the ice sheet to the outlet. The pond outlets may also be partially blocked by ice and snow accumulation and outflow vs storage height relations are altered. Frozen outlets can result in either complete retention, or in the case where outflows bypass the outlet, a shorter retention time for the incoming stormwater.

The dry pond appeared to retains ability to detain stormwater and to allow the effective settling of sediment even in the winter and spring. Apparently it retains a substantial effectiveness as a grassed filter.

Unit loadings on an annual basis were calculated for

comparison with previous research. The calculations were based on an 830 mm annual average rainfall for Guelph, a runoff coefficient of 0.45 for the residential areas and 0.85 for the commercial area. Mass loadings of tss were based on the average concentration measured at the pond inlets.

A comparison of the estimated annual loading of suspended solids from the two residential and the commercial catchments showed an estimated annual loading similar to research previously completed in Ontario for residential areas as shown in Table 14.1. Calculations of the ss loading from the commercial area were found to be substantially higher than that previously measured while calculations for the residential area resulted in loadings similar to previously reported values.

A grid survey of the wet pond's sediment found an average accumulated sediment depth of 5.7 cm. Using the above unit area loading for this study, a measured specific gravity of 1.23 and 5 1/2 years of pond use resulted in an estimated sediment accumulation of 3.4 cm. Development of individual lots during the course of 5 1/2 years are the likely sources of the balance of sediment accumulation.

Table 14.1: Unit area loads of suspended solids.

Study	Annual Unit Loads $(kg \cdot ha^{-1} yr^{-1})$	
	Residential	Commercial
Singer (1977)	619	825
Marsalek (1978)	388	233
This Study	840	1993

Measurements of volatile suspended solids (vss) at the street gutter and pond inlets averaged 34 mg/L or 15% of total suspended solids while 5 mg/L (42% of tss) of vss was measured at the pond outlets as shown in Figure 14.2.   Reduction of volatile suspended solids during retention for the two ponds averaged 86%.   However, the dissolved volatile solids after retention increased by 58%, on average for the two ponds, resulting in only a 12% overall reduction in total volatile solids during detention in the ponds.

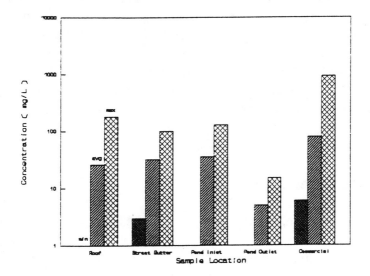

Figure 14.2:   Variation   in   volatile   suspended   solids concentration by location.

Volatile suspended solid concentrations measured at the commercial site were 2.3 times higher than those measured in the residential catchments at the inlets to the ponds and at the street gutter.   A possible explanation for this is that oil and grease concentrations are associated with the suspended solids and contribute much of the volatile portion.   Three studies cited in Stenstrom (1984) found between 81-96 % of oil and grease was adsorbed or otherwise attached to suspended solids and that this oil and grease was three times higher in runoff from commercial and parking lot areas than from runoff in residential areas (Stenstrom, 1984).

14.5.2 Dissolved Solids

There was a consistent increase in the concentration of dissolved solids, from the street gutter (63 mg/L) to the pond inlets (104 mg/L) and finally the pond outlet (252 mg/L). Between the pond inlet and pond outlet there was on average a 60% increase in volatile dissolved solids and a 142% increase in non-volatile dissolved solids. This increase is unexplained at this time and is likely due either to a sampling effect created from outlet samples bearing pre-event concentrations of pond water or the presence of undetected groundwater inflow between events. Evaporation rates of up to 12.5 mm/day were measured at the wet pond and are responsible for a portion of the increased concentrations of dissolved solids in the ponds.

14.5.3 Total Phosphorus

Total phosphorus concentrations were highest at the pond inlets, averaging 0.29 mg/L, 38% higher on average than those phosphorus values measured at the street gutter (0.18 mg/L) (Figure 14.3). The likely explanation for this is a first flush effect of catch basin sumps which tend to accumulate organic matter during periods of low flow. Septic conditions often occur in the sump with an accompanying release of nutrients from the vegetation. A mean annual concentration of 0.35 mg/L of total phosphorus was reported by Waller (1980) for surface runoff from urban areas in Ontario. The median event-mean concentration measured during the NURP (EPA, 1983) study was 0.33 mg/L.

Since phosphorus is often chemically bound to sediment and colloidal material, the removal of suspended solids in detention ponds should also remove some phosphorus. The average percent reduction of phosphorus levels across the two ponds was

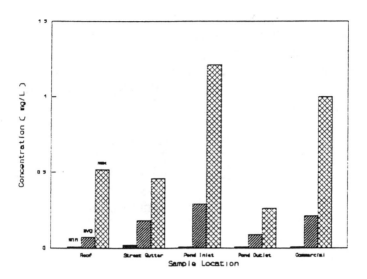

Figure 14.3:   Variation in total phosphorus concentration by
               location.

approximately 70 percent.   Average removal rates for total
phosphorus from wet detention ponds monitored by US EPA
(1986) ranged from 3% to 79%. Monitoring data from detention
ponds monitored in Ontario cited phosphorus removal efficiencies
from between 25% to 86% (Ontario, MOE 1991).

Whether the phosphorus removal during pond detention
obtained in this study and in others is real or apparent is not,
however, clear. The availability of phosphorus for biological
growth is   normally   associated   with   the   dissolved   or
orthophosphate fraction of the total phosphate, which is unlikely
to be affected by a short pond retention time. Additionally, algae
which often grow in wet ponds are quite capable of extracting
and using the phosphorus deposited with sediment (G. van Eck,
1982).   This phosphorus could later be released as dissolved
phosphorus during decay of the algae. Additionally, conditions
in a wet pond such as low oxygen levels and elevated

temperatures may also enhance the release of dissolved phosphorus from the sediment for use in biological processes.

Comparison of the data on an annual unit load basis using the method of calculation described earlier for suspended solids shows good agreement for the phosphorus loading rates from the residential areas as shown in Table 14.2. Phosphorus loadings from the commercial area were less than half of that found in previous research.

Table 14.2: Unit area loads of total phosphorus.

| Study | Annual Unit Loads $(kg \cdot ha^{-1}yr^{-1})$ | |
	Residential	Commercial
Marsalek (1978)	1.5	3.5
Waller (1980)	1.1[1]	
This Study	1.1	1.5

[1] Urban area in total.

Total phosphorus was one of two parameters which exhibited different concentrations between the two residential catchments. The newer catchment had substantially higher concentrations of total phosphorus, almost three times more than the older area at both the street gutter and at the pond inlet.

14.5.4 Biochemical Oxygen Demand (BOD$_5$)

Pond inlets and roof downspouts were both characterized by high first-flush BOD$_5$ values which quickly diminished. The commercial catchment had the highest concentrations of BOD$_5$ as shown in Figure 14.4.

Reductions in BOD$_5$ concentrations in the detention ponds were due to the settling of suspended organic material as indicated by the measured reduction of volatile suspended solids (86%) between the pond inlet and outlet.

The average BOD$_5$ reduction in the two ponds was about 30%

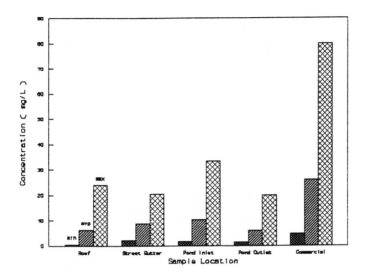

Figure 14.4: Variation in BOD$_5$ concentration by location.

This represented a BOD$_5$ reduction from 10.4 mg/L to 6.1 mg/L. The average reduction of BOD$_5$ across the dry pond was substantially higher than that for wet pond: 50% vs 10 %.

For this study, an annual unit BOD$_5$ load of 38.8 kg·ha^{-1}yr^{-1} was calculated. Unit load data reported by Marsalek (1978) for urban test catchments in Ontario ranged from 18.9 to 35.3 kg·ha^{-1}yr^{-1}.

Data on BOD$_5$ obtained from measurements in storm sewers in the United States indicated loadings ranging from 28 to 43 kg·ha^{-1}yr^{-1} (Heaney el al., 1977). An annual unit loading for the commercial catchment was calculated to be 184 kg/ha^{-1}yr^{-1}.

14.5.5 Fecal Coliform

Fecal and total coliform are indicators of the presence of pathogenic organisms in water. For water bodies which receive stormwater and have a recreational use that involves body contact this is a concern. Discharges from all three catchments sampled in this study went directly to a wetland and micro-biological constituents would not be expected to have an adverse impact in this environment.

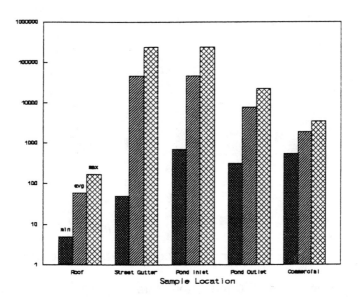

Figure 14.5: Variation in fecal coliform count by location.

The lowest average counts (an arithmetic mean was used in expressing the counts for coliform) recorded for fecal coliform were from the roof downspout samples (59 MPN/100 mL) and the samples collected from the commercial catchment (1880 MPN/100 mL). Highest coliform values where recorded from the street gutter and the inlet to the ponds. Both of these locations had similar average counts for coliform (46,000 MPN/100 mL). Although the outlets from the ponds showed a reduced coliform count the absolute magnitude of coliform counts in water leaving the ponds remained high.

Typical concentrations of fecal coliform counts in street gutters and from rooftops reported previously found fecal coliform counts for these sample locations of 10,000 and 100 MPN/100 mL respectively (Ontario Ministry of the Environment, 1983).

Sediment and catch basin sump water were found in the same study to have low fecal coliform counts, 20/g solids and 20 MPN/100 ml respectively. Catch basin sumps, in particular, would not be expected to contribute significant amount of coliform to the stormwater.

14.5.6 Sodium and Chloride

Pond inlet and outlet sampling, as would be expected, resulted in a similar average concentration for both measurements of chloride and sodium. The normal ratio of Cl/Na (obtained from solution of NaCl) was observed. A lower average for both sodium and chloride at the street gutter as compared to the pond inlets was due to the sampling point for gutter flow being on a minor road that did not receive as much salt as did the main road. Higher concentrations of sodium and chloride would likely have been recorded had sampling been completed immediately after road salt application which was not the case for most of sampling done.

The highest concentration of chloride recorded during this study was from the wet pond outlet (493 mg/L). Concentrations of chloride from residential areas have been recorded at concentrations as high as 786 mg/L (Waller, 1986).

14.6 Conclusions

The following conclusions can be drawn:

1. Results of the sampling program in Guelph indicated that the concentrations and mass loadings of BOD_5, suspended solids, total phosphorus, and fecal coliform were similar in magnitude to previous studies. Published annual loadings are a reasonable base for the design of quality control structures.

2. The concentration and mass loading of BOD_5, and the mass loading of suspended solids from the commercial area differed substantially from that found in previous research and warrants further investigation.

3. Most pollutants were washed off the rooftops within the first 2-3 mm of rainfall, within 6-7 mm of rainfall for the commercial area and within 10-15 mm for the residential areas.

4. Both the wet and dry quantity control structures can be effective at reducing suspended solids and associated pollutants. Reductions of 95% for suspended solids , 70% percent for total phosphorus and 30% for BOD_5 were measured for the wet and dry ponds in this study.

5. Although reductions of fecal coliform and BOD_5 were observed, neither were effectively removed by pond detention.

6. Incorporating a grassed-swale type filter into wet ponds could be an effective method of improving the performance of the these structures during winter and spring seasons.

14.7 References

Bascombe, A.D., Ellis, J.B., Revitt, D.M. and Shutes, R.B.E.(1990). The development of ecotoxicological criteria in urban catchments. *Water Science and Technology*, 22(10/11):173-179.

Bissonnette P. (1986) Bellevue's urban storm water permit and program, in *Urban runoff quality-impact and quality enhancement technology*, eds. B. Urbonas and L.A. Roesner. American Society of Civil Engineers.

Canada Mortgage and Housing Corporation (1980a). *Evaluation of stormwater impoundments in Winnipeg.*

Canada Mortgage and Housing Corporation (1980b). *Stormwater management technology systems demonstrations in the city of St. Thomas.*

Canada Mortgage and Housing Corporation (1981). *Stormwater runoff treatment by impoundment: Barrhaven Pilot Study.*

Collins, M.A., Roller K. and Walton G. (1991). Management of toxic pollutants in storm water runoff. Paper #AC91-056-004, Water Pollution Control Federation, 64th Annual Conference, Toronto, Ontario, Oct. 7-10.

Environmental Protection Agency (1983). Results of the Nationwide Urban Runoff Program: Water Planning Division, Washington, D.C., vols 1-3.

Environmental Protection Agency (1986). Methodology for Analysis of Detention Basins for Control of Urban Runoff Quality, Nonpoint Source Branch. Office of Water, Washington, D.C.

Garrie, H.L. and A. McIntosh. (1986). Distribution of benthic macroinvertebrates in a stream exposed to urban runoff. *Water Resources Bulletin*, 22(3):447-458.

Gast, H.F., Suykerbuyk, R.E.M. and Roijackers R.M.M. (1990). Urban storm water discharges: effects upon plankton communities. *Water Science and Technology*, 22(10/11):155-162.

Heaney, J.P., et al. (1977). Nationwide evaluation of combined sewer overflows and urban stormwater discharges. EPA-600/2-77064 Environmental Protection Agency, Washington, D.C. 1977.

Huber W. (1986) Deterministic Modeling of Urban Runoff Quality. *Urban Runoff Pollution*, ed. H.C. Torno, J. Marsalek and M. Desbordes, NATO ASI Series, Springer-Verlag Berlin Heidelberg. pp.39-57.

Livingston, E.H. (1986). Stormwater Regulator Program in Florida in *Urban runoff quality-impact and quality enhancement technology*, eds. B. Urbonas and L.A. Roesner. American Society of Civil Engineers.

Marsalek, J. (1978). Pollution due to urban runoff: unit loads and abatement measures. Environmental Hydraulics Section, Hydraulics Research Division, National Water Research Institute, Canadian Centre for Inland Waters. Burlington, Ontario.

Marsalek, J. (1986). Toxic contaminants in urban runoff: a case study. *Urban Runoff Pollution*, ed. H.C. Torno, J. Marsalek and M. Desbordes, NATO ASI Series, Springer-Verlag Berlin Heidelberg. pp.39-57.

Marsalek J. (1991). Urban drainage in cold climate: problems, solutions and research needs. Rivers Research Branch, National Water Research Institute, Canadian Centre for Inland Waters, Burlington, Ontario.

O'Neill J. E. (1979). *Pollution from urban land use in the Grand and Saugeen watersheds*. Ontario Ministry of the Environment.

Ontario Ministry of the Environment, Laboratory Services Branch (1977). Microbiological characteristics of stormwater runoffs at East York (Toronto) and Guelph separate storm sewers.

Ontario Ministry of the Environment, Pollution Control Branch (1982). Physical-chemical treatment and disinfection

Ontario Ministry of the Environment (1983). Rideau River Stormwater Management Study.

Ontario Ministry of the Environment (1991). Stormwater Quality Best Management Practices.

Payne, J.A., and P. D. Hedges (1990). An evaluation of the impacts of discharges from surface water sewer outfalls. *Water Science and Technology*, 22(10/11):127-135.

Ross, W.C. and S.K. Lior (1991). Impacts of municipal stormwater management programs on industrial dischargers. AIChe Conference, 1991 Summer National Meeting, August 20, 1991, Pittsburgh, PA, Session no.30 Paper no. 30B.

Sartor, J.D., G.B. Boyd and E.M. Archibald. (1974). Water pollution aspects of street contaminants. *Journal Water Pollution Control Federation*, 46:458-467.

Stenstrom, M.K., Silverman G.S. and Bursztynsky. (1984). Oil and grease in urban stormwater. *Journal of Environmental Engineering*, 110(1):58-72.

Shaver, E.H. (1986). Infiltration as a stormwater management component, in *Urban runoff quality-impact and quality enhancement technology*, eds. B. Urbonas and L.A. Roesner. American Society of Civil Engineers.

Singer, S. (1977) Unpublished technical memorandum, Ontario Ministry of the Environment, Toronto, cited in O'Neill, J. E. (1979), *Pollution from urban land use in the Grand and Saugeen watersheds*, Ontario Ministry of the Environment.

Standard Methods for the Examination of Water and WasteWater, 16th ed., American Public Health Association, 1985.

Waller D.H. and W.C. Hart (1986). Solids, Nutrients and Chlorides in Urban Runoff. *Urban Runoff Pollution*, eds. H.C. Torno, J. Marsalek and M. Desbordes, NATO ASI Series, Springer-Verlag Berlin Heidelberg. pp.59-85.

Waller D.H. and Z. Novak (1980). Pollution loadings to the great lakes from municipal sources in Ontario. *Journal of Water Pollution Control Federation*, pp. 387-395.

Wanielista, M.P. (1977). Manual of Stormwater Management Practices. Final report submitted to the East Central Florida Regional Planning Council cited in Livingston, E.H. (1986), Stormwater Regulator Program in Florida, in *Urban runoff quality-impact and quality enhancement technology*, eds. B. Urbonas and L.A. Roesner. American Society of Civil Engineers.

Willemsen, G.D., Gast, H.F., Franken, R.O.G. and Cuppen J.G.M. Urban storm water discharges: effects upon communities of sessile diatoms and macro-invertebrates. *Water Science and Technology*, 22(10/11):147-154.

Zeno D.W. and C.N Palmer (1986). Stormwater management in Orlando, Florida, in *Urban runoff quality-impact and quality enhancement technology*, eds. B. Urbonas and L.A. Roesner. American Society of Civil Engineers.

Chapter 15

Simulation of Stormwater Management Pond Configurations

Ronald L. Droste, A. Charles Rowney, and Craig R. MacRae
Gore and Storrie Ltd.
857 Northwest Road, Kingston, Ontario, K7P 2N2

Treatment of stormwater in detention facilities is central to stormwater management practice. The performance of these facilities, however, is contingent on the mode of operation and a number of design variables. This chapter investigates the sensitivity of the performance of batch and continuous plug flow (PF) operation modes to variations in these design variables on a fractional removal and exceedance basis for fecal coliforms (FC) and suspended solids (SS).

15.1 Introduction

Detention of stormwater in ponds, possibly incorporating physical or chemical treatment, is the most commonly employed method for achieving effluent quality objectives for urban runoff. Spatial, financial and other engineering constraints, however, place an upper limit on the size of these facilities. Consequently, as water quality objectives become more stringent, the performance of

0-87371-898-4/93/$0.00 + $.50

these facilities, as measured by removal efficiency and number and duration of exceedances, has become a critical design constraint. Performance, however, is a function of mode of operation (batch or continuous), and pond configuration. The latter consist of a number of design parameters relating to flow dynamics within the facility and the temporal duration of treatment processes acting on the body of water contained within the facility as a whole.

Inefficiencies associated with short-circuiting, dead storage zones, and re-entrainment of settleable materials are typically observed in practice. A continuous hourly runoff quantity and quality simulator with treatment algorithms was used to evaluate the sensitivity of pond performance to variations in selected design parameters. The analyses was carried out for FC and SS for a long term precipitation series. The model was calibrated for quantity and quality to two typical urban catchments in the Ottawa area.

These analyses formed the basis of a systematic evaluation of the performance of a range of pond configurations in terms of removal efficiency and number and duration of exceedances. Parameters examined were: mode of operation (continuous and batch); pond volume; number of cells in the pond system; permanent pool (PP) size; discharge pipe diameter; batch mode detention time (TD); catchment runoff and quality characteristics; pollutant removal rate coefficient; and dry weather flow routing/by-passing. Four pond types considered representative of the likely range of pond configurations and modes of operation were evaluated. These were: continuous mode, PF (5 cells); continuous mode, completely mixed (CM) flow (1 cell); batch mode, 24 hour detention; and, batch mode, 72 hour detention. All ponds were simulated with and without a PP.

15.2 Simulation Methodology

The overall treatment performance of a pond will be influenced by the ratio of influent volume to active storage volume, interevent time, shape of the hydrograph, baseflow rate, and quality of the influent. To account for the random variation of these input factors, a long term simulation approach was adopted. A historical 29 year rainfall time series (1960-1988), with actual variation of wet and dry periods for the area of interest was obtained for a gauge at Ottawa International airport.

The analysis of the large number of combinations of pond configurations, modes of operation, and other variables of interest would generate large, cumbersome data files. To facilitate the analysis, statistical methods were used on the generated flow quantity-quality time series to derive a minimum precipitation database required to produce representative simulations of the long term performance of any configuration (Droste et al., 1992; RMOC, 1992). This representative precipitation file which consisted of six key years ranging from wet-to-average-to-dry years, was then used in all subsequent analyses.

The continuous stormwater quantity-quality model QUAL-HYMO (Rowney and MacRae, 1992(a)), was selected as the simulator because of its versatility and ease of use. Excess precipitation is generated in the upland zone using a modified SCS-CN method for pervious areas and a volumetric runoff coefficient for impervious areas. Overland flow routing is performed using either the Williams' or Nash Unit Hydrograph methods. Baseflow was determined using a modified baseflow recession constant method. The simulations were performed over the swimming season, June through September of each year. Consequently, soil freeze-thaw and snowmelt processes, although accounted for in the model, were not utilized.

Pollutant loadings were determined using a buildup-washoff method for FC and a rating curve approach for SS in five separate size fractions (NURP distributions were assumed, EPA, 1986). Flow routing through a storage facility is achieved using the storage indication method. First order decay processes are applied to simulate bacterial dieoff and discrete particle settling is used to model sedimentation processes. A schematic of the POND routine in QUALHYMO is shown in Figure 15.1. The algorithms are described in detail by Rowney and MacRae

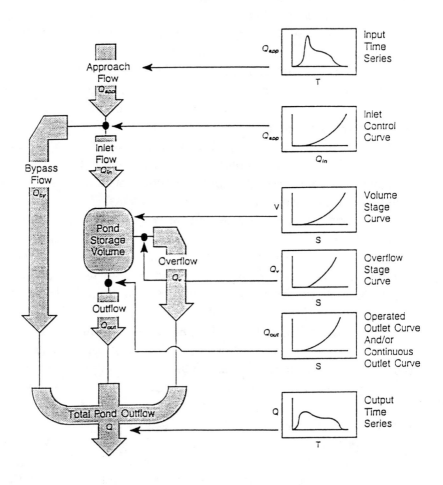

Figure 15.1: Control pond routing in QUALHYMO.

(1992b).

The model was calibrated to two Ottawa area watersheds, the Coventry and Merivale Trunk Storm Sewer (MTSS) catchments. Discrete events were sampled for FC and SS concentrations during 1981 (three events) and 1985 (three events) in the Coventry and MTSS catchments, respectively. The calibrated QUALHYMO runoff quantity-quality parameters for these catchments are described in RMOC (1992).

Land use in the Coventry catchment is predominantly single residential housing while in the MTSS woodland and industrial land uses predominate. The impervious fractions are approximately the same in each catchment (21% for Coventry and 19% for MTSS) but the Coventry catchment has much shorter times to peak. The major difference between the two catchments is the considerably lower quality of runoff from the Coventry catchment compared to the MTSS system. RMOC (1992) contains significantly more detail on calibrations and catchment characteristics.

15.3 Pond Design Variables

The number of pond parameters that could have a significant effect on treatment were categorized as follows:

- Mode of operation: batch and continuous flow (CF)
- Pond active storage volume
- Effluent pipe diameter
- TD in batch mode
- Number of cells in the pond system
- PP size
- Storm influent quantity and quality characteristics
- Pollutant removal rate coefficients
- Routing dry weather flow (DWF) into or around the pond

These parameters were systematically varied in over 7,000 simulations to develop performance and design curves.

Performance was measured in terms of removal efficiency (fractional removal) and number and duration of exceedances (FC at 100 and 200 counts/dl and SS at 20 and 50 mg/L).

15.4 Simulation Results

It is not necessarily true that a configuration change will produce a consistent change in performance from dry to wet years because of all the variables affecting treatment, particularly event volume distributions and separation times. Therefore results from the six years were generally averaged.

15.4.1 Batch Treatment

The results of the simulations were somewhat unexpected for the batch mode of treatment. It was found that most of the variables except pond volume, DWF and the dieoff rate coefficient did not have any significant influence on the degree of treatment or the number of exceedances within the ranges of parameter values examined. Results for batch and CF modes of operation for percent removal, number and hours of exceedances are shown in Table 15.1. The effects of DWF and variations in the dieoff rate coefficient are discussed separately. The average results of all treatment parameters and the upper and lower lines that are one standard deviation away from the mean value are plotted as a function of pond volume in Figures 15.2 (fraction removed), 15.3 (number of exceedances), and 15.4 (duration of exceedances) for FC only.

Active Storage Volume

The active storage volume was varied from 3 to 20 mm (depth over the watershed). The standard deviations of the average treatment (whether percent removal, number or duration of exceedances) as shown in Figures 15.2 to 15.4 is primarily due to variation in the meteorological input parameters between years.

For FC, pond performance increased 38% for removals, number of exceedances decreased by 68% to 80%, and duration of exceedances decreased by 82% to 88% over the 3 to 20 mm range in volumes. Efficiency with respect to SS increased by approximately 26% for fraction removed, from 90% to 96% for number of exceedances, and on the order of 95% for duration of exceedances. The treatment performance curves follow a classical diminishing returns pattern with the inflection point occurring between 6 and 9 mm for fraction removed, and at approximately 9 mm for number and duration of exceedances. The standard deviation as a percent of the average value ranges from approximately 16% to 2.6% (fraction removed), 28% to 3.2% (number of exceedances), and 53% to 109% (duration of exceedances) for 3 to 20 mm pond volumes respectively. With the exception of the influence of DWF and variation in the dieoff rate coefficient, the standard deviation at any volume exceeds the variation in performance attributed to all other design variables. The rationale for this observation is discussed in the following sections.

Effluent Pipe Diameter

The discharge pipe diameter did not have an influence on treatment in a batch pond because the large discharge pipe sizes used for the batch ponds drained the pond quickly. Pond drain times for pipe sizes used in the simulations are shown in Table 15.2 for the MTSS system.

For 610 mm diameter pipes or larger the maximum draining time for ponds with a volume of 12 mm or less (for 90% of the volume) is 21 hours which is considerably less than the average interevent time. The time difference for draining lower fractions of the pond volume as a function of the minimum and maximum discharge pipe diameter is much less. Therefore a discharge pipe size of 610 mm or greater does not have a significant influence on the performance of a batch pond.

Table 15.1: Removal statistics for MTSS ponds.

Operating Mode[***]	Det. Time (hr)	No. of Cells[*]	PP[**] (%)	Pond Volume (mm)				
				3	6	9	12	20
FC Fractional Removal								
BATCH	24	1/5	0/20	0.724	0.876	0.935	0.960	0.981
	72	1/5	0/20	0.717	0.865	0.925	0.952	0.980
CONTINUOUS	–	1	100	0.612	0.749	0.817	0.857	0.915
	–	5	100	0.732	0.878	0.933	0.959	0.983
CONTINUOUS	–	1	50	0.565	0.728	0.811	0.859	0.921
	–	5	50	0.673	0.848	0.918	0.951	0.981
SS Fractional Removal								
BATCH	24	1/5	0/20	0.734	0.823	0.865	0.890	0.925
	72	1/5	0/20	0.747	0.838	0.882	0.907	0.943
CONTINUOUS	–	1	100	0.701	0.791	0.838	0.867	0.912
	–	5	100	0.773	0.862	0.905	0.929	0.959
CONTINUOUS	–	1	50	0.682	0.775	0.825	0.856	0.904
	–	5	50	0.744	0.835	0.881	0.909	0.947
FC No. of Exceedances								
BATCH	24	1/5	0/20	17.1	12.2	9.2	7.2	5.4
	72	1/5	0/20	10.0	6.1	4.3	2.7	2.0
CONTINUOUS	–	1	100	22.5	21.7	20.6	19.4	16.6
	–	5	100	22.3	20.4	19.3	17.9	13.4
CONTINUOUS	–	1	50	20.8	18.9	18.3	16.9	14.2
	–	5	50	20.7	18.4	17.1	15.7	13.2
SS No. of Exceedances								
BATCH	24	1/5	0/20	14.4	5.5	2.6	1.3	0.5
	72	1/5	0/20	9.7	5.9	3.4	1.6	0.9
CONTINUOUS	–	1	100	21.4	14.3	10.9	9.0	4.8
	–	5	100	15.6	9.9	7.5	5.5	4.1
CONTINUOUS	–	1	50	20.0	15.6	12.8	11.0	7.9
	–	5	50	14.2	9.7	7.3	6.6	6.3
FC Duration of Exceedances								
BATCH	24	1/5	0/20	273.9	194.9	125.2	83.5	47.9
	72	1/5	0/20	278.5	201.8	130.0	83.3	34.2
CONTINUOUS	–	1	100	428.5	462.8	471.6	435.7	352.9
	–	5	100	325.0	254.0	186.4	140.8	79.3
CONTINUOUS	–	1	50	583.8	616.7	600.8	558.3	471.3
	–	5	50	476.3	346.8	259.6	195.9	103.8
SS Duration of Exceedances (hr)								
BATCH	24	1/5	0/20	232.0	134.5	64.8	33.0	9.6
	72	1/5	0/20	246.9	148.4	71.5	38.9	12.2
CONTINUOUS	–	1	100	308.3	186.9	115.0	70.2	18.7
	–	5	100	237.4	122.2	55.8	35.0	11.8
CONTINUOUS	–	1	50	357.8	239.2	150.3	101.1	36.4
	–	5	50	269.6	148.8	65.1	33.7	14.9

```
*    Batch run data with 1 and 5 cells were pooled together.
**   Batch run data with PP's of 0 and 20% were pooled together.
***  Batch ponds had discharge pipe diameters of 610 and 1220 mm.
     Continuous ponds had discharge pipe diameters of 304 and 610 mm.
```

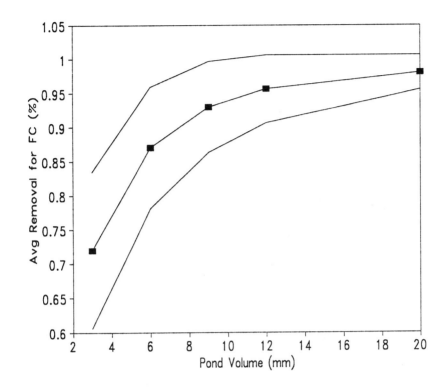

Figure 15.2: FC removal in 24 and 72 hr batch ponds for MTSS.

Detention Time

TD in a batch pond was expected to have a significant influence. Although the overall percent removal did not change significantly when the TD was changed from 24 to 72 hours, the number of exceedances dropped significantly as shown in Table 15.1.

Benefits gained in terms of fractional removal by increasing the detention of pond contents are somewhat offset by an increase in the number of events that do not receive full treatment. At longer TDs the pond volume is full for greater periods of time and the likelihood of an event occurring and overflowing the pond increases. Mixing of fresh influent with pond contents that have been treated for some time will deteriorate the overflow

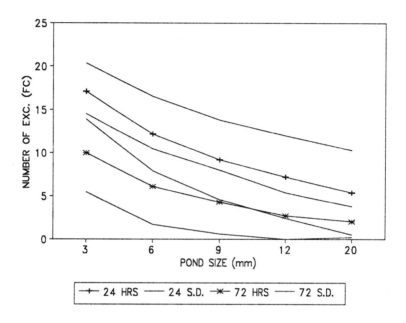

Figure 15.3: Number of FC exceedances for MTSS batch ponds.

quality and probably result in an exceedance. This is the primary explanation why overall removals did not change to a significant extent as TD was increased from 24 to 72 hours as shown in Table 15.1.

The reasons for the relatively large decrease in the number of exceedances with an increase from 24 to 72 hours TD (Table 15.1), is due to complex interactions of many factors. The main reason is that the 24 hour detention period is not sufficient for dieoff to reduce FC levels below 100/dl. Consequently, normal operation of the pond will often result in an exceedance. Effluent that is discharged from the 72 hour batch pond frequently approaches 100% removal. Overflow from a pond operated at either TD results in an exceedance. Since the most common

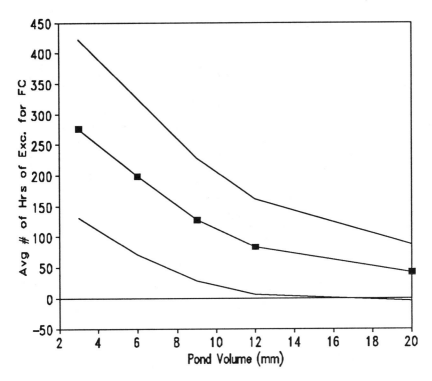

Figure 15.4:   Duration of FC exceedances for MTSS batch
ponds.

output from a pond is effluent that has resided in the pond for the
specified TD, the number of exceedances decreases for the 72
hour batch pond.

Permanent Pool

Increasing the PP by up to 20% of the total pond volume did not
cause a significant change in treatment indices.   Although the
number of overflows will increase to some extent, the water in
the PP zone has resided in the pond for a long time and will have
experienced higher degrees of treatment.   This will cause some
dilution of fresh influent during an event.   For events that cause
an overflow, the quality of the overflow will be slightly improved

Table 15.2: Drain times of MTSS test ponds.

Pond Volume (mm)	3	6	9	12	20
Volume Removed	Time to Drain Indicated Volume (hr)				
Diameter 304 mm					
100	45	90	235	280	300
90	22	44	66	88	147
75	17	32	49	65	108
Diameter 610 mm					
100	16	32	48	64	107
90	6	11	16	21	35
75	4	7	11	15	24
Diameter 914 mm					
100	9	18	27	36	60
90	3	6	8	11	18
75	2	4	6	7	12
Diameter 1220 mm					
100	4	12	18	24	41
90	2	4	7	9	15
75	1	2	4	5	8

due to dilution. For events that do not overflow the available volume, quality will again be improved while the contents reside in the pond and when the effluent is finally discharged.

In this analysis total design volume was divided into active and PP storage. As PP size increased, active storage volume was decreased in order to maintain a constant total pond volume. Consequently, there will be a breakpoint as PP size increases where the overall performance of the pond significantly decreases because of the increased volume of overflows.

Influent Quantity-Quality

The lower quality runoff generated in the Coventry catchment was responsible for a deterioration in pond performance for 24 hour detention for both FC and SS. The removal mechanics in QUALHYMO will not result in large differences in overall percent removal but the number and duration of exceedances will significantly increase as influent concentrations of pollutant rise. Influent concentration was not a significant factor influencing performance for a 72 hour detention batch pond for either FC or SS.

15.4.2 Continuous Flow Treatment

In this case the data were separately grouped according to the configuration parameters of PF, CM and PP of 50 and 100% (four different combinations) to determine average removals and exceedances over the six years. Each of these configuration changes resulted in a significant change in one or more of the treatment indices. The data for CF configurations are given in Table 15.1 for ease of comparison with batch treatment results.

The data are plotted in Figures 15.5 (fraction removed), 15.6 (number of exceedances), and 15.7 (duration of exceedances) for FC data only. The plots for the CF data do not show lines of one standard deviation above and below the average because of the amount of information displayed.

Active Pond Storage

The CF mode of operation was also evaluated over a 3 to 20 mm range of active storage volumes. Pond performance increased on the order of 34% to 63% in fraction removed, 26% to 40% in number of exceedances, and 17% to 78% in the duration of

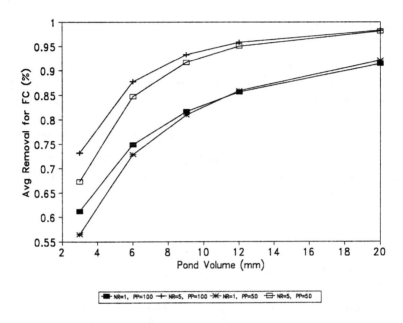

Figure 15.5:   FC removal for CF ponds for MTSS.

exceedances for FC over this volume range. Performance for SS increased from 27% to 32% for fraction removed, 55% to 77% in number of exceedances, and 90% to 95% in duration of exceedances.

Similar to the batch ponds, the performance curves follow a classical diminishing returns form. The variation in performance over the range of volumes simulated is attributed to the influence of other design parameters such as number of cells, and the size of the PP.

Effluent Pipe Diameter

Under CF operation, pond contents begin to discharge at the beginning of an event and continue to discharge at a rate

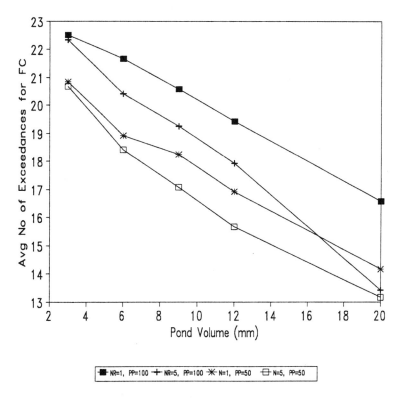

Figure 15.6: Number of exceedances for CF ponds for MTSS.

proportional to the depth of water in the pond.   Maximum
treatment will be obtained by limiting the discharge rate to
maximize the detention of the influent.   In this analysis pipe
diameters of 304 and 610 mm were evaluated. Pipes smaller than
304 mm in diameter are subject to clogging.

Number of Cells

Not surprisingly, a PF (five cell) pond improves treatment
performance over a CM reactor by 15% to 17% for fraction of
pollutant removed, and approximately 14% and 58% in number
and duration of exceedances, respectively for FC over the range
of volume examined.   Pond effectiveness for SS removal

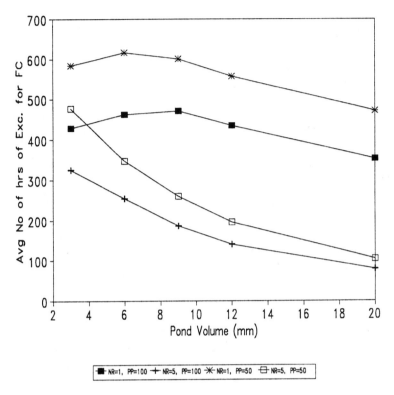

Figure 15.7: Duration of FC exceedances in CF ponds for MTSS catchment.

increased by 6% to 11% in fraction removed, 4% to 5% in number of exceedances, and 2% to 6% in duration of exceedances. However, a PF reactor is only marginally superior in performance in terms of fraction removed to a batch pond. But the number of exceedances is on the order of 50% higher for the PF CF pond compared to the batch configurations.

The explanation of why removals are higher in a PF CF pond but the number of exceedances is also higher compared to batch configurations, is that although the five cell pond is a close approximation of an ideal PF reactor, there is still some dispersion. A portion of the influent does disperse through the five cell CF pond and escapes in the effluent with very little treatment. In contrast, batch mode operation ensures that no effluent is released before the specified retention time provided there are no overflows. These occurrences are enough to raise the number of exceedances significantly for the five cell CF pond relative to the batch configurations. The fractional removal rate of the five cell CF pond remains at a slightly higher value because its volume utilization is higher than an equivalent batch pond.

Permanent Pool

Increasing the PP of a CF pond from 50% to 100% increased the performance due to an increase in volume utilization. Fractional removal increased on the order of 1% to 8%, number of exceedances actually increased by 1% to 17%, while duration of exceedances decreased by 25% to 32% for FC over the range of volumes simulated. Performance for SS increased from 1% to 4% in fraction removed; there was a 7% to 39% decrease in the number of exceedances, and 11% to 49% decrease in duration of exceedances. The effect of the increase in PP size generally increased with increasing active storage capacity and the number of cells.

Influent Quantity-Quality

Fractional removals increase by 7.6% to 14.6% for FC between MTSS and Coventry for a 3 mm pond implying that poorer water quality degrades pond performance. This effect diminishes with increasing volume (-0.1% to 1.7% increase in FC removal for a 20 mm pond). The same trend was observed for SS which showed a 3.2% to 5.7% increase for a 3 mm pond and no effective change for a 20 mm pond. Number of exceedances also declined for FC by 3% to 9% for a 3 mm pond and 5% to 10% for a 20 mm pond. However, performance with regard to SS deteriorated, with number of exceedances increasing from 1.5 to 38.7% for a 3 mm pond and -1% (decrease) to 720% increase for a 20 mm pond. The behaviour of duration of exceedances was also complex. The higher quality influent recorded an increase in duration of exceedances of between 33% and 71% for a 3 mm pond and a 4.6% to 21% decrease for a 20 mm pond for FC over the lower quality influent. The same trend was observed for SS where an increase in duration of exceedances between 23% to 85% was recorded for the higher quality influent for a 3 mm pond and a range of -43% (decrease) to a 20% increase was noted over the lower quality influent in a 20 mm pond. The differences were somewhat due to the different hydrographs and pollutographs generated in each catchment.

The rank of configurations with respect to each other in terms of any performance index was consistent regardless of the influent quality.

15.4.3 Sensitivity Analysis of FC Dieoff Rate

Gietz,(1983) measured FC dieoff rates that ranged rom 0.9 to 2.4 d^{-1} (base e). All previous runs were made at a FC dieoff rate of 2.5 d^{-1} (t_{90} = 22 hr) to determine the best performance of the ponds. The influence of FC dieoff rate on treatment was examined by using FC dieoff rates at the lower (k_d = 0.9 d^{-1}; t_{90} = 62 hr) and average (k_d = 1.8 d^{-1}; t_{90} = 30 hr) values reported by Gietz (1983).

The analysis was performed by examining a batch pond receiving MTSS simulated runoff and a continuous mode pond receiving Coventry simulated runoff. The pond characteristics were as follows:

Batch (1):   24 hr TD; 0% PP; 1 cell; pipe diameter- 610 mm

Batch (2):   72 hr TD; 0% PP; 1 cell; pipe diameter- 610 mm

PF: 100% PP; 5 cells; pipe diameter- 304 mm

CM: 50% PP; 1 cell; pipe diameter- 304 mm

These pond configurations bracket the range of performance.

All pond volumes and all six key years were included in the analysis. The plots of fraction removed and number of exceedances for FC are shown in Figures 15.8 to 15.11 for the PF and 72 hour batch mode operating conditions.

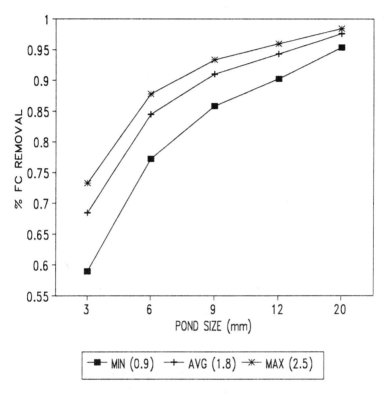

Figure 15.8: FC removal in a PF CF (MTSS) pond as a function of dieoff rate coefficient.

From Figures 15.8 and 15.9 it is observed that the influence of the variation in dieoff rate coefficient in FC removal decreases as pond volume increases. Referring to Figure 15.8 for example, the difference between FC removal using a dieoff rate coefficient of 0.9 (minimum) and 2.5 (maximum) was 15% at a pond volume of 3 mm and only 8% at a volume of 20 mm. The same trend was observed for 72 hour batch mode operation as shown in Figure 15.9.

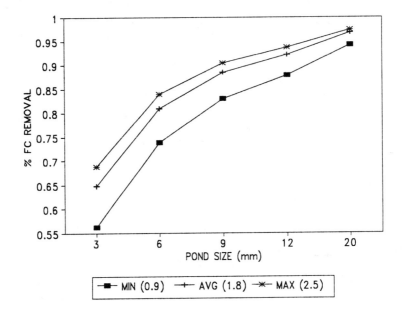

Figure 15.9:   FC removal in a 72 hr batch (MTSS) pond as a
function of dieoff rate coefficient.

In terms of number of exceedances the variation in dieoff rate
coefficient increases in significance with increasing storage
volume. Referring to Figure 15.10 for a PF CF pond the variation in performance was <1% at 3 mm volume to approximately
10% for 20 mm volume for a range of 0.9 (minimum) to 2.5
(maximum) in the dieoff rate coefficient. For 72 hour batch
operation an increase from 0.9 to 2.5 d^{-1} in the dieoff rate
coefficient produced decreases of 47% and 75% in number of
exceedances at 3 and 20 mm volumes respectively. In the CM
CF pond the number of exceedances did not exhibit any
sensitivity to dieoff rate changes.

The 72 hour batch pond is the best treatment option in terms
of removal and number of exceedances regardless of the dieoff

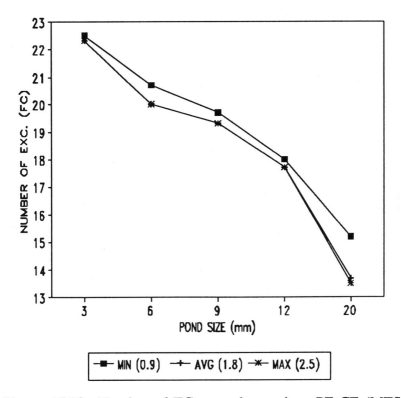

Figure 15.10: Number of FC exceedances in a PF CF (MTSS) pond as a function of dieoff rate coefficient.

rate. It is also interesting to note that for either 6 or 12 mm ponds, there is a very large increase in the number of exceedances when the FC dieoff rate drops to its minimum value but that there is no difference in the number of exceedances at average or maximum dieoff rates. The exceedances were primarily due to excess volume of runoff over the pond volume. Removal rates for the lower dieoff rate were not as satisfactory for the 72 hour batch pond. Consequently, the variability in dieoff rate may not produce effluent that reliably achieves the target goals of 100 FC/dl at 4 or less times per swimming season unless a pond approaching 20 mm of runoff volume is designed.

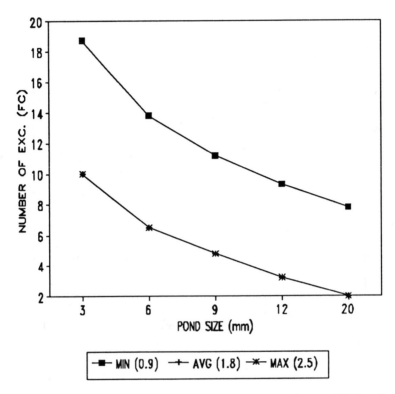

Figure 15.11: Number of FC exceedances in a 72 hr batch (MTSS) pond as a function of dieoff rate coefficient.

15.4.4 The Effect of Dry Weather Flow on Treatment

DWF will cause pond contents to be released sooner than desired, decreasing the time for treatment. Although pond contents will be diluted to a degree by DWF, the lower detention period will result in deterioration of treatment as DWF rate increases. An exercise was performed to determine the critical DWF rate where the rate of deterioration of treatment begins to accelerate. The DWF rate was specified in terms of the time required for the DWF to fill the pond volume.

For this analysis only two pond volumes were examined: 6

and 12 mm and the average year (1971) runoff input series was used. Imposing a uniform DWF input to the ponds will not add any significant variability to the results. The following configurations were analyzed:

Batch: TD- 24 hr; 5 cells; 0% PP; pipe diam.- 610 mm
PF: 5 cells; 100% PP; pipe diameter- 610 mm

Both MTSS and the Coventry catchments were included in the analysis. Typical results for MTSS catchment are shown in Figures 15.12 and 15.13 (fraction removed), and Figures 15.14 and 15.15 (number of exceedances) for FC and SS respectively. The results for the Coventry runoff simulations were similar.

Figure 15.12 for FC removal indicates that pond performance will deteriorate rapidly for a DWF rate that fills the pond in less than 5 days. A similar inflection point occurs on Fig. 15.13 for SS removal. To avoid undue deterioration of pond performance by DWF bypass, measures may be required when the DWF rate fills the pond in five days or less and a bypass may be recommended when the DWF rate will fill the pond in ten days or less. The same general trend is observed in Figures 15.14 and 15.15 for the effect of DWF on FC and SS number of exceedances, respectively.

Selecting a base value for the DWF is difficult given seasonal variation of DWF. If the value selected is too high then significant amounts of runoff will be bypassed to the receiving water without treatment. This is a particular concern for events of long duration. Although careful evaluation of DWF variation is required to select the optimum value, fortunately there is some leeway in the allowance for DWF inflow into a pond since the simulations have shown that a DWF that fills the pond in five days or more will not significantly deteriorate pond performance. Therefore this amount of DWF can be allowed to enter the pond at all times. When DWF fluctuations will rise above this value the challenge is to design a DWF bypass that will not violate the five day pond fill up rule and to assess if the bypassed flow will not result in a significant amount of untreated runoff.

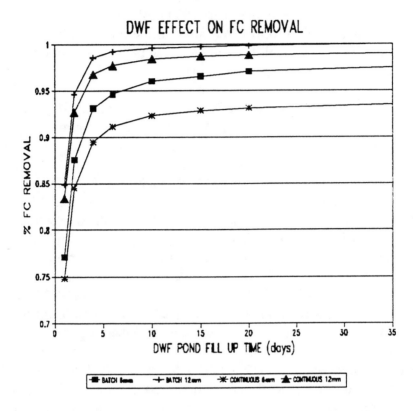

Figure 15.12: DWF effect on FC removal in MTSS ponds.

If the amount of untreated runoff is high when maintaining the five day rule, there are two solutions: pond volume can be increased which increases the DWF inflow allowance or a more sophisticated control structure can be designed that directs all flow into the pond during a runoff event and bypasses DWF at other times.

15.5 Summary of Simulation Results

The results of the simulations have shown that a PF CF pond under ideal conditions with no short-circuiting or dead volume provides the highest degree of treatment on an overall removal

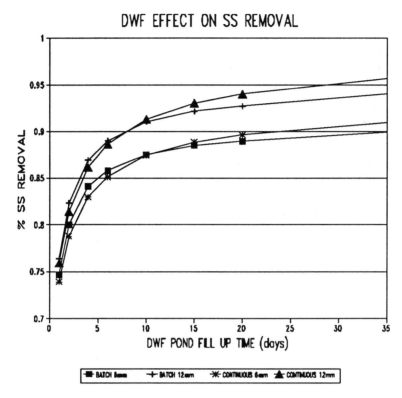

Figure 15.13: DWF effect on SS removal in MTSS ponds.

basis for the parameters of concern. The size of the active storage volume, the number of cells, and the size of the PP were found to be significant design parameters. Although small outlet pipe diameters are preferred, there was no significant treatment variation when the pipe diameter of CF ponds was less than 610 mm.

The batch mode of operation is only marginally inferior in performance to a PF reactor in terms of removal (within 2%) while providing a 50% decrease in the number of exceedances. This observation holds over a likely range of dieoff rate coefficients. For the batch mode of operation, discharge pipe sizes between 610 and 1220 mm and a PP of 0 or 20% did not cause any significant change in treatment parameters. The

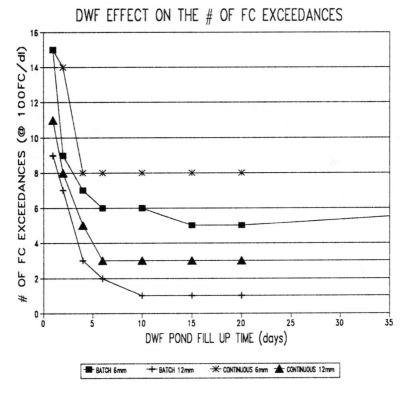

Figure 15.14: DWF effect on the number of FC exceedances for MTSS ponds.

number of cells (between one and five) in the batch reactor did have a tendency to increase removals by up to 5% but the number of exceedances did not change significantly with respect to this configuration change.

DWF rates which will fill a pond in less than five days will rapidly accelerate the degradation of pond performance. Consequently, it may be mandatory to bypass DWF in those situations where DWF rates exceed the five day criterion. DWF rates that fill a pond in ten days or longer will not deteriorate the treatment in any pond.

Pond volumes between 9 and 12 mm of runoff for the entire catchment provide optimal performance based on the treatment

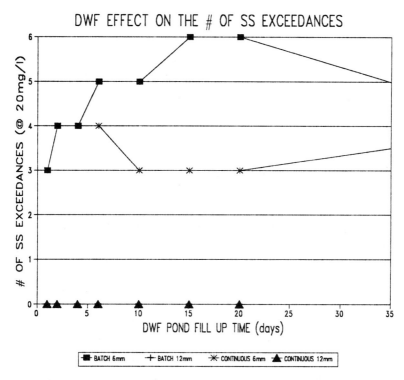

Figure 15.15: DWF effect on the number of SS exceedances for
              MTSS ponds.

indices. However, variability in the FC dieoff rate will not allow
the target value of four exceedances per summer to be
consistently achieved even in the most efficient pond configuration (72 hour batch mode) without disinfection.

15.6 References

Droste, R.L., Whittaker, J., and D'Aoust, S.G. (1992). Database requirements for design of stormwater treatment
facilities. Submitted to the 1992 ASCE Stormwater
Management Conference.

EPA (1986). Methodology for analysis of detention basins for control of urban runoff quality. US Environmental Protection Agency, EPA440/5-87-001, Washington, DC.

Gietz, R.J. (1983). Urban runoff treatment in the Kennedy-Burnett settling pond. Rideau River Stormwater Management Study, Regional Municipality of Ottawa-Carleton, Ottawa.

RMOC (1992). Rideau river stormwater management study. Report to the Regional Municipality of Ottawa-Carleton by Gore & Storrie Limited.

Rowney, A.C. and MacRae, C.R. (1992(a)). QUALHYMO user's manual: A continuous hydrologic simulation language, Version 2.1. Technical University of Nova Scotia, Dept. of Civil Engineering, Halifax, Nova Scotia.

Rowney, A.C. and MacRae, C.R. (1992(b)). QUALHYMO technical reference manual: A continuous hydrologic simulation model, Version 2.1. Technical University of Nova Scotia, Dept. of Civil Engineering, Halifax, Nova Scotia.

Urbonas, B., Guo, J.C.Y., and Tucker, L.S. (1989). Optimization of stormwater quality capture volume. In Urban Stormwater Quality Enhancement: Source Control, Retrofitting and Combined Sewer Technology, H.C. Torno (ed.), ASCE, 1989, pp. 94-110.

Chapter 16

A Continuous Simulation Technique for the Assessment of Infiltration Basins

Neil R. Thomson
Department of Civil Engineering
University of Waterloo
Waterloo, Ontario, N2L 3G1

This chapter provides an overview of a simulation technique that can be used to assess the performance of an infiltration basin facility. The developed model provides the necessary quantity and quality linkage between the surface and groundwater regime, and allows for the continuous simulation of the basin which is required for a long-term site-specific design assessment. Simulation results from a conceptual infiltration basin design are presented and discussed for both a single and a multiple event hydrograph and pollutograph.

16.1 Introduction

Over the last few years, a wide range of *Best Management Practices* has been developed in order to aid in the protection of the water quality of receiving water bodies in need of assistance due to actual or, potential, impacts of stormwater discharges.

0-87371-898-4/93/$0.00 + $.50

359

Some of the more desirable structural *Best Management Practices* are the induced infiltration alternatives (e.g., infiltration basins, infiltration trenches, and porous asphalt pavements) since they are capable of mitigating the stormwater impact of flooding, streambank erosion, and receiving water quality problems (Md WRA, 1984). More importantly, however, induced infiltration may maintain, or perhaps enhance, stream baseflow thereby allowing a stream to increase its ability to assimilate contaminants, and thus strengthen the potential for aquatic life. Unfortunately, the implementation of infiltration alternatives is only possible for areas where the underlying hydrogeological conditions are favourable.

The most promising of the induced infiltration alternatives for servicing large drainage areas or developments is the infiltration basin. An infiltration basin is constructed by excavation of a large dugout, or formation of earth berms or embankments in order to create a relatively large open surface storage area. A typical infiltration basin which may be used to infiltrate high-frequency runoff events is presented in Figure 16.1 (Schueler, 1987).

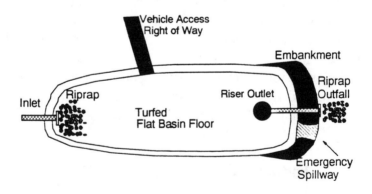

Figure 16.1: Typical infiltration basin.

For this configuration, stormwater runoff enters the basin through the inlet structure and passes over the riprap apron which reduces the energy capacity of the runoff. This reduction in runoff velocity is required to reduce the potential for erosion of the basin floor. The basin floor is turfed and is as level as physically possible in order to distribute the storm runoff volume evenly over the entire basin floor area. The basin stores a specified volume of water which is allowed to infiltrate into the surrounding subsurface regime through the floor and side walls of the basin. An outlet structure is provided to allow less frequent storms to be released, while major events can be discharged over the emergency spillway.

Infiltration basins are thought to be very effective in removing fine particulate matter and soluble contaminants present in urban stormwater runoff; however, little field data exists to substantiate the effectiveness of this removal process (Schueler, 1987). The quantity of sediment build-up within an infiltration basin is problematic since it reduces the maximum infiltration rate of the underlying soil strata. Sediment build-up rate is a function of the storage capacity of the infiltration basin, the volume of inflow into the basin, and the sediment loading. It has been recommended by Md WRA (1984) that an infiltration basin be designed to drain completely within 3 days (72 hours) following the occurrence of a storm event. The primary purpose for this design requirement is to encourage aerobic conditions for bacteria growth within the upper portions of the soil profile.

16.2 Potential Problems

A serious concern with the use of infiltration basins is the potential for groundwater and soil contamination. This concern has arisen because of the potential impact on aquifers which are used as a primary water supply source. An illustration of this situation is presented in Figure 16.2 and 16.3. In Figure 16.2, an infiltration basin has developed a zone of groundwater

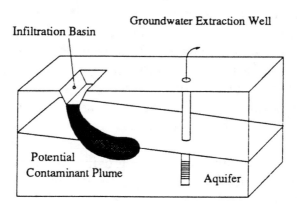

Figure 16.2: Components of the continuous simulation model.

contamination that threatens a water supply source, while in Figure 16.3 the zone of groundwater contamination will eventually migrate into a surface water body thus degrading this aquatic ecosystem.

Unfortunately, to date, data relating to the long-term impact of infiltration basins on the subsurface regime is not very extensive. Several monitoring studies have been conducted in the U.S. (e.g., Nightingale 1987a,b; Yousef *et al.*, 1987), and have shown that the concentration of heavy metals quickly decreased with soil depth. However, it should be noted that these studies are site specific and variations between land use, soil type, runoff water quality characteristics, and the groundwater flow regime have not been investigated.

The impact of infiltration basins on the subsurface regime is extremely important since the remediation of groundwater contamination problems can be very expensive, and groundwater contamination problems are usually considered long-term.

16.3 Continuous Simulation Model

In order to assess the impact of a proposed, or existing, infiltration basin on the subsurface regime, a continuous

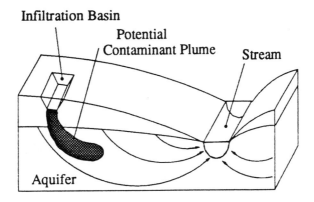

Figure 16.3: Subsurface configuration.

simulation model has been proposed. This model provides the necessary linkage between the infiltration basin and the subsurface regime for the interchange of both fluid and contaminants which is required for a long-term site specific performance assessment. This long-term performance assessment allows for the simulation of actually observed multiple storms where the sequence of events is important, rather than basing the infiltration basin performance on a generic or hypothetical set of conditions (Marshall Macklin Monaghan Ltd., 1991).

16.3.1 Conceptualization

Conceptually, the developed continuous simulation model can be illustrated as shown in Figure 16.4. For a given drainage area and meterological conditions, a continuous profile of the surface runoff hydrograph and pollutograph can be generated. These profiles are routed into the infiltration basin using a variable time stepping scheme, and the appropriate overflow and infiltration rates are determined. Since the time frame for each storm event is, in general, small compared to the infiltration time frame (hours as compared to days), the portion of inflow over the design

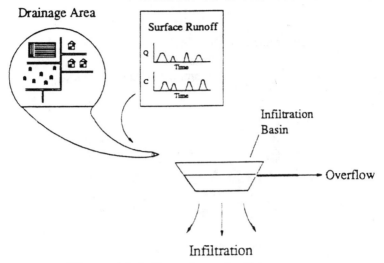

Figure 16.4: Boundary conditions.

volume of the infiltration basin (if any) is removed as overflow. The remaining volume within the infiltration basin is available for infiltration. The actual infiltration rate is determined by the depth of water in the basin and the transient pressure head distribution within the subsurface. For simplicity, the subsurface has been conceptualized as a two-dimensional spatial domain which is assumed to coincide with the regional groundwater flow direction. Although a fully three-dimension representation would capture more completely the actual performance of the basin, the two-dimensional representation should be adequate for most situations (Fipps *et al.*, 1988).

In general, the following steps are involved for each simulation time step:

1. determine added fluid and pollutant mass to the infiltration basin;

2. remove the overflow portion (if required);

3. determine the infiltration rate into the subsurface; and

4. determine the mass transport of contaminant in the

subsurface regime.

The calculation of the infiltration rate into the subsurface (Step 3) and the subsequent loss of fluid and contaminant mass from the infiltration basin is perhaps the most computationally burdensome step. The is due in part to the nonlinear flow equation that governs the transient pressure head distribution in the subsurface, and also due to the iterative procedure required to determine the depth of water in the basin at the end of each time step.

A basin model and a subsurface model are the two basic components of this continuous simulation model and are described in the following two sections.

16.3.2 Basin Model

The basin model is essentially a level pool routing algorithm (Chow *et al.*, 1988). The fluid mass balance equation for an infiltration basin is given by

$$\frac{dS(h_w^b)}{dt} = I(t) - O(t,h_w^b) - E(t,h_w^b) \qquad (16.1)$$

where $S(h_w^b)$ is the storage within the pond, h_w^b is the depth of fluid in the basin, $I(t)$ is the continuous fluid inflow, $O(t, h_w^b)$ is the basin overflow, and $E(t, h_w^b)$ is the infiltration rate from the basin into the subsurface. The basin storage is determined from a basin stage-storage curve which is input into the developed model as a set of tabular data. The basin overflow is determined based on a stage-discharge curve which is also input into the model as a set of tabular data. Finally, the infiltration rate $E(t, h_w^b)$ is determined implicity as part of the calculation of the pressure head distribution within the subsurface.

A similar mass balance equation can be formulated for the mass of contaminants in the infiltration basin. In this formulation it is assumed that the contaminant of concern is completely mixed within the basin.

16.3.3 Subsurface Model

Within the subsurface domain two distinct zones can be defined. The first zone is the saturated zone where all of the available pore space is occupied by water, while the second zone is the unsaturated zone where all the available pore space is occupied by water and soil air. Assuming that the soil air within the unsaturated zone is infinitely mobile, then the governing equation which describes the continuous two-dimensional transient pressure distribution in the subsurface may be expressed as:

$$S_w S_s \frac{\partial h_w}{\partial t} + \phi \frac{\partial S_w}{\partial t} = \frac{\partial}{\partial x_i}\left[k_{rw}K_{ij}\frac{\partial}{\partial x_j}(h_w + e_j)\right] \quad (16.2)$$

where S_w is the water saturation, S_s is the specific storage, t is time, h_w is water pressure head, ϕ is porosity, K_{ij} is the saturated hydraulic conductivity tensor, k_{rw} is the relative permeability with respect to water, and e_j is the unit vector in the direction of the vertically upward component of the coordinate system. Equation (16.2) represents a modified version of Richard's equation (Richard's 1931). Associated with (16.2) is an appropriate set of initial and boundary conditions. The initial conditions define the water pressure head distribution prior to the start of simulation, while the boundary conditions are defined around the external boundary of the spatial domain. There are two types of boundary conditions that apply to (16.2): the first denoted as Type-1, represents a known water pressure head along a portion of the boundary, and the second denoted as Type-2, represents a known Darcy flux along a portion of the external boundary (Huyakorn and Pinder, 1983). To complete the statement of the governing equation (16.2), constitutive relationships relating water saturation to capillary pressure, and relative permeability to water saturation must be provided. One of the most widely used set of constitutive relationships are those developed by van Genuchten (1980). Assuming that water saturation is a single valued function of the capillary pressure head, van Genuchten developed the following empirical relationship

$$S_{we}(h_w) = \left(1 + (\alpha h_{aw})^n\right)^{-m} \tag{16.3}$$

with

$$S_w = (1 - S_{wr})\, S_{we} + S_{wr} \tag{16.4}$$

and

$$h_{aw} = h_a - h_w \tag{16.5}$$

where S_{we} is the effective water saturation; S_{wr} is the residual water saturation; h_{aw} is the capillary pressure; h_a is the soil air pressure in equivalent units of water pressure head; and α, n, and m are empirical constants. Based on the theoretical work of Mualem (1976), van Genuchten (1980) derived the following empirical relationship for relative permeability using (16.3)

$$k_{rw}(S_{we}) \quad S_{we}^{\frac{1}{2}}(1 - [1 - S_{we}^{\frac{1}{m}}]^m)^2 \tag{16.6}$$

To represent the mass transport of a contaminant within the subsurface domain, the following governing equation is employed in this model

$$\frac{\partial}{\partial t}(\phi S_w C_w + \rho_s(1-\phi)C_s) = \frac{\partial}{\partial x_i}\left(\phi S_w D_{ij}\frac{\partial C_w}{\partial x_j}\right) - v_i\frac{\partial C_w}{\partial x_i} \tag{16.7}$$

where C_w is the concentration of the dissolved constituent of concern in the water phase, ρ_s is the soil particle density, C_s is the concentration of the sorbed constituent in the solid phase, D_{ij} is the hydrodynamic dispersion tensor and accounts both for mechanical mixing and diffusion, and v_i is the Darcy velocity determined from the water pressure distribution. Equation (16.7) accounts for the transport processes of advection, dispersion, diffusion, and sorption (de Marsily, 1986). Also associated with (16.7) is a set of initial and boundary conditions. The initial conditions define the spatial distribution of both the water phase concentration and soil phase concentration prior to the start of simulation. Three types of boundary conditions are possible: the first, Type-1, defines a known concentration along a portion of the domain boundary, the second, Type-2, defines a specified diffusive mass flux along a portion of the domain boundary, while the third, Type-3, defines a known total mass flux along a portion of the domain boundary. All three of these boundary conditions are used to represent various portions of the spatial domain boundary for the infiltration basin configuration discussed in Section 16.4.

The solution of the governing equations (16.2) and (16.7) in conjunction with their appropriate constitutive relationships, and boundary and initial conditions where solved numerically by the finite element method. Full advantage was taken of the influence coefficient approach in forming the element matrix equations. Fluid mass was conserved in the solution of (16.2) due to the direct application of the finite element method to the mixed form of Richard's equation instead of the conventional pressure head form. A full Newton-Raphson iterative scheme was employed in order to converge to the required solution at each timestep. The linear system of equations was solved by an iterative, preconditioned orthomin solution method (Mendoza, 1991). To reduce numerical oscillations that are inherent in the solution of mass transport equation (16.7), an upwind finite element scheme was implemented (Huyakorn and Pinder, 1983). A variable timestepping scheme was also implemented in the model to improve the computational efficiency. This scheme allows the time step increment to be automatically selected depending on the convergence criteria, the maximum change in water pressure head and/or concentration, and the time to the next inflow event.

16.4 Example Results

To demonstrate the applicability of the developed continuous simulation model the results from a single event simulation and a multiple event simulation are presented. For each of these simulations, the basin geometry and subsurface configuration are identical.

The infiltration basin considered in these analyses has a bottom width of 25 m, a length of 60 m, side slopes of 3:5 (v:h), and a maximum ponding depth of 2.4 m. The subsurface configuration consists of the 35 m deep by 700 m long two-dimensional spatial domain as presented in Figure 16.5. This spatial domain was discretized into 3864 triangular elements and 2030 nodes. Sand was used as the soil medium and assigned a saturated hydraulic conductivity of 0.001 cm/s, a porosity of 0.3, a longitudinal dispersivity of 5 m, and a transverse dispersivity of 2 m. For the constitutive relationships, (16.3)-(16.6), a value of α of 0.2/m, n of 2.2, $m=1-1/n$, and S_{wr} of 0.2 were employed. Although in this example one soil zone was employed, the developed methodology can be applied to subsurface domains which are comprised of more than one soil type.

The boundary conditions for the subsurface configuration are presented in Figure 16.6. A no-flow condition was assigned along the bottom of the spatial domain, and along the portions of the left and right boundaries corresponding to the unsaturated zone. For the remaining portions of the left and right boundaries a prescribed water pressure head corresponding to a hydrostatic pressure profile was assigned. The water table was positioned approximately 3 m below the base of the infiltration basin at a one percent slope. Across the top of the spatial domain, a prescribed flux of 2.5 cm/year with a concentration of zero was

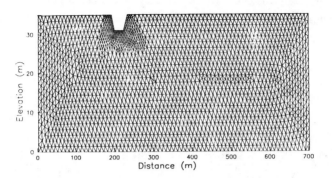

Figure 16.5: Subsurface configuration.

prescribed, except for the portion of the boundary corresponding
to the location of the infiltration basin which was handled in the
following iterative fashion for each time step:

1) The water elevation in the basin is estimated based on the
   added fluid mass to the basin,

2) The submerged portion of the boundary is prescribed a water
   pressure head equal to the hydrostatic profile in the basin
   while the remaining portion is prescribed a zero water flux
   condition,

3) Using this set of boundary conditions the transient water
   pressure head distribution is determined and the total
   infiltration rate $E(t, h_w^b)$ in (16.1) is calculated,

4) This infiltration rate is used in conjunction with (16.1) to
   determine the updated water elevation in the infiltration basin
   and then this process is repeated until convergence is attained.

Once convergence has been attained for the pressure head
distribution, the concentration of the pollutant in the fluid
contained within the infiltration basin is assigned to the
submerged portion of the basin boundary, while the remaining

portion of the basin boundary is assigned a zero diffusive flux value. The concentration in the fluid entering the spatial domain through the left boundary is assigned a value of zero, indicating that this fluid is devoid of the pollutant.

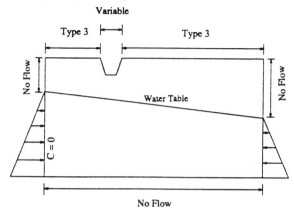

Figure 16.6: Boundary conditions.

The initial condition for the water pressure head was determined from a steady state flow analysis using the above defined boundary conditions for flow, and a prescribed uniform flux of 2.5 cm/year across the top of the spatial domain. It was assumed the initial water and soil phase concentrations are zero throughout the spatial domain.

16.4.1 Single Event

This single event simulation consists of one event routed into the infiltration basin and the subsequent water pressure head and concentration distribution determined within the subsurface regime. The hydrograph and pollutograph which are used as input into the infiltration basin are presented in Figure 16.7. For simplicity, a triangular shape and time base of 3 hours was used for each of these profiles. Over the 3-hour period a total of 4000 m^3 of water and 3000 kg of a conservative contaminant (e.g., chloride) were routed into the infiltration basin.

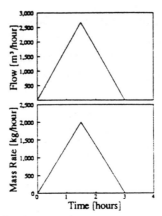

Figure 16.7:   Input hydrograph and pollutograph for the
single event simulation.

The pressure head and concentration distributions are
presented in Figures 16.8 through 16.11 for a simulation time of
0.25, 15, 36, and 72 hours, respectively.   Note that the
concentration contour increment in all of these figures is 200
mg/L. In Figure 16.8 (0.25 hours) it can be observed that the
water table (which is approximated by the zero water pressure
head contour) is responding to infiltration from the basin, and that
a contaminant plume has been formed below the infiltration basin.
At 15 hours into the simulation (Figure 16.9) the subsurface
regime below the infiltration basin is saturated and a substantial
plume exists. By 36 hours (Figure 16.10) into the simulation the
infiltration basin is almost empty, while at 72 hours (Figure
16.11) the water table is returning to its initial position. Notice
that the morphology of the contaminant plume below the basin is
almost identical at 36 and 72 hours, which suggests that the early
portion of the infiltration event contributes predominantly to the
migration of the plume.

A single event simulation approach can also be used
effectively to analyze a proposed infiltration basin during the
hydraulic design phase. A brief study of the variations in
infiltration rate with maximum ponding depth, antecedent soil
moisture conditions, the location of the water table, and the
physical soil characteristics of the underlying soil has been
reported elsewhere by Graham and Thomson (1991).

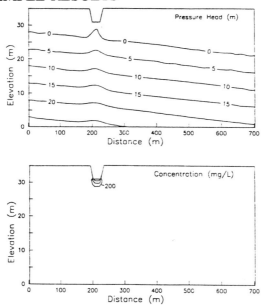

Figure 16.8:   Pressure head and concentration distribution at
0.25 hours.

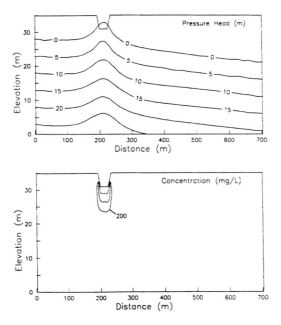

Figure 16.9:   Pressure head and concentration distribution at
15 hours.

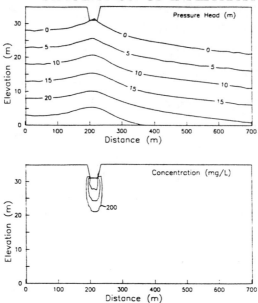

Figure 16.10: Pressure head and concentration distribution at
36 hours.

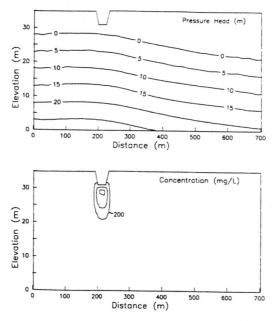

Figure 16.11: Pressure head and concentration distribution at
72 hours.

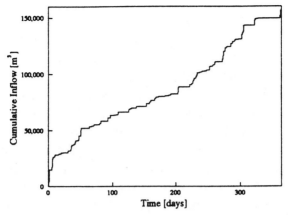

Figure 16.12: Cumulative inflow to infiltration basin.

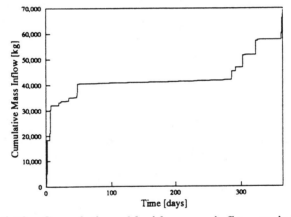

Figure 16.13: Cumulative chloride mass inflow to infiltration
            basin.

16.4.2 Multiple Event

The multiple event simulation consists of a continuous hydrograph
and pollutograph routed into the infiltration basin for a period of
one year. The hydrograph and pollutograph employed in this
analysis were determined from STORM (HEC, 1977) using the
coefficient method for a 20 hectare drainage area. Hourly
precipitation and daily temperature data were used to generate
rainfall/runoff or snowmelt/runoff information as shown in Figure
16.12. To indicate the upper-end member of the migration

potential for stormwater runoff constituents, chloride was selected
and assigned a value of approximately 1200 mg/L to the
snowmelt-runoff and a value of 25 mg/L to the rainfall/runoff
(Oberts, 1990). Figure 16.13 presents the cumulative mass inflow
to the infiltration basin. The chloride concentration distribution at
the end of one year of simulation is presented in Figure 16.14.
The resulting plume has a very curious morphology due in part
to the high chloride loading which occurs at the start and end of
the one-year simulation period (see Figure 16.13). This increase
in chloride loading is a result of the high concentration of
chloride in the snowmelt/runoff. The leading edge of the chloride
plume which has broken away from the portion of the plume
which is contiguous with the infiltration basin is related to the
initial high loading rate at the start of the simulation period.

16.5 Summary

A continuous simulation technique for the assessment of
infiltration basin performance has been developed. This technique
accounts for the quality and quantity linkage between the
infiltration basin and the subsurface regime.

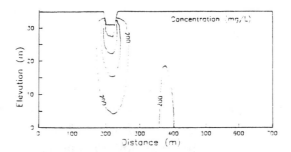

Figure 16.14: Chloride   distribution   after   one   year   of
            simulation.

The proposed approach appears to be useful for both single and multiple event simulations; however, a comprehensive validation of the developed model using actual field observations should be undertaken to ensure that the necessary processes have been taken into consideration.

Additional research into soil clogging, soil freezing and a fully three-dimensional representation of the infiltration basin need to be addressed.

16.6 References

Chow, V.T., Maidment, D.R. and Mays, L.W. (1988). Applied Hydrology. McGraw-Hill, New York.

de Marsily, G. (1986). Quantitative Hydrogeology, Academic, San Diego, Calif.

Fipps, G., Chescheir, G.M., and Skaggs, R.W. (1988). Modelling seepage from stormwater infiltration ponds. Symposium on Coastal Water Resources. AWRA.

Graham, E.I., and Thomson, N.R. (1991). Analysis of transient infiltration basin operation, in Proceedings of the 1991 CSCE annual conference, Vancouver, British Columbia, Vol 1, 315-323.

Huyakorn, P.S., and Pinder, G.F. (1983). Computational methods in subsurface flow. Academic Press, Inc. pp. 473.

Marshall Macklin Monaghan Limited (1991). Stormwater quality best management practices. Report prepared for Ontario Ministry of the Environment, pp. 177.

Maryland Water Resources Administration (Md WRA 1984). Standards and specifications for infiltration practices. Maryland Department of Natural Resources, Annapolis, MD.

Mendoza, C. (1991). Orthofem User's Guide. Waterloo Centre for Groundwater Research, University of Waterloo, Waterloo, Ontario.

Mualem, Y. (1976). A new model for predicting the hydraulic conductivity of unsaturated porous media. Water Resour. Res., 12:513-522.

Nightingale, H.I. (1987a). Water quality beneath urban runoff water management basins. Water Resour. Bull., 23(2):197-205.

Nightingale, H.I. (1987b). Accumulation of As, Ni, Cu, and Pb in retention and recharge basin soils from urban runoff. Water Resour. Bull., 23(4):663-672.

Oberts, G.L. (1990). Design considerations for management of urban runoff in wintry conditions. Proceedings of the International Conference-Urban Hydrology Under Wintry Conditions, Narvik, Norway.

Richards, L.A. (1931). Capillary conduction of liquids through porous mediums. Physics,1:318-333.

Schueler. T.R. (1987). Controlling urban runoff: A practical manual for planning and designing urban BMPs. Dept. of Environmental Programs. Metropolitan Washington Council of Governments.

van Genuchten, M. Th. (1980). A closed form equation for predicting the hydraulic conductivity of unsaturated soils. Soil Sci. Soc. Am. J., 44:892-898.

Yousef, Y.A., Wanielista, M.P., and Harper, H.H. (1986). Design and effectiveness of urban retention basins. Proceedings of Urban runoff quality-impact and quality Enhancement Technology. Eds. B. Urbonas and L.A. Roesner, ASCE., 338-349.

Chapter 17

Water Quality Modelling of a Proposed Reservoir

Lawrence H. Woodbury and G. Padmanabhan
Department of Civil Engineering
North Dakota State University
Fargo, North Dakota 58105

This chapter introduces modelling techniques that have been found to be successful in evaluating future water quality conditions in proposed reservoirs. An example of an actual reservoir modelling study is used to illustrate the approach.

17.1 Introduction

Reservoirs are valuable surface water resources formed by the construction of dams or other hydraulic structures. These facilities demand minimum standards of water quality in order to attain desired benefits for recreation, water supply, irrigation, fish and wildlife uses. Since reservoirs generally become semi-permanent facilities, careful planning is required to insure that they do not become long-term liabilities to society. Once constructed and filled, reservoirs offer little opportunity for water

0-87371-898-4/93/$0.00 + $.50
© 1993 by Lewis Publishers

resource professionals to manage water quality degradation. All natural and artificial lakes (reservoirs) undergo a natural process of aging called eutrophication. Eutrophication is essentially an increase in biological productivity, and is usually measured by changes in certain water quality parameters. It is essential that any proposed reservoir planning study include an assessment of potential water quality conditions within the reservoir.

Too often, reservoir projects are sold during the planning stage with overly optimistic estimates of future water quality conditions, which are based on little or no rational technical analysis. This in turn has resulted in inflated benefit-cost ratios and flawed economic studies. Decision makers deserve much better information on projected water quality conditions, which in turn translates into realistic estimates of benefits to be realized over the project life.

The following discussion uses an actual water quality planning study for a proposed run-of-the-river reservoir in a case study approach to describing a successful modelling technique. The problem is approached by first conducting a water quality data collection program. The water quality data base is then used as input to one of two different water quality models. The first is a computer program called FLUX, to simulate nutrient loadings. The output of FLUX (nutrient loadings) is then used as input to the second model, BATHTUB, which produces trophic state indices as output. The trophic state indices are then used to assess and estimate future eutrophication potential of the proposed reservoir. Modelling results are easily comprehended by laymen and professionals for decision making purposes.

17.2 Background Information

Winger Dam and Reservoir is a proposed water resources development project in northwestern Minnesota, USA. The project, which is to be located just south of the City of Winger on the Sand Hill River, consists of a permanent pool for recreation and a temporary surcharge pool capacity for flood storage. The permanent pool extends in a northeasterly direction upstream from the proposed dam site for a distance of approximately six miles.

During the course of project development, the Minnesota Department of Natural Resources (MDNR) prepared a Draft Environmental Impact Statement (DEIS) which was published in June of 1987. This document contained the first preliminary assessment of anticipated water quality conditions in the proposed reservoir, based upon field data collected by MDNR personnel at various times between the months of April and September of 1986. This data included water temperature, dissolved oxygen, biochemical oxygen demand (BOD), fecal coliform bacteria, total suspended solids, total dissolved solids, total alkalinity, total phosphorous, and sulfate.

After completion of the DEIS, the U. S. Environmental Protection Agency (USEPA) and the Minnesota Pollution Control Agency (MPCA) raised several water quality and wetland protection concerns to the project sponsor, the Sand Hill River Watershed District. In order to adequately address the water quality issues raised by USEPA and MPCA, the District commissioned a modelling study to investigate the issues and develop a suitable response.

There are several water quality models available which could be utilized for assessing the long-term water quality response of the proposed Winger reservoir. All of these models vary with many degrees of sophistication. Some require only regionalized data estimates, while others require detailed water quality sampling for data input. The water quality modelling effort conducted by the District essentially consisted of three phases:

1. development of a water quality and flow rate data base;
2. modelling of reservoir nutrient loadings; and
3. modelling of reservoir response with respect to water quality.

A water quality monitoring plan was developed and approved by the MPCA for implementation in 1989. The data base generation was conducted on a continuous basis between the months of August 1989 and July 1990. The data base served as input for modelling reservoir loadings of various nutrients having water quality ramifications. For this purpose, the District utilized the computer program FLUX (U. S. Army Corps of Engineers, 1987). The reservoir loadings, as modeled with FLUX, then served as input data to the reservoir water quality response model BATHTUB (U. S. Army Corps of Engineers, 1987). The BATHTUB model produces results which serve as an aid in predicting the eutrophic status of the proposed reservoir.

17.3 Reservoir Hydrology

17.3.1 Watershed Characteristics

The Sand Hill River is a major tributary of the Red River of the North, which flows north into the Canadian Province of Manitoba. The Sand Hill River basin, which can be characterized as long and narrow, has an average width of about eight miles and a length of approximately 55 miles. The total size of the watershed is about 426 square miles at the river's outlet at Climax, Minnesota. In comparison, the drainage area of the proposed reservoir at Winger is about 91.9 square miles, or 21% of the total basin drainage area.

The Sand Hill River originates in Sand Hill Lake, located about four miles south of the City of Fosston, Minnesota. In the most easterly portion of the watershed, the elevation rises to nearly 1,350 feet above sea level. The river flows generally in a westerly direction, but zigzags north and south over a range of five to seven miles. At the proposed dam site, the flood line elevation is proposed to be 1165 feet above sea level. This represents a drop from the highest point of the upstream watershed of about 185 feet.

The contributing watershed of the proposed reservoir is glacial in origin and its soils support agricultural uses. It is mostly gently rolling terrain with numerous potholes, the majority of which have been drained. A large portion of the watershed has been cleared of original tree growth to allow farming. The soils for the contributing drainage area exhibit rather complex relief, in which slopes range from gently sloping to steep. The soils are mostly loam.

17.3.2 Climate

The watershed is located near the center of the North American continent and has a continental climate. Characteristics of the climate are warm summers and cold winters. The mean temperature for winter months is 10.0 F, and 68.0 F for the summer months. On the average, there are 15 days above 90.0 F in the summer, and 55 days below 0.0 F in the winter. The average annual precipitation for the Sand Hill River basin is 22 inches with extremes varying from a minimum of 10 inches to a maximum of 34 inches. About 75%, or nearly 17 inches, of the annual precipitation falls during the period of April through September. Approximately 15% of the precipitation normally occurs as snow.

Drought can be a problem in July and early August in the entire basin, especially in the center portion where soils are thin and light. Rainfall is normally adequate in May and June for agricultural purposes. Heavy rainfall during the spring and early summer months contributes to flooding problems in the basin (Houston Engineering, Inc., 1991).

17.3.3 Reservoir Morphometry

For the purposes of this discussion, the reservoir morphometry will be described by mean reservoir depths, water residence or flow-through characteristics, and stratification potential. Taken as a whole, the reservoir at a permanent pool elevation of 1190 feet above mean sea level (MSL) is about 5.75 miles long. At its peak flood pool elevation of 1196 MSL, the reservoir is 6.8 miles long. The permanent pool surface area 1217 acres and its volume is 11,131 acre-feet. The maximum permanent pool depth at the dam site is 24.8 feet and the overall mean depth throughout the reservoir is 9.1 feet. As was indicated previously, the watershed area contributing to the reservoir is 91.9 square miles. Average annual precipitation over the contributing basin is 22 inches, and the average annual evaporation is 32 inches (Houston Engineering, Inc., 1991).

The proposed Winger reservoir is typical of most manmade impoundments. The long, narrow configuration of the reservoir reflects the existing river valley topography. For the purposes of water quality modelling, the reservoir can be divided into three separate and distinct segments. The lower segment, just upstream from the dam, can be thought of as a relatively deep water basin with a very narrow littoral zone along its periphery. If there is any chance for stratification within the proposed reservoir, it will occur within this lower segment.

The middle segment of the reservoir, approximately 2.35 miles in length, has a maximum depth of 13 feet and mean depth of 7.7 feet. It is assumed that this portion of the lake will maintain a polymictic, or a completely mixed state. As such, no stratification is anticipated. The littoral zone around the periphery of the middle segment can be considered to be more extensive than the lower segment.

The upper segment is approximately 1.25 miles in length. With a maximum depth of 7 feet and a mean depth of 3 feet, this reservoir segment can only be characterized as a completely mixed shallow lake. In fact, this can be considered to border on wetland status in consideration of its extensive littoral and vegetative zones.

A last item to be discussed with respect to reservoir morphometry concerns hydraulic residence time. From the standpoint of water quality modelling, there are two types of residence times. The first type of residence time is a long-term, steady-state residence time which is established after the reservoir has filled and reached maturity. Using an average annual flow rate (28 cubic feet per second), a steady-state hydraulic residence time of 0.55 years was determined for the proposed Winger reservoir. Since residence time is simply the reservoir volume divided by average annual flow rate, and since reservoir volume is relatively constant, the annual average volumetric flow rate becomes of prime importance. The lower the flow rate, the higher the residence time, and vice versa. Therefore, when flows are extremely low, residence times can be expected to increase dramatically. Flow data recorded at the Winger site between August 1989 and July 1990 resulted in an average annual flow of 7.3 cubic feet per second (CFS), corresponding to a residence time of 2.1 years.

Residence times are also affected by precipitation, evaporation, and ground water inflow/outflow from the reservoir. Assuming a negligible ground water inflow contribution, it is noted that average annual evaporation exceeds average annual precipitation by about 10 inches per year. The resulting net deficit to the reservoir water budget can significantly increase the hydraulic residence time. For the 1989-90 data period at the Winger site, the average annual flow would be reduced to 5.9 CFS, with a corresponding residence time of 2.6 years.

From the standpoint of water quality and the ability of in-lake nutrients to support aquatic growth, it can generally be assumed that a lake with a very high residence time will enhance the eutrophication process. On the other hand, when assessing the impact of a lake's mixing characteristics on eutrophication, it can be assumed that a well mixed lake (polymictic) will enhance internal recycling of nutrients, resulting in higher nutrient values and accelerated eutrophication. Only the lower segment of the proposed Winger reservoir seems to have any potential for reducing mixing and maintaining a dimictic or stratified state. The middle and upper segments can be expected to be polymictic and enhancers of the eutrophication process.

17.4 Water Quality Monitoring Program

The purpose of the water quality monitoring plan was to develop additional water quality data on watershed inflow to the proposed reservoir area. This data can then be used as input to the reservoir water quality models, which in turn can provide estimates of future reservoir states of eutrophication and water quality. The objective of the water quality monitoring program was to collect enough data to enable the mathematical modelling of the proposed Winger reservoir.

In order to insure that the effects of seasonal variation are included in the samples, it is usually recommended that the sampling period be conducted over at least a three year period (Minnesota Pollution Control Agency, 1990). However, in the case of the Winger project, time constraints dictated a one year sampling period. Sampling frequency was designed to be more intense during spring runoff and summer storm runoff events. The intent of the increased sampling frequency during peak runoff events was to capture the water quality parameter variation in relation to the flow rate variation.

The Sand Hill River Watershed District and the MPCA mutually agreed on a specific set of parameters to be included in the water quality monitoring plan. These parameters are ones normally utilized as input to the FLUX and BATHTUB models. Seven of the parameters are water quality related, and one is a hydraulic parameter. The parameters include: total phosphorous, ortho phosphorous, total suspended solids, total volatile suspended solids, total Kejedahl nitrogen, nitrate, nitrite, and instantaneous flow rate (hydraulic parameter).

17.5 Water Quality Modelling

17.5.1 Model Formulation

The procedures utilized for conducting a water quality modelling study of the proposed Winger reservoir involved the following steps:

1. problem definition;
2. data collection and compilation;

3. model implementation of reservoir loadings;
4. model implementation of reservoir response to loadings; and
5. interpretation of results.

The first step in any modelling process is problem definition. In the case of the proposed Winger reservoir, it was necessary at this stage to describe the reservoir and watershed characteristics. It was also appropriate at this stage to determine the study type and model type. In the case of this project, the study type was determined to be predictive in nature and the model type selected was a eutrophication response model.

17.5.2 Modelling Reservoir Loadings - FLUX

In the data reduction phase, tributary water quality data are reduced or summarized in a form which can serve as model input. Since the models generally deal with conditions averaged over a growing season within defined reservoir areas (segments), data reduction involves the averaging or integration of individual measurements, sometimes with appropriate weighting factors. The FLUX computer program is designed to facilitate reduction of tributary inflow monitoring data. Using a variety of calculation techniques, FLUX estimates the average mass discharge or loading that passes a given tributary monitoring station, based upon grab-sample concentration data and a continuous flow record. In the case of the proposed Winger reservoir, FLUX made possible the estimation of average mass loadings for a variety of chemical/nutrient parameters. The results of the FLUX analysis for the 1989-90 data at the Winger site are given in Table 17.1.

17.5.3 Modelling Reservoir Responses - BATHTUB

Given the reservoir mass loadings for various nutrient constituents, the model implementation phase proceeds with

Table 17.1: Winger Reservoir Loadings - FLUX.

Parameter	Flux (kg/yr)	Conc. (ppb)
Ortho Phosphorous	447.7	68.84
Total Phosphorous	722.8	111.13
Total Kejedahl Nitrogen	25348.2	3897.16
Nitrate	8829.0	1357.42
Nitrite	308.9	47.49
Total Suspended Solids	78196.9	12022.40

developing reservoir responses from accumulated mass loading. The BATHTUB computer program permits application of empirical eutrophication models to morphometrically complex reservoirs. The program performs water and nutrient balance calculations in a steady-state, spatially-segmented hydraulic network which accounts for advective transport, diffusive transport, and nutrient sedimentation. Eutrophication related water quality conditions (expressed in terms of total phosphorous, total nitrogen, chlorophyll-a, transparency, organic nitrogen, particulate phosphorous, and hypolimnetic oxygen depletion rate) are predicted using empirical relationships previously developed and tested for reservoir applications. In addition, BATHTUB calculates Carlson's Trophic State Indices (TSI), including the phosphorous TSI, the chlorophyll-a TSI, and the secchi disc TSI.

A key option of the BATHTUB computer program allows the modeller to reduce a given reservoir into a selected number of segments for more detailed analysis on a segment-by-segment basis. Figure 17.1 shows various types of possible segmentation schemes provided for in BATHTUB. In the case of Winger reservoir, this capability was put to good use because of the unique nature of various portions of the reservoir. As was previously described, the Winger reservoir was divided into three unique segments, similar to Scheme 3 in Figure 17.1. The specific hydraulic and geometric characteristics of each segment are compiled in Table 17.2.

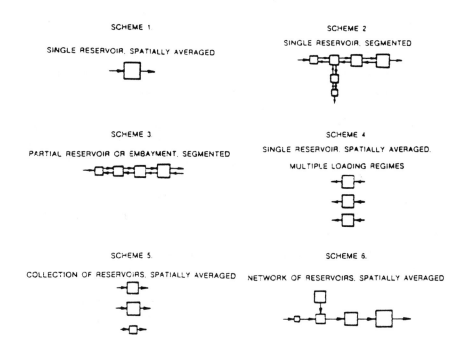

Figure 17.1: BATHTUB segmentation schemes.

Table 17.2: Reservoir Segment Characteristics.

Length (mi.)	Area (ac.)	Surface Volume (AF)	Max. Depth (ft.)	Mean Mix Depth (ft.)	Hypo Depth (ft.)
	Upstream Segment				
1.25	196	584	7	3	NA
	Middle Segment				
2.35	404	3107	13	7.7	NA
	Lower Segment (Near Dam)				
2.15	617	7440	24.8	12.1	NA
				4.66	10.5

17.6 Water Quality Modelling Results

The BATHTUB computer model was utilized to examine the effects of nutrient loadings on in-lake nutrient concentrations for data collected between August of 1989 and July of 1990. The hydrologic conditions during this year resulted in a low flow year by any standard of measurement. Therefore, the results of this modelling exercise are on the conservative side. Flow rates vary from as low as 0.6 CFS to a high of about 35 CFS. The mean annual flow for the year was about 7.3 CFS, averaged over the whole data collection period. The corresponding hydraulic residence time was about 2.6 years, which includes net evaporation losses.

Using the FLUX output as input to the BATHTUB model, in-lake values of various parameters were predicted for each of the three segments of the proposed reservoir. A summary of the BATHTUB output is reproduced in Tables 17.3, 17.4, 17.5, and 17.6. Units for each parameter value are defined as follows:

$$\text{Total P TSI} = 4.2 + 33.2 \log P$$
$$\text{Secchi TSI} = 60 - 33.2 \log (\text{Secchi})$$
$$\text{Chl-a TSI} = 30.6 + 22.6 \log (\text{Chl-a})$$

where:
ppb = parts per billion or mg/m^3 m = meters
TSI is dimensionless   P is in ppb
Secchi is in m   Chl-a is in ppb

In reviewing the predicted in-lake total phosphorous concentrations for each segment of the reservoir, it is noted that values range from 88 mg/m^3 in the upper segment, 79 mg/m^3 in the middle segment, to 58 mg/m^3 in the lower segment near the dam. Total phosphorous concentrations appear to decrease with distance from the upper end of the reservoir to the lower end of the reservoir at the dam. This could be substantiated by increased total phosphorous entrapment within the shallower vegetated areas of the reservoir. Nevertheless, all total phosphorous values predicted are significantly higher than the minimum threshold concentration of 50 mg/m^3 used as a guideline by the Minnesota

Table 17.3: BATHTUB output - upper segment.

Parameter	Units	Value
Total Phosphorous (Total)	ppb	87.71
Total Nitrogen (Total N)	ppb	2108.79
Organic Nitrogen (Org. N)	ppb	1498.01
Chlorophyll-a (Chl-a)	ppb	55.51
Secchi Disk (Secchi)	m	0.55
Chl-a TSI	-	70.0
Secchi TSI	-	68.7
Total P TSI	-	68.7

Table 17.4: BATHTUB output - middle segment.

Parameter	Units	Value
Total P	ppb	78.50
Total N	ppb	1870.88
Organic N	ppb	1068.73
Chl-a	ppb	36.69
Secchi	m	0.65
Chl-a TSI	-	65.9
Secchi TSI	-	66.3
Total P TSI	-	67.1

Table 17.5: BATHTUB output - lower segment.

Parameter	Units	Value
Total P	ppb	58.40
Total N	ppb	1364.43
Organic N	ppb	752.40
Chl-a	ppb	22.81
Secchi	m	0.75
Chl-a TSI	-	61.3
Secchi TSI	-	64.2
Total P TSI	-	62.8

Table 17.6: BATHTUB output - reservoir.

Parameter	Units	Value
Total P	ppb	69.79
Total N	ppb	1652.41
Organic N	ppb	977.47
Chl-a	ppb	32.68
Secchi	m	0.68

Pollution Control Agency (MPCA,1990).

Another measure of recreational lake acceptability is the secchi disk reading, also known as transparency. The MPCA has indicated that it would be "desirable" for the secchi disk reading to remain at above 1.5 meters for at least 75% of the summer to avoid the perception of an "impaired" or "no swimming" lake. The MPCA has also correlated this secchi disk reading to an average summer surface-water phosphorous concentration of 40 mg/m^3. This correlation is apparently the basis for the recommendation that total phosphorous concentrations remain below 50 mg/m^3 (MPCA,1990).

From a review of the BATHTUB computed secchi disk readings, it is apparent that the transparency also increases with distance downstream from the upper end of the reservoir to the lower end. Specifically, the secchi disk values predicted varied from 0.5 meters in the upper end to 0.7 meters near the dam. These values seem to correlate quite well with total phosphorous predictions within each segment. However, again the predicted exceeded the standard suggested by this MPCA.

The BATHTUB modelling output also includes a calculation of Carlson's Trophic State Index for chlorophyll-a, secchi disk, and total phosphorous. All three TSI's register between 69 and 70 for the upper segment of the reservoir. Proceeding further downstream into the deeper waters of the reservoir, the TSI is reduced to a value between 66 and 67. The lower segment of the reservoir predicts TSI's ranging from 61 to 64. From the standpoint of predicting eutrophication, the following criteria have been established for establishing the trophic status of lakes and reservoirs (Reckhow, 1983):

Oligotrophic	TSI less than 40
Mesotrophic	$41 < TSI < 50$
Eutrophic	$51 < TSI < 70$
Hypereutrophic	TSI greater than 70

From reviewing the above criteria, it appears that the BATHTUB modelling effort has predicted eutrophic to hypereutrophic conditions in the upper more shallow segment of the reservoir. The middle and lower segments could be classified as eutrophic.

17.7 Summary

This study focused on predicting the water quality of a proposed reservoir.   All indicators seemed to predict a reservoir with eutrophic characteristics.   Hydraulic residence times are sufficiently long to insure significant entrapment of nutrients, such as nitrogen and phosphorous.   Although it is feasible that the lower end segment of the lake could stratify and aid in controlling nutrient dispersion, the middle and upper segments will likely be polymictic and completely mixed, which will probably enhance internal cycling of nutrients. This would in turn result in higher nutrient concentrations in the upper two segments of the reservoir. By themselves, these physical characteristics of the reservoir would forecast a probable eutrophic condition.

Water quality modelling efforts utilizing both FLUX and BATHTUB have yielded predictions which again substantiate a potential eutrophic condition. Total predicted phosphorous levels in all segments of the reservoir significantly exceeded the MPCA suggested maximum value of 50 mg/m^3.   In addition, the predicted secchi disk values also fall short of the MPCA 1.5 meter standard. All computed values of Carlson's Trophic State Index (TSI) predicted a reservoir with eutrophic characteristics.

17.8 References

Cooke, G. D., Welch, E. B., Peterson, S. A.., and Newroth, P. R. (1986). Lake and Reservoir Restoration. Butterworth Publishers - Ann Arbor Science, Stoneham, Massachusetts, U.S.A.

Henderson-Sellers, B. (1984).   Engineering Limnology. Pitman Advanced Publishing Program, Boston, Massachusetts, U.S.A.

Houston Engineering, Inc. (1991). Assessment of Reservoir Water Quality, Project No. 4, Winger Dam, Sand Hill River Watershed District, Fertile, Minnesota, U.S.A.

Minnesota Department of Natural Resources (1987), Draft Environmental Impact Statement - Project No. 4, Winger Dam, Sand Hill Watershed District. St. Paul, Minnesota, U.S.A.

Minnesota Pollution Control Agency (MPCA, 1990). Personal communication. St. Paul, Minnesota, U.S.A.

Reckhow, K. H., and Chapra, S. C. (1983). Engineering Approaches For Lake Management, Volume 1: Data Analysis and Empirical Modelling. Butterworth Publishers - Ann Arbor Science, Stoneham, Massachusetts, U.S.A.

Thornton, K. W., Kimmel, B. L., and Payne, F. E. (1990). Reservoir Limnology - Ecological Perspectives. John Wiley and Sons, Inc., New York, New York, U.S.A.

U. S. Army Corps of Engineers (1987). Empirical Methods For Predicting Eutrophication In Impoundments, Report 4, Environmental Laboratory, Waterways Experiment Station, Vicksburg, Mississippi, U.S.A.

Chapter 18

Interactive Computer Aided Infrastructure Design and GIS - the Future?

A. R.V. Ribeiro
 GWN Systems Inc.
 #200, 11133-124 Street, Edmonton, Alberta, T5M 0J2

This chapter provides an overview of interactive computer aided infrastructure design, or more generally computer aided engineering (CAE) and geographic information systems (GIS), how the technology has developed to today, and what its future holds.

18.1 Introduction

In the last ten years the personal computer industry has experienced an explosive growth in power and capability at a corresponding exponential decline in cost per megahertz of the central processing unit (CPU). The industry has gone from 8 bit processing in the early eighties to 32 bit processing today. In terms of processing capability we have gone from PC machines with under one million instruction sets per second (MIPS) to machines now with up to 15 MIPS, which brings the technology

0-87371-898-4/93/$0.00 + $.50
© 1993 by Lewis Publishers

close to that of workstations of a few years ago and exceeding mainframes of a decade ago.

The area of geographic information systems (GIS) has evolved from the earliest civilization's need for maps for trade and military purposes (Hodgkiss, 1981) and in Roman times, the agrimensores, or land surveyors for government taxation (Dilke, 1971) to modern uses of resource, urban, and infrastructure development.

The evolution of the PC has been matched with a parallel progress in surveying and mapping technologies, such as laser survey instruments and digital storage of survey data. On the mapping side we have the development of highly accurate digital analytical stereo plotters and orthophoto technology; for remote sensing we have Landsat and SPOT imagery which can provide raster pixel resolution to 10 by 10 m.

Similarly we have seen development of computer aided engineering (CAE) software which complements computer aided drafting (CAD) packages.

In many ways the GIS and the CAE industry has evolved independently with few of the systems being integrated. What is needed is an integration of raster/vector techniques in GIS and an integration of infrastructure design and GIS.

18.2 Geographic Information Systems (GIS)

18.2.1 Maps and Spatial Information

With the industrial revolution and the resultant increase in demand for resources, the demand for topographic and thematic maps accelerated greatly. In the twentieth century, stereo aerial photography and remotely sensed imagery have allowed photogrammetrists to map large areas with greater and greater accuracy. Geologists, soil scientists, ecologists, urban planners and engineers now have a source of useful information for resource extraction and management. In more recent times the same data is now seen as an important source for environmental impact assessments (e.g. Brinkman and Smyth, 1973, FAO, 1976).

The mathematical techniques for spatial analysis was not developed until the 1930s and 1940s in parallel with developments in statistical methods and time series analysis. However the lack of computing tools hindered any practical progress in the field until the 1960s with the advent of the digital computer. At this point the conceptual methods for spatial analysis and the potential of quantitative thematic mapping and spatial analysis began to flourish (Cliff and Ord, 1981; Ripley, 1977; Webster, 1977).

Prior to the use of computers for mapping the spatial database was a hard copy drawing on paper or film. The information comprised lines, points, areas and symbols which are explained in the legend. Additional information which could not be described directly on the sheet were referenced to external printed material, reports, memoirs, appendices, etc. Therefore most hard copy maps have the following in common:

1. the original data had to be simplified (classified) or reduced;

2. maps had to be drawn accurately;

3. large areas could only be represented on a number of separate map sheets (depending on scale);

4. it was difficult and expensive to extract hard copy data for use with other spatial data; and

5. they were static qualitative documents which were difficult to update and almost impossible for presenting quantitative information.

Because of the prohibitive costs of generating hard copy maps, they are normally used and are considered relevant for 20 years or more.

18.2.2 Computed Aided Mapping and Analysis

With the development of new trends for resource assessment, land evaluation and planning in the 1960s and 1970s, people began to demand an integrated, multidisciplinary method to analyze the mapped data. There are two approaches: (i) the "*gestalte*" method (Hopkins, 1977) which attempts to find "naturally occurring" environmental units that can be recognized, described and mapped in terms of the total interaction of the attributes under study, and (ii) the more conventional "monodisciplinary" resource maps, such as those of geology, landform, soil, vegetation and land use. The former has proven to be too general and difficult for retrieval of specific information about attributes. So there has remained a market for the more conventional monodisciplinary maps.

Engineers and planners in particular, in North America, have realized that data from several monodisciplinary surveys can be combined and integrated by overlaying transparent copies of resource maps to derive resultant polygons of specific combinations (McHarg, 1969). One of the first computer packages to use this approach was SYMAP (Fisher, 1978), short for SYnagraphic MAPping (Greek origin synagein, meaning to bring together), which includes a set of modules for analyzing data, manipulating them to produce choropleth or isoline inter-polations, and allows results to be printed with overprinting of line printer characters to produce gray scales. A number of raster or grid cell mapping programs followed this development.

The raster-based maps did not receive acceptance by cartographers who continued with their high-quality hard-copy maps till 1977, by which time the experience of using computers in map-making had advanced enough that Rhind (1977) was able to present the following list of reasons for using computers for cartography:

1.   to make existing maps more quickly;

2.   to make existing maps more cheaply;

3.   to make maps for specific user needs;

4.   to make map production possible in situations where skilled staff are unavailable;

5.   to allow experimentation with different graphical representations of the same data;

6.   to facilitate map making and updating when the data are already in digital form;

7.   to facilitate analysis of data that demand interactions between statistical analysis and mapping;

8.   to minimize the use of printed maps as a data store and thereby to minimize the effects of classification and generalization on the quality of the data;

9.   to create maps that are difficult to make by hand, e.g., 3D maps or stereoscopic maps;

10.   to create maps in which selection and generalization procedures are explicitly defined and consistently executed; and

11.   to allow for savings and improvements through a review of the whole map-making process as a result of automation.

By the late 1970s, in North America, a considerable investment in the development and application of computer-aided cartography had been undertaken by government and private agencies (Tomlinson et al., 1976; Teicholz and Berry, 1983). Unfortunately this did not result in immediate and direct savings in costs. Hardware was expensive and trained staff was and still is in short supply. As a result many problems and issues are still prevalent today. Technological development, particularly in hardware, has often outstripped management's ability to keep up. Indeed many cartographers miss the opportunity that digital mapping has provided them and instead have used the technology merely to produce maps more accurately and quickly. A comment which may be applied not only to cartographers but to engineers as well, was made by Poiker (1982) *"computer cartography is like a person with the body of an athlete in his prime time and the mind of a child"*.

Computerized cartography has been primarily directed to automating existing manual methods rather than exploring new ways to handle spatial data, in spite of the cost and many man years of development. The results from the 1960s and 1970s development in mapping can be summarized as an automation of existing tasks with emphasis on accuracy and visual quality and spatial analysis (raster) at the expense of good graphical results. This is changing today albeit too slowly for many.

18.2.3 Geographic Information Systems (GIS)

A number of fields have grown concurrently with computerized mapping and spatial analysis. These are cadastral and topographic mapping, thematic cartography, civil engineering, geography, mathematics of spatial variation, soil science, surveying and photogrammetry, rural and urban planning, utility networks, remote sensing and image analysis. Although initially there has been a duplication of effort for the different applications, the concurrent development results in the possibility of linking many kinds of spatial data processing together into a truly general-purpose GIS (Figure 18.1).

Workers in all disciplines wish to develop a powerful set of tools for collecting, storing, retrieving, transforming and displaying spatial data from the real world for their particular purposes. This set of tools constitutes a *"Geographical Information System"* (or GIS). Geographical data describe objects from the real world in terms of: (a) their position with respect to a known coordinate system, (b) their attributes that are unrelated

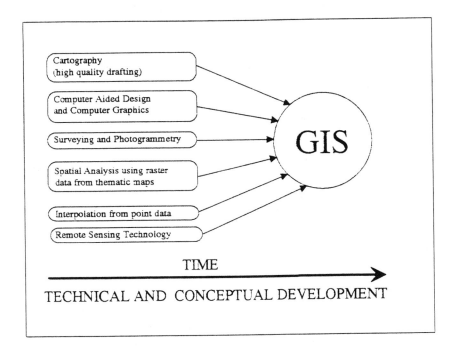

Figure 18.1: GIS linking parallel developments.

to position (such as soil type, cost, pH, etc.) and (c) their spatial interrelations with each other (topological relations), which describe how they are linked together or how one can travel between them (Burrogh, 1986).

Although there are a lot of similarities between CAD and GIS they are not the same. The major difference is the need of GIS to handle a diversity of data and the analysis methods used. Because of this, many original GIS systems were designed independently with their own graphics engine. More recently, advanced CAD systems have allowed third party application developers to build a GIS system around their CAD engine. The efficiency and functionality vary from one CAD system to another.

A GIS system is more than a mapping system because data

can be accessed, transformed and manipulated interactively. They can be used to study environmental processes or to analyze results of trends, or to anticipate results of different planning decisions. The GIS system can be used to model the consequences of a course of action before the mistakes have been irrevocably made in the landscape itself. With concerns today of environmental impacts of development, the GIS ability to analyze becomes more and more relevant.

18.2.4 The Components of a GIS

A GIS has three important components: (i) computer hardware, (ii) sets of application modules, and (iii) a proper organizational context. These need to be balanced if the system is to function properly.

The basic hardware components are: (i) the central processing unit (CPU), (ii) the visual display unit (VDU or monitor), (iii) the input device or digitizer (or mouse), (iv) the storage device or hard disk drive, (v) the backup device or tape drive, and (vi) the output device or plotter.

The software package for a GIS system consists of five basic technical modules (Figure 18.2):

1. data input and verification;
2. data storage and database management;
3. data analysis and transformation;
4. data output and presentation; and
5. interaction with the user.

Data input is the process of transforming data captured from existing maps, field observations, aerial photography and satellite imagery into a graphical database (Figure 18.3). Data storage and database management concerns the linkage (topology) of the graphical elements with the attribute database via the database management system (DBMS). Data transformation contains two types of operations: (i) transformations to data to bring them up

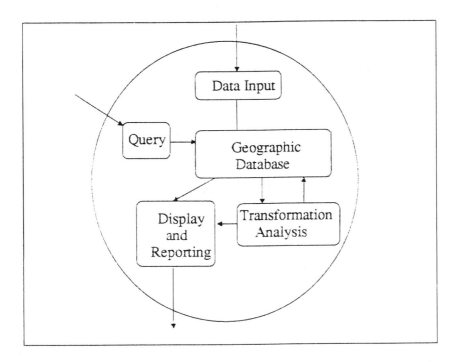

Figure 18.2: Main software components of a GIS.

to date, match other data and remove errors, and (ii) data analyses applied to achieve answers to queries (such as polygon overlays, unions, etc.). Transformations can operate on the spatial and non-spatial aspects of the data, either separately or in combination. Data output is the process of displaying results for presentation in the form of maps, tables and figures. The final module or interaction with user - query input - is one of the more critical components. Today query input is generally made through a menu driven environment.

The effectiveness of a successful GIS system is governed by its organizational structure. It is not sufficient to purchase hardware and software, it is most critical that personnel and managers be retrained to use the new technology in an organizational context as shown in Figure 18.4.

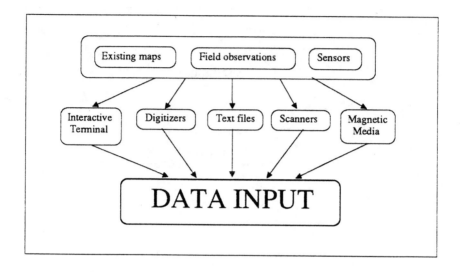

Figure 18.3: GIS data input.

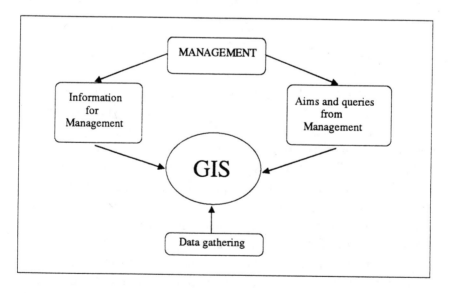

Figure 18.4: Organization aspects of GIS.

18.3 Infrastructure Design

Only twenty years ago engineers were just starting to use hand held calculators for design. Any programming required main frame or mini-computers with batch processing. Up and until PC-based computer-aided-drafting (CAD), much of the infrastructure design was done manually. Even today a large segment of engineering is not automated. There has been a dramatic increase in the last five years because of the increase in the availability of computer-aided-engineering (CAE) software for infrastructure design. Additional benefits have accrued due to the increasing computing power of the PC hardware systems.

Unfortunately many of the CAE packages have been designed to merely duplicate the old design process, similar to the evolution in digital cartography. This decreased the production process and cost, but did not add to the digital information base. There is still a large amount of duplication in the design office. Instead of having manually-drawn as-built plans there are digital CAD plots and disks in the archive. Computer printouts together with computerized files of input data and reports replaced manual design sheets.

18.3.1 Computer Aided Engineering and Design

CAE and CAD technology and development flourished in the mid 1980s together with that of the PCs. However, its initial emphasis was automation of the manual process. This is partly because integration was not possible due to the immaturity of the technology and secondly because of the conservative nature of the design community.

Take the case of civil site design; engineers still compute volumetrics for many earthworks projects by cross-sections and the average-end-area method. With digital terrain models (DTM), earthworks can be computed more accurately with the prismoidal method using the triangulated irregular network (TIN) surfaces of the original ground and final design. With the availability of

DTM software packages on PCs, more consultants are now taking advantage of the improvements in technology.

Similarly survey instruments have advanced in technology. Now many in the industry are using electronic and laser surveying equipment. With a total station, a site can be surveyed easily with the information stored digitally on a data collector. A digital elevation model (DEM) of random masspoints and breaklines for the site can be collected instead of a set of xy grid points. Similarly DEMs can be collected for road and drainage projects instead of the traditional traverse and level survey for cross-sectional information.

18.3.2 Case Example: Subdivision Design

In order to demonstrate change in the technology of infrastructure design a case study using a typical subdivision development process will be used.

Past and Current Approach - Manual Process

Subdivision development was and is today still being designed in a sequential, monodisciplinary process as shown in Figure 18.5. The main sequence for a typical subdivision development is:

1. a subdivision project is initiated;

2. a mapping company generates a hard copy topographic map of the site, or a grid survey is undertaken;

3. the planner lays out the subdivision plan using the map;

4. the surveyor determines the legal description of the lots, etc. (cadastre);

5. the engineer does the grading plan;

6. the engineer/designer lays out the storm network;

7. the engineer/designer calculates the runoff and designs the pipe sizing, manually or with simulation models;

8. the designer/draftsman plots the plan and profile information;

9. the bill of material is extracted for the tender documents;

10. the process is repeated for sanitary, water, etc;

11. the project is tendered and constructed; and

12. on completion the as-built information is prepared and submitted to the local agency.

Current and Future Approach - Automated Process

Using computer aided interactive infrastructure design modules and GIS the process can be improved by using the multidisciplinary approach with the GIS as the central focal point for the design team. Geographic data can be readily accessed by all members and inter-disciplinary communication and feedback between team members will assist in the optimization of the design process.

Similarly the CAE application software must also work together in an integrated manner so that relevant information can pass from one module to another with minimal modifications and conversion as shown in Figure 18.6. COGO, DTM, GIS, PLAN and STORM are CAE application modules (GWN Systems Inc., 1987-1992), for coordinate geometry design, digital terrain modelling for site works, geographic information system, subdivision baseplan design and storm drainage design respectively. The system is summarised:

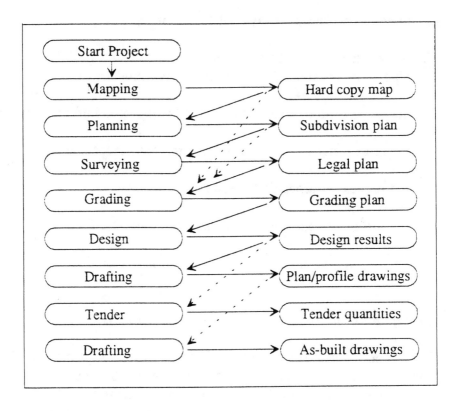

Figure 18.5: Sequential monodisciplinary design process.

1. A subdivision project is initiated.

2. A mapping company or surveyors generate a DEM of the site. A DTM model is used to generate a TIN model and contours of the site. With a digital terrain model contour plots can be generated at any time, scale or contour interval.

3. The planner reviews the site. Because the consultant is using a DTM the planner can visually analyze the site in 3D, check aspect, undertake slope analysis, sight lines, etc. By working with others on the design team he can very quickly layout the lots, roads, etc., taking into account other design concerns.

The design can be done digitally with PLAN or similar packages.

4. The surveyor uses COGO to define the legal base and make adjustments as required. Data from PLAN can be used directly, thereby minimizing time required to generate the legal plan.

5. The engineer/designer does grading with PLAN and DTM.

6. The engineer/designer lays out the storm drainage network, manhole locations, drainage areas, etc. with PLAN.

7. The engineer/designer calculates the pipe sizing requirement with STORM using the Rational method or simulation models.

8. The engineer/designer automatically plots the plan and profile drawings using STORM.

9. The engineer/designer automatically extracts the bill of materials from the attribute database with STORM.

10. The process is repeated for sanitary, water, systems, etc.

11. The project is tendered and constructed.

12. On completion the as-built information is prepared digitally and submitted to the local agency together with the database information for inclusion in their GIS system.

18.3.3 A typical Interactive Infrastructure Design Package

GWN-STORM is one module of a series of integrated interactive computer aided design software developed by GWN Systems Inc (1992). It is designed to run on a number of platforms using

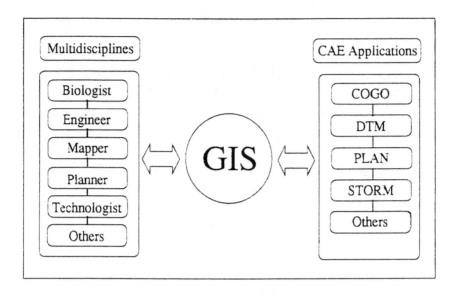

Figure 18.6: Multidisciplinary design process with GIS.

UNIX, DOS and the Apple Macintosh operating systems. The user interface is through interactive dialogue boxes within the graphics environment. Its uses a knowledge-based design approach with user definable parameters.

The dialogues within the Intergraph MicroStation CAD are event-driven. With event-driven programming, hydraulic and hydrologic calculations can be displayed as interactive dialogues. With event-driven programming, it is possible to present the user with options whereby three variables can be displayed and with a toggle switch select the variable to be re-calculated if one of the two input variables are modified. For the case of the hydraulic calculator, if the diameter is toggled, then any changes to the slope or flow variables will result in a new diameter being re-calculated as shown in Figure 18.7.

The graphical elements in the CAD system, such as manholes, pipes and drainage areas are linked to database records (attributes) which contains fields with design information. As the design progresses, results are automatically stored in their respective database fields. This approach is taken to allow for automatic transfer to the GIS system.

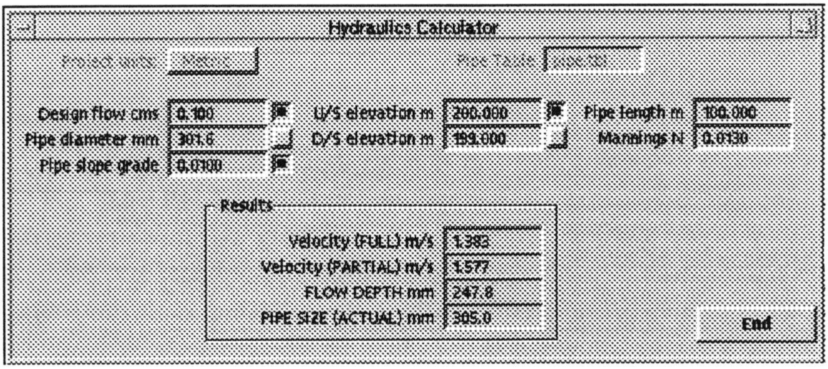

Figure 18.7: Event driven hydraulic calculator.

18.4 The Future: Closing the Loop

GIS and integrated computer aided infrastructure design applications for PCs exist today. The hardware and software technology has improved exponentially in the last ten years and will continue to improve in the future. What is needed is a better understanding of the advantages these technologies can provide and the development of better interaction between the multidisciplines to maximize the synergy and benefits.

With a GIS system political, social and environmental issues can be addressed in a holistic manner. Simulations can be undertaken to assess the environmental impacts of development prior to construction. Engineers, scientists and planners can be proactive instead of reactive.

By linking simulation models to GIS and infrastructure design one can now:

1. Assess the impacts of storm runoff on structures;

2. Model pollution loading in streams, etc;

3. Assess the impact of infiltration on sanitary systems based on ground water seepage, soil conditions, etc;

4. assess impact of development on wetlands; and

5. site solid waste disposal areas based on polygon analyses.

All the above examples can and are being done without GIS or computer modelling, but these are done at a high cost and as one-off type studies. Without GIS it is very difficult to reassess a different scenario if there are changes in methodology or legal requirements. In addition all the data collected is in hard copy and is usually unavailable, or difficult to transpose for use by others.

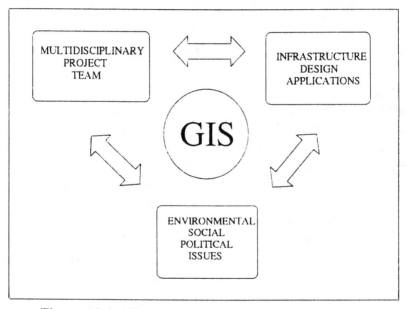

Figure 18.8: The future GIS approach to design.

18.4.1 Conclusions

Cost-effective PC-based GIS and integrated interactive infrastructure design systems are available today. These systems can be used in a multidisciplinary way to minimize data collection costs and maximize economic, social, political and environmental benefits for society as shown in Figure 18.8.

The technology for GIS and integrated CAE design is available now. Let us all work together to close the loop and move ahead into the future.

18.5 References

Brinkman, R. and Smyth, A.J. (1973). Land evaluation for rural purposes. Summary of an expert consultation, Wageningen, The Netherlands, 6-12 October 1972. International Institute for Land Reclamation and Improvement, Wageningen, Publication No. 17.

Burrough, P.A. (1986). Principles of Geographical Information Systems for Land Resources Assessment. Clarendon Press, Oxford.

Cliff, A.D. and Ord, J.K. (1981). Spatial processes: models and applications. Pion, London.

Dilke, O.A.W. (1971). The Roman land surveyors. An introduction to the agrimensores. David and Charles, Newton Abbot, U.K.

Fisher, H.T. (1978). Thematic cartography - what it is and what is different about it. *Harvard papers in theoretical cartography*. Laboratory for Computer Graphics and Spatial Analysis, Harvard.

GWN Systems Inc. (1987-1992). GWN-COGO, GWN-DTM, GWN-GIS, GWN-PLAN and GWN-STORM - Computer aided engineering applications. Edmonton, Alberta, Canada.

Hodgkiss, A. G. (1981). Understanding maps. Dawson, Folkestone, U.K.

Hopkins, L.D. (1977). Methods for generating land suitability maps: a comparative evaluation. *Am. Inst. Plan. J.* October, 386-400.

McHarg, I.L. (1969). Design with nature. Doubleday/ Natural History Press, New York.

Poiker, T.K. (formerly Peuker) (1982). Looking at computer cartography. GeoJournal 6, 241-9.

Rhind, D. (1977). Computer aided cartography. *Trans. Inst. Br. Geogrs* (N.S.) 2, 71-96.

Ripley, B.D. (1981). Spatial statistics. Wiley, New York.

Teicholz, E. and Berry, B.J.L. (1983). Computer graphics and environmental planning. Prentice Hall, Englewood Cliffs, NJ.

Tomlinson, R.F. (1984). Geographic information systems - a new frontier. *In Proc. Int. Symp. on Spatial Data Handling.* 20-24 August, Zurich, pp. 1-14.

Tomlinson, R.F., Calkins, H.W., and Marble, D.F. (1976). Computer handling of geographical data. UNESCO, Geneva.

Webster, R. (1977). Quantitative and numerical methods for soil survey and classification. Oxford University Press, Oxford.

Chapter 19

GIS Based Hydraulic Model Pictures the Interceptor Future

Uzair M. Shamsi
The Chester Engineers
P.O. Box 15851
Pittsburgh, PA, 15244

Albert A. Schneider
Allegheny Co. Sanitary Authority
Pittsburgh, PA

This chapter assesses the hydraulic capacity of the Chartiers Creek Interceptor of the Allegheny County Sanitary Authority (ALCOSAN) to convey dry and wet weather flows under the existing and anticipated ultimate future development conditions. The assessment was performed by developing a hydraulic model of the interceptor system using U.S. EPA's computer program Storm Water Management Model (SWMM). The model consisted of 56 tributary sewer subareas, 75 sewer segments, 96 manholes, 34 orifices, and 34 weirs/outfalls. The input data for the model were extracted from a Geographic Information System (GIS) of the watershed. The watershed GIS consisted of data layers of digitized subareas, soil associations, and census tracts; present and future land use; and USGS Digital Elevation Models (DEMs).

0-87371-898-4/93/$0.00 + $.50

19.1 Introduction

Established in 1946, The Allegheny County Sanitary Authority (ALCOSAN) is by far the largest of all the sewerage systems in Allegheny County, Pennsylvania and serves more than half of the local municipalities, through a series of interceptor lines and a sewage treatment plant. The Chartiers Creek watershed, shown in Figure 19.1, is one of the largest watersheds in Allegheny County, enveloping all or portions of 23 municipalities, including the City of Pittsburgh, twelve boroughs and ten townships. More than 90% of the watershed is served by Chartiers Creek Interceptor. Of the 56 tributary areas shown in Figure 19.1, 22 have combined sewers. Most of the separate sewer tributary areas contribute wet weather inflow to the interceptor. The service area includes urban, suburban, and rural areas. The area is approximately 148 square miles of rolling hills, whose development ranges from very dense commercial, residential, and industrial areas; sparsely developed rural and agricultural areas; lightly to extensively strip mine disturbed areas; to varying size tracts of undeveloped forested areas. The elevation within the six 7.5 minute USGS quadrangles covering the area ranges from 690 to 1,500 feet above mean sea level.

The interceptor ranges in size from 24 inches to 45 inches. The ALCOSAN sewage treatment plant is located on the east bank of the Ohio River. The interceptor flow is conveyed to the treatment plant through a 54-inch underground tunnel that passes under the Ohio River. Most of the service areas tributary to the interceptor have diversion chambers. If operating correctly, a diversion chamber should permit all dry weather flows and a limited amount of wet weather flows to enter the interceptor.

The watershed is expected to undergo significant development in the future. Recently, a new interceptor serving several developing communities has been connected to the ALCOSAN interceptor, raising concerns about the interceptor's hydraulic capacity to convey the future flows. This paper addresses a case study (Shamsi, 1991) that assessed the hydraulic capacity of the Chartiers Creek Interceptor of ALCOSAN to convey the dry and

Chartiers Creek Interceptor Study
Existing Service Areas

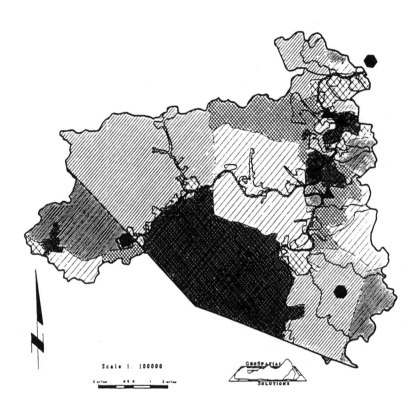

Municipal Boundaries

County Boundary

Subcatchments

Interceptor

Watershed Boundary

⬢ Raingage

▲ Flowmeter

Figure 19.1: Existing service areas.

wet weather flows under the existing and anticipated ultimate future development conditions.

19.2 Watershed Geographic Information System

This study developed and analyzed a GIS for the Chartiers Creek watershed in order to provide input to the watershed SWMM. The geographic data set that was created consisted of digital components that would allow preprocessing, analysis, and display using techniques commonly referred to as remote sensing and GIS analysis.   An analysis of the terrain, hydrography, soil associations, land use, census properties, and locations of major trunk and interceptor sewers was conducted to develop the watershed GIS.

19.2.1 Technical Approach

The software used in this project was primarily ARC/INFO, a vector based GIS, and ERDAS, an image processing and raster based GIS. Additional programs were written whenever needed for data format changes or the creation of a product for which the methodology was not available in either of the commercial packages.   This included the determination of percent slope averages, the creation of the percent impervious and developable/non-developable images, and the construction of tabular output.

The vector data format is a topologically constructed set of points, nodes, lines, and polygons which define locations, boundaries, and areas.   A raster format is a regular grid of uniform size cells, with a data value associated with each cell. The reason for using both kinds of data processing and handling formats is to take advantage of the best features of each. Data entry of vector information is more easily performed using a file digitized into the vector format.   All GIS layers except the elevation data were initially digitized in vector format from their respective base maps utilizing ARC/INFO. The resulting polygon topology was then converted to raster format for GIS analysis.

Satellite, elevation imagery, and overlays for GIS cross-tabulation are appropriately dealt with in raster format wherein every cell of a given layer/image is registered to the corresponding cell of every other layer. The GIS analysis of the watershed and subareas was done in raster format utilizing ERDAS software. The SPOT image and the DEMs were initially in raster format. Each data set or information layer was co-registered to common coordinates, so that every raster grid cell matched its corresponding cell in other layers.

19.2.2 GIS Analysis and Results

Land use classes were derived from a manual interpretation of the SPOT image. Each land use class was assigned a percent impervious value. These value were based on the Soil Conservation Service (SCS) estimates for similar land use types. Certain open space land use types were assigned a zero percent impervious value. Table 19.1 summarizes the watershed land use classes and their percent impervious values.

19.2.3 Summary of Themes by Subarea

The final GIS analyses were performed on raster information layers with the following themes: subareas, percent slope, slope greater than 20 percent, slope less than or equal to 20 percent, land use, impervious area, soils, and census tracts. The area of each subarea was the basic theme against which the various other themes were summarized. The ERDAS "Summary" function, as well as an in-house written function called "Sumave," were utilized.

Table 19.1: Watershed land use and percent imperviousness.

Land Use	% impervious	% of study area
High Density Residential	52	2.5
Medium Density Residential	28	21.5
Low Density Residential	16	4.2
Commercial	85	0.6
Industrial	72	2.9
Open Wooded	0	33.2
Urban Disturbed	0	0.9
Rural Disturbed	0	6.8
Strip Mine, Quarry, Oil Well	0	5.2
Open Space	0	3.0
Open, Non-wooded	10	18.1
Major Roadway	100	1.1

The following tables were produced:

19.1. A subarea table showing area in acres, percent slope and mean percent imperviousness of each subarea. These values were used in SWMM to compute subarea wet weather flows.

19.2. A soil association table showing percent of various soil associations in each subarea prepared by digitizing SCS soil association maps. This table was used to determine subarea infiltration parameters for stormwater runoff modelling.

19.3. A census tracts table showing percent of various census tracts in each subarea prepared by digitizing U.S. Census Bureau census tract maps. Census tract data for population, number of dwellings, number of persons per dwelling, market value of housing units, and family income were obtained from the 1980 census data. Subarea percentage of census tracts and census tract percentages in subareas were applied to total tract values to estimate weighted mean subarea values. Mean subarea values computed in this manner were used in the watershed SWMM to estimate subarea dry weather flows.

The following color, raster, ARC/INFO maps were produced:

1. Existing service areas (Figure 19.1)
2. Present land use (Figure 19.2)
3. Census Tracts (Figure 19.3)
4. Soil Associations (Figure 19.4)
5. Future land use (Figure 19.5)

19.2.4 Determination of Future Developable Land

The present land use map (Figure 19.2) shows that substantial watershed area is still undeveloped, especially to the west of Chartiers Creek. In order to model ultimate future development in the watershed, six additional subareas were outlined as shown in Figure 19.5. These subareas were based on topography roughly following the natural surface water drainage basins on undeveloped land. By overlaying the vector land use layer and the vector "greater than/less than or equal to 20 percent slope" layer, a map coverage resulted showing existing developed areas, and future developable and non-developable areas. Based on the criterion that future development can occur only on 20 percent or less slopes of non-developed and non-strip mined areas, the nonqualifying classes were eliminated resulting in a vector layer of future developable land shown in Figure 19.5.

The six future subareas were superimposed on the Watershed GIS future land use map to compute for each future subarea the area with slopes steeper than 20 percent and strip mines. This area and area already developed was subtracted from the total future area to compute future developable area in each future subarea. Future subareas outlined in this manner, therefore, are based on the assumption that each and every developable parcel of land will be ultimately developed. Table 19.2 shows how the developable land area of future subareas was calculated.

Table 19.2: Future subareas.

Sub Area ID	Area (Acres)	Area w/ steep slopes & strip mines (Acres)	Area already developed (Acres)	Developed Area (Acres)
600	1,415.6	613.4	444.8	357.4
610	19,775.4	3,632.5	2,254.8	13,888.1
620	1,240.4	87.0	73.4	1,080.0
630	2,399.3	539.6	46.3	1,813.4
640	7,245.0	1,698.1	1,037.5	4,509.5
650	3,920.6	511.7	1,663.7	1,745.2

19.3 Hydraulic Model of the Interceptor

The U.S. EPA's SWMM (Huber and Dickinson, 1991) was developed in the early 1970's and has been continually maintained and updated. It is perhaps the best known and most widely used of the available urban runoff quantity/quality models. SWMM simulates dry weather sewage production and wet weather flows on the basis of land use, demographic conditions, the hydrologic conditions in the watershed, meteorological inputs and conveyance/treatment characterizations. Based upon this information, SWMM can be used to predict combined sewage flow quantity and quality values. SWMM also provides the ability to perform detailed analyses of conveyance system performance under a wide range of flow conditions. As such, it is well suited to this study and is the model of choice for use in most combined sewer overflow feasibility studies.

The use of SWMM to model the Chartiers Creek watershed to assess the available capacity of the Chartiers Creek interceptor will be particularly advantageous for the following reasons:

1. SWMM represents the best means of producing estimates of current dry and wet weather flow rates from a service

Chartiers Creek Interceptor Study
Present Land Use

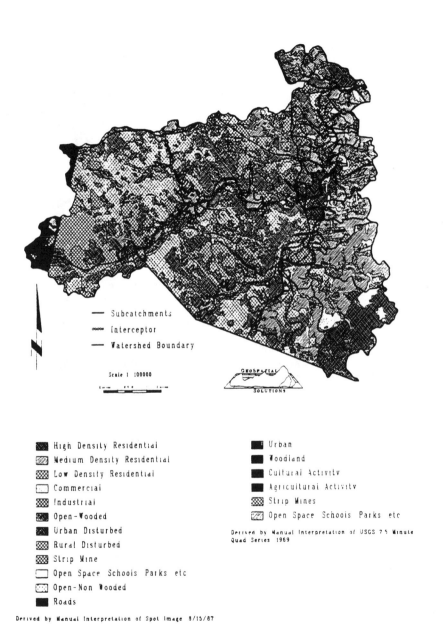

Figure 19.2: Present land use.

Chartiers Creek Interceptor Study
Census Tracts

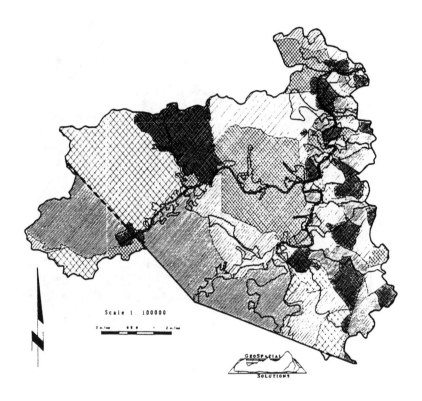

Census Tracts
County Boundary
Subcatchments
Interceptor
Watershed Boundary

Figure 19.3: Census tracts.

Chartiers Creek Interceptor Study
Soil Associations

Gilpin-Upshur-Atkins

Culleoka-Weikart-Newark

Dormont-Guernsey-Culleoka

Urban Land-Philo-Rainsboro

Urban Land-Rainsboro-Alleg Var

Urban Land-Dormont-Culleoka

Strip Mines-Guernsey- Dormont

Dormont-Culleoka

Dormont-Culleoka-Newark

Udorthents-Culleoka-Dormont

— Allegheny/Washington County Boundary

— Subcatchments

∞∞ Interceptor

— Watershed Boundary

Figure 19.4: Soil associations.

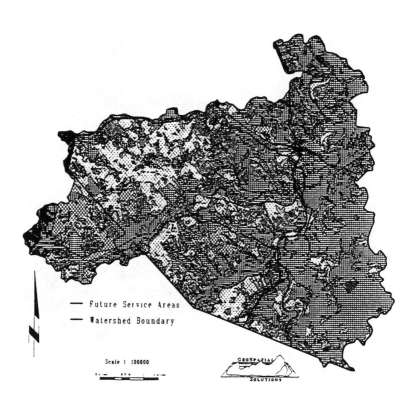

Chartiers Creek Interceptor Study
Potential Development

— Future Service Areas
— Watershed Boundary

Scale 1 100000

Developable Lands
Present Land Use and Slope < 20%

Open-Wooded
Urban Disturbed
Rural Disturbed
Open Non-Wooded
Woodland
Agricultural Activity

Non-Developable Lands

Presently Developed,
Strip Mines, or
Slopes > 20%

Figure 19.5: Future land use.

area as large and diverse as the Chartiers Creek watershed. Flow estimates can be prepared based upon current land use conditions, topography, interceptor sewer characteristics, and selected meteorological conditions. The model can be calibrated against measured flow rates.

2. SWMM represents the best means of modelling the performance of the interceptor conveyance system under a range of dynamic flow conditions.

3. The use of SWMM provides a ready means of accounting for anticipated future development characteristics in an assessment of available capacity.

4. Using SWMM, it will be possible to assess capacity in response to wet weather input. This characteristic can be very useful for subsequent analyses related to abatement of combined sewer overflows (CSOs).

5. The watershed SWMM developed under this study can be expanded to model water quality and assess the effectiveness of a range of CSO abatement or treatment options.

6. Future CSO and system monitoring data expected to be collected in compliance with the Pennsylvania Department of Environmental Resources' CSO strategy can be used to improve calibration of the model.

19.3.1 Modelling Strategy

SWMM is flexible enough to allow different modelling approaches to the same area. An approach which adequately describes the service area and accurately computes and routes the flows at reasonable computing time and effort should be adopted.

After a review of the data, maps, and literature available to complete this study the following modelling strategy was adopted:

1. Run SWMM's Transport Block to generate sanitary flows (dry weather flows) in all the subareas.

2. Use Transport block to model flow division (an approximate way to separate dry weather flows from combined flows) from the areas where detailed modelling of diversion chambers in the Extran Block is not feasible.

3. Enter the entire amount of sanitary sewage flow to the diversion chambers.

4. Run SWMM's Runoff Block to generate stormwater runoff (wet weather flow) in all the subareas.

5. For combined sewer subareas, enter the entire amount of runoff to the diversion chambers. For separate sewer subareas, enter only a fraction of total runoff to the diversion chambers to account for uncontrolled infiltration and inflows (I/I flows). This fraction will be determined by comparing the modeled and measured flows from separate subareas.

6. Import wet weather flows from Runoff Block into the Transport Block and combine them with the dry weather flows. Enter the combined flows to diversion chambers.

7. Use SWMM's Extran Block to regulate the dry weather flows entering the interceptor and route them through the interceptor.

8. Study interceptor response (capacity, surcharging, manhole flooding, etc.) from the Extran output.

19.3.2 SWMM Schematic Diagram

It is highly recommended that a sewer system schematic diagram be prepared prior to model building. A schematic diagram is developed by discretization, a procedure for the mathematical abstraction of the physical drainage system. For the computation of hydrographs, the drainage basin may be conceptually represented by a network of hydraulic elements, i.e. subareas, sewers, manholes, and diversion chambers. Hydraulic properties of each element are then characterized by various parameters, such as size, slope, and roughness coefficient. The schematic diagram for the study watershed is shown in Figure 19.6.

19.4 Precipitation and Flow Monitoring

Site selection for monitoring rainfall and flow was made in conjunction with the general arrangement of the watershed, delineated subareas, and the interceptor sewer system in order to maximize the extent and value of the data collected for use in calibration of watershed SWMM. Flow monitoring is necessary for several reasons. One reason is that hydraulic characteristics of the collection system may need to be confirmed or corrected. Another reason is that the flow monitoring provides necessary calibration and verification data to develop a mathematical model of the interceptor. Rainfall duration and intensity is so variable that parts of the collection system may be dry while others are deluged. Knowing that rain occurred only in a part of the collection system can have major implications on the hydraulic loading of the interceptor. Rainfall data coupled with flow monitoring data close the loop and establish cause and effect.

The determination of feasible locations for installing raingauges was based upon a combination of factors such as watershed topography, shape of the watershed, and availability of space. The objective was to distribute the raingauges in such a manner that most of the watershed storms can be captured. This objective requires that each raingauge cover approximately an equal watershed area without leaving any watershed area uncovered. Since the eastern half of the watershed accounts for most of the service area, three raingauges were uniformly distributed in this part. The western half of the watershed was covered by one gage because it covers a relatively small service area. These locations are shown in Figure 19.1. Measurements were taken for four months.

The determination of feasible locations for installing flow monitors was based upon a combination of factors such as type of flow (separate or combined), location with respect to the interceptor, and access. It was decided to install three flow monitors to sample a typical combined sewer service area, a typical separate sewer service area, and the interceptor. The flow monitoring locations are shown in Figure 19.1. Flow monitoring was also conducted for four months.

19.5 Model Calibration

Model calibration consists of adjusting model parameters (for example, imperviousness, or roughness) until the predicted output agrees with measured observations. For example, the modeled hydrograph may be adjusted to agree with the measured hydrograph. Parameter estimates should fall within reasonable ranges of known values in order to enhance confidence in model results. This is easier for physically measurable parameters such as area, elevation and slope and harder for more abstract parameters such as imperviousness, depression storage, and roughness. As a result, the former are often fixed during the calibration and the latter are varied. It is often necessary to account for an unmodelled effect by varying a parameter beyond its normal range. In general, rainfall/runoff parameters and dry weather flow base infiltration rate estimating coefficients are adjusted to calibrate model output against measured conditions.

SWMM calibration was performed by comparing model outputs from Runoff, Transport, and Extran blocks to measured flow rates. All the three SWMM blocks Runoff, Transport, and

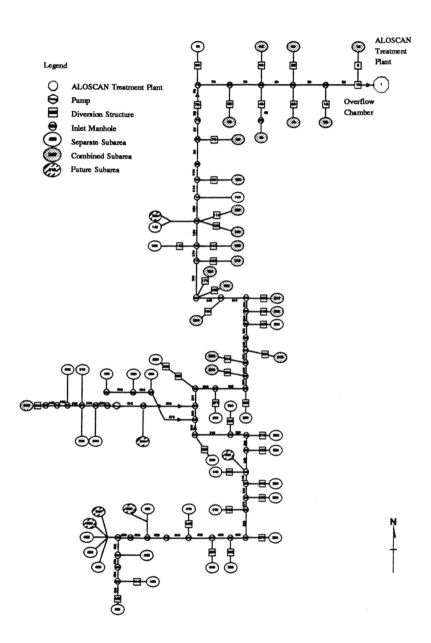

Figure 19.6: SWMM schematic diagram.

Extran were calibrated separately. Runoff and Transport block input parameters were calibrated against measured wet and dry weather subarea flows, respectively. Extran block input parameters were calibrated against measured interceptor flows both under dry and wet weather flow conditions. Only the Extran block calibration is described here.

Extran block is based on physical properties of the interceptor conveyance system including sewers, manholes, and diversion chambers. Accurate information about the length, slope, and size of the sewers was obtained from the construction drawings. However, Manning's roughness coefficient for the sewers cannot be determined with full certainty and may be adjusted during the calibration process. The diversion chambers' geometry was determined from the construction or as-built drawings. However, SWMM's capability to model flow division in a diversion chamber is based on simplifying assumptions. For example, diversion chambers are treated as storage elements where water level rises or drops depending on the size of orifice opening for dry weather flow to the interceptor or the width and height of the overflow weir for combined sewer overflows. Conveyance within a chamber due to sloped bottom troughs is not accounted for. A diversion chamber is assumed to have a flat bottom with uniform cross section throughout its depth. Last but not least, partial openings due to orifice plates cannot be modeled. Thus, discrepancies between modeled and observed interceptor flows are likely. These discrepancies can be reduced by adjusting the parameters controlling the dynamics of flow in diversion chambers such as orifice coefficients, weir coefficients, and cross sectional area of the chambers.

19.5.1 Dry Weather Flow Calibration

Figure 19.7 shows the interceptor calibration for dry weather flows. The modeled interceptor dry weather flows were compared against the measured mean flows of May 28 and June 25 flows. The flow averaging was performed to account for two

typical types of interceptor dry weather flows. The first type of
dry weather flows (June 25) are found during prolonged dry
weather periods and peak at about 24 cfs. The second type of dry
weather flows (May 28) are found after substantial rainfall events
and peak at about 28 cfs. The latter type of flows include effect
of wet weather infiltration on dry weather flows.

It can be seen from the plots in Figure 19.7 that modeled dry
weather flows match the observed flows quite satisfactorily. The
difference between the modeled and observed average daily flow
and volume is less than two percent, which indicates a
satisfactory calibration of the SWMM for dry weather interceptor
flows. This degree of calibration was attained by the following
SWMM parameter values:

·	Manning's "n"	= 0.013
·	Diversion chamber orifice coefficient	= 0.65
·	Diversion chamber weir coefficient	= 3.00
·	Manhole/diversion chamber area	= 50 ft^2

19.5.2 Wet Weather Flow Calibration

The interceptor tributary subarea flows are regulated in the
diversion chambers. The diversion chambers allow a limited
quantity of flow to enter the interceptor and divert any flows in
excess of this amount to the receiving waters. If properly
designed and maintained, this limited quantity is approximately
3.5 times the average dry weather flow rate. Improper design and
lack of maintenance may cause the diversion chambers to allow
more or less wet weather flows to enter the interceptor than set
by this standard. Furthermore, some separate sewer tributary
subareas have no diversion chambers and are referred to as having
direct connections with the interceptor. These subareas are not
always 100 % water-tight and some quantities of wet weather
flow are able to infiltrate into the interceptor. Thus, interceptor
flows increase during the storm events and corresponding
interceptor flows are called wet weather flows. In conclusion, the

TIME	OBSERVED FLOW			MODELED
(HOURS)	June 25, 91	May 28, 91	MEAN	FLOW
	(CFS)	(CFS)	(CFS)	(CFS)
0	21.05	25.63	23.34	22.30
1	20.60	25.63	23.12	21.96
2	19.09	24.46	21.78	21.00
3	18.25	22.85	20.55	19.41
4	15.85	21.49	18.67	17.51
5	14.39	21.05	17.72	15.91
6	13.21	18.88	16.05	15.80
7	13.05	18.25	15.65	16.81
8	13.70	17.83	15.77	18.55
9	16.82	18.88	17.85	21.77
10	18.88	22.85	20.87	27.07
11	21.27	25.63	23.45	29.10
12	23.08	27.75	25.42	27.36
13	23.77	27.98	25.88	25.77
14	23.31	27.98	25.65	25.09
15	23.08	27.51	25.30	24.13
16	21.49	26.80	24.15	23.04
17	20.83	25.63	23.23	22.58
18	20.60	25.16	22.88	22.59
19	20.60	24.46	22.53	22.57
20	20.60	24.46	22.53	22.20
21	20.38	24.93	22.66	22.50
22	20.60	24.93	22.77	22.90
23	20.60	25.63	23.12	23.03
TOTAL	465	577	521	531
VOLUME	1,674,360	2,075,940	1,875,150	1,911,420
MEAN	19.38	24.03	21.70	22.12
% DIFFERENCE IN VOLUME				1.93
% DIFFERENCE IN MEAN FLOW				1.93

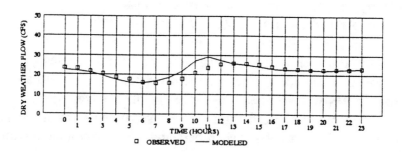

Figure 19.7: Dry weather flow calibration.

interceptor wet weather flows include only that portion of the total subarea wet weather flow that is allowed by the diversion chamber to enter the interceptor, and any wet weather contribution of direct connection separate sewer subareas.

Figure 19.8 shows the interceptor calibration for wet weather flows. The modeled interceptor wet weather flows were compared against the measured flows of June 11, 1991. It can be seen from the plots in Figure 19.8 that modeled wet weather flows match the observed flows satisfactorily. The difference between the modeled and observed peak flow and volume is 5.5% and 7.7% respectively, which indicates a satisfactory calibration. This degree of calibration did not require any further adjustment of the model parameters established under dry weather flow calibration.

19.6 Interceptor Capacity Analysis

The following five type of flows were considered under present and anticipated future development conditions:

- Dry weather flows,
- 2-year, 1-hour wet weather flows,
- 25-year, 1-hour wet weather flows,
- 2-year, 24-hour wet weather flows, and
- 25-year, 24-hour wet weather flows.

The capacity analysis was performed by comparing the maximum simulated flows in various sewer segments of the interceptor to their full flow capacities. The full flow capacity is defined as the maximum flow that can be conveyed by a sewer without surcharging, and it is estimated from sewer size, slope, and roughness. If a sewer is surcharged, the flow becomes pressurized and the sewer can convey more flow than its full flow capacity with or without manhole overflows and street flooding. Surcharged sewers causing manhole overflows and street flooding are unacceptable.

TIME	RAINFALL	OBSERVED		MODELED	
(HOURS)	(INCHES)	FLOW	VOLUME	FLOW	VOLUME
		(CFS)	(CF)	(CFS)	(CF)
19	0.00	22.17	39,906	24.32	43,776
20	0.75	22.17	79,812	24.06	87,084
21	0.10	32.65	98,676	32.77	102,294
22	0.25	45.44	140,562	44.15	138,456
23	0.00	46.78	165,996	49.36	168,318
0	0.00	46.41	167,742	49.17	177,354
1	0.15	45.44	165,330	48.85	176,436
2	0.05	41.00	155,592	48.17	174,636
3	0.00	43.48	152,064	48.64	174,258
4	0.20	44.85	158,994	48.60	175,032
5		44.85	161,460	48.49	174,762
6		45.96	163,458	48.06	173,790
7		44.20	162,288	40.19	158,850
8		43.48	157,824	38.15	141,012
9		43.48	156,528	36.79	134,892
10		42.71	155,142	32.71	125,100
11		42.71	153,756	30.92	114,534
12		41.88	152,262	27.27	104,742
13		41.00	149,184	24.69	93,528
14		34.91	136,638	23.49	86,724
15		31.50	119,538	22.88	83,466
16		29.16	109,188	22.59	81,846
17		27.98	102,852	22.40	80,982
18		27.98	100,728	22.40	80,640
SUM	1.50		3,305,520		3,052,512
% DIFFERENCE IN VOLUME					−7.7
% DIFFERENCE IN PEAK FLOW					5.5

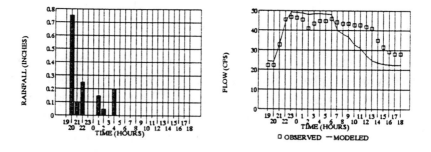

Figure 19.8: Wet weather flow calibration.

19.6.1 Present Conditions

The present flows were simulated by running the calibrated SWMM with the existing tributary subarea characteristics. Figure 19.9 shows a plot of maximum dry weather flows versus interceptor capacity for all the interceptor segments modeled in SWMM. The maximum sewer flows reported here represent the peak flows simulated by the watershed SWMM. Figure 19.9 shows that the peak interceptor dry weather flow from the entire watershed is 35.7 cfs. Since the peak flows are less than the full flow capacity in all the interceptor segments, it is demonstrated that the interceptor has adequate capacity to transport the present day dry weather flows from the watershed.

Figures 19.9 also shows plots of maximum wet weather flows under the 2-year/1-hour, 25-year/1-hour, 2-year/24-hour, and 25-year/24-hour storm conditions. The peak interceptor wet weather flows for the four design storms are 54.5, 62.7, 59.8, and 63.3 cfs, respectively. Peak wet weather flows are greater, under certain flow conditions, than the full flow capacity in the first three interceptor segments located at the watershed outlet. Table 19.3 summarizes the interceptor capacity deficits with respect to types of wet weather flows.

Table 19.3:   Interceptor capacity deficits under present wet weather flows.

Design Flow	Deficient Interceptor Segments	Capacity Deficit (cfs)
2-year/1-hour	30	3.9
25-year/1-hour	20	4.0
	30	9.2
2-year/24-hour	20	1.1
	30	6.0
25-year/24-hour	20	4.6
	30	9.1
	50	0.1

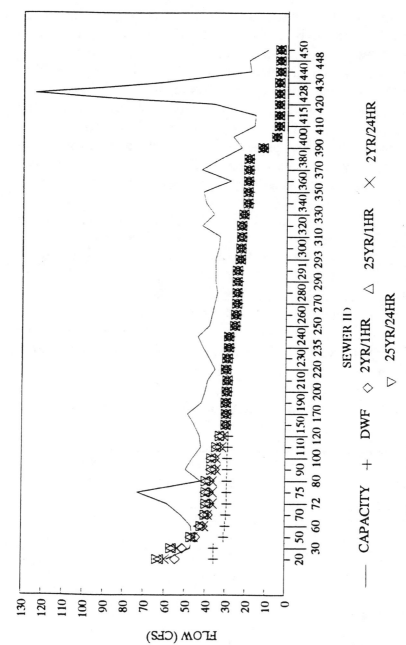

Figure 19.9: Interceptor flows under present conditions.

The previous table shows that flows resulting from the short duration (1 hour) storms are as critical as the long duration (24 hours) storms since both types of design flows cause approximately the same amounts of capacity deficits and the same maximum wet weather flows. This phenomenon can be explained by the fact that high rainfall intensity associated with the short duration storms is known to cause more severe flooding conditions. It can be seen that, under all four types of wet weather flows, sewer 30 will have the maximum capacity deficit of 9.2 cfs. Sewers 20 and 50 will have capacity deficits of 4.6 cfs and 0.1 cfs, respectively. Although interceptor capacity deficits due to surcharge interceptor segments did occur as described above, no manhole overflows (street flooding) were noted in the SWMM output. Thus, it can be concluded that the four types of wet weather flows surcharged the interceptor near the watershed outlet, but these flows were not large enough to cause manhole overflows. In other words, while the full pipe capacity of the interceptor for the above sewers is exceeded, surcharging is not severe enough to produce overflows.

19.6.2 Future Conditions

The future flows were produced by loading the calibrated SWMM with subarea data descriptive of anticipated future land development conditions in the six future subareas (600, 610, 620, 630, 640, and 650). Figure 19.10 shows a plot of future maximum dry weather flows versus interceptor capacity. It can be seen that the peak interceptor dry weather flow is 45.5 cfs, which is 27.5 percent greater than the present day dry weather flows. The peak flows are less than the full flow capacity in all the sewers except sewer 220. The capacity deficit based on full pipe flow for this sewer is 3.7 cfs. However, the model indicates that while the full pipe capacity in this segment is exceeded, no manhole overflows would be produced. Thus, despite a capacity

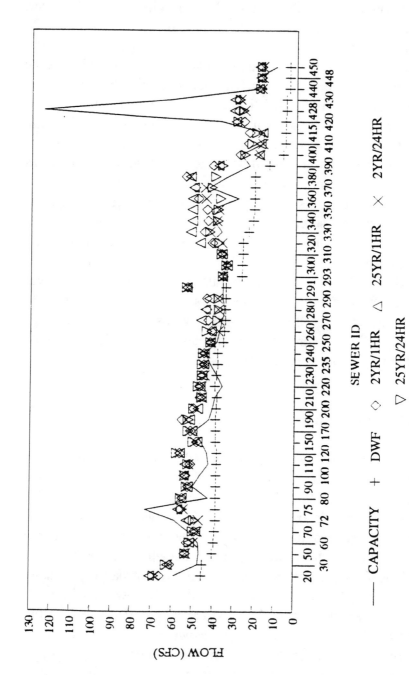

Figure 19.10: Interceptor flows under future conditions.

deficit, the surcharge head of the interceptor will enable sewer 220 to convey future dry weather flows. It is therefore concluded that the interceptor is adequate to handle future dry weather flows from the potential future service areas with the possibility of dry weather surcharging, provided that no major industrial development will take place and commercial and institutional growth will continue at the existing rate.

Figures 19.10 also shows plots of maximum future wet weather flows. The peak wet weather flows for the four design storms are 66.0, 70.6, 68.0, and 68.6 cfs, respectively. Peak wet weather flows are projected to be greater than the full flow capacity of the interceptor in all but a few interceptor segments where capacity was quite high already as indicated by the spikes on the capacity curve. A total of twelve modeled interceptor manholes overflowed during the course of future wet weather flows associated with 25-year/1-hour rainfall events. Since the other three types of wet weather flows also demonstrate a similar pattern, it can be concluded that the interceptor does not have sufficient capacity to convey the future wet weather flows.

19.7 Conclusions

The study concludes that under the existing development conditions the interceptor is adequate to convey the dry weather flows, but will result in a slight surcharge during wet weather conditions.

Under the ultimate future development conditions, the interceptor is expected to convey the dry weather flows with slight surcharging, and wet weather flows with moderate to severe interceptor surcharging, manhole overflows, and street flooding.

19.8 References

Huber, W.C., and Dickinson, R.E. (1991).   Storm Water
Management Model. User's manual, Version 4, Environmental
Research Laboratory, U.S. Environmental Protection Agency,
Athens, Georgia.

Shamsi, U.M. (1991).   Chartiers Creek Interceptor Study.
Draft report submitted by The Chester Engineers to the
Allegheny   County   Sanitary   Authority,   Pittsburgh,
Pennsylvania.

Chapter 20

Application of a GIS to Assess Changes in Flood Hydrology of Urbanizing Watersheds

Ivan Muzik
Department of Civil Engineering
The University of Calgary, Calgary, Alberta, T2N 1N4

This chapter describes an application of a geographic information system to support hydrologic simulation of runoff on watersheds undergoing urban development. Described are modifications of the Soil Conservation Service runoff curve number method, for prediction of a new flood regime of watersheds undergoing land use and land cover changes such as urban development.

20.1 Introduction

Hydrologists and engineers are concerned with estimating runoff volumes and peaks from rainfall. In areas undergoing rapid development a special concern is how the frequencies of peak flows will change due to urbanization. Predictive techniques have been developed by various authors based on regression of flood frequencies against basin characteristics. For example, Sauer and

Thomas (1983) of the U.S. Geological survey reported the results of a nationwide study in which a regression-based estimating procedure is used to adjust the equivalent rural peak discharge to the urban condition by means of an index of urbanization called the basin development factor. This factor and basin size, channel slope, basin rainfall, basin storage and impervious surfaces were used in multiple regression analysis to develop prediction equations for urban peak flows of various return periods. A similar study was conducted by Sherwood (1987). Veenhuis and Gannett (1987) made an assessment of the effects of urbanization on floods by analysis of the 25 years of streamflow gauging record in the Austin area. Peak discharges for the 2- and 100-year return periods were found to be 33% and 10% greater in the later period of the record representing higher degree of urbanization. A regression equation they tested generally overpredicted the peak discharges by 20% to 45%.

The major drawback of a regression-based method is the need for discharge data covering both prior and post urbanization periods to develop not only the regression equations but also the flood frequency curves for the two periods. Such data are not readily available in many parts of the world. Furthermore, regression equations have coefficients reflecting local conditions and may not have general validity. An alternative to this empirical approach is to use a physically-based rainfall-runoff model sensitive to changes in watershed parameters caused by urbanization.

The Soil Conservation Service (SCS) runoff curve number method (SCS, 1972) in combination with a dimensionless unit hydrograph is an example of such a model in the domain of lumped systems simulation. The SCS method has gained a wide acceptance among hydrologists because it is relatively simple, requires input data which are usually readily available, and yet retains a significant measure of physical relevance through the use of such parameters as soil type, land use and land cover. In principle, the method can be expanded to distributed or semi-distributed simulation, such as the HEC-1 model.

In recent years, there has been a growing interest in the use of geographic information systems as the means of integrating hydrologic models with spatial and temporal data, thus producing a time and cost efficient simulation system.   Moreover, geographic information systems can readily accept remotely sensed data by satellites.   Indications are (Schultz, 1988) that remote sensing is going to play a major role in hydrologic modelling in future.

The following sections describe a geographic information system and hydrologic simulation software originally developed to support flood hydrograph simulation on natural watersheds using the SCS runoff curve number and dimensionless unit hydrograph methods.   Because of the sensitivity of the runoff curve number method to land use/cover changes the software has been found well suited to studies involving watershed changes from natural to more developed conditions accompanying the spread of urban settlements.   The standard SCS method has been modified to include stochastic components of the rainfall-runoff process.   Regional stochastic parameters were derived from analysis of fifty-five watersheds in Alberta.   The software was expanded to include a Monte Carlo simulation technique capable of deriving synthetic flood frequency curves corresponding to various stages of development in a watershed.   A simplified flowchart indicating the major steps in generation of a synthetic flood frequency curve for a watershed, utilizing a GIS as the source of spatial and temporal data, is shown in Figure 20.1.

20.2 Rainfall Runoff Model

The following discussion focuses on the hydrologic components of the GIS supported simulation of flood hydrographs and synthetic flood frequency curves. This includes the selection of rainfall input, determination of rainfall abstractions, and transformation of excess rainfall into direct runoff by means of a unit hydrograph.

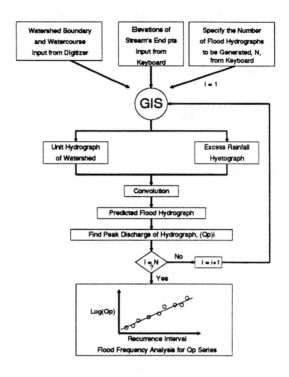

Figure 20.1: Simplified flowchart of simulation procedure.

20.2.1 Rainfall Input

The simulation system allows the user to specify rainfall hyetographs of any shape. However, to generate synthetic flood frequency curves, the following procedure is used: 1. The duration of rainfall is randomly determined from its probability distribution. An example of log-Pearson type III distribution fitted to storm durations in southern Alberta is shown in Figure 20.2. 2. The return period of the rainfall T is chosen by drawing a random number from a uniform distribution of numbers between zero and one. 3. The depth of the rainfall P is computed from:

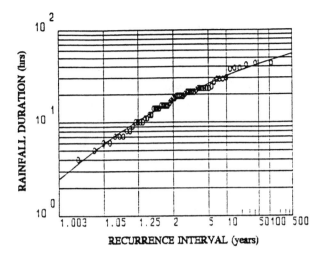

Figure 20.2: Probability distribution of rainfall duration fitted by log Pearson type III distribution.

$$P = \bar{x} + K_T \, s \qquad (20.1)$$

where x, s and K_T are the mean, standard deviation, and frequency factor, respectively, of the Gumbel distribution of the depth of rainfall, automatically provided by the GIS for the analyzed watershed. 4. The time distribution of the rainfall is determined by a random selection of one of the nine time distribution curves of rainfall in heavy storms described by Huff (1967). The resulting hyetograph is then modified by means of the SCS method for abstractions before it is finally converted into the direct runoff hydrograph. This procedure is repeated several thousand times during the Monte Carlo simulation.

20.2.2 Rainfall Abstractions

Application of the unit hydrograph method for prediction of a flood hydrograph requires the determination of excess rainfall. Excess rainfall is that part of the observed total rainfall which is neither retained on the land surface nor infiltrated into the soil. Assuming Hortonian overland flow (Chow, 1988), excess rainfall becomes direct runoff at the watershed outlet. The difference between the observed total rainfall and the excess rainfall is called abstractions. Abstractions include interception (i.e. precipitation caught and retained on vegetation above the ground), and infiltration of water into the soil. Interception and depression storage are usually lumped together and termed the initial abstraction.

Various approaches exist for determining the rate of abstractions from rainfall. In this chapter, the Soil Conservation Service (1972) method for abstractions has been adopted. The method, known as the runoff curve number method, has been modified to take into account stochastic properties of abstractions.

Standard SCS Method

The Soil Conservation Service (1972) procedure for computing abstractions from storm rainfall relates the depth of excess rainfall or direct runoff P_e to the total rainfall depth P, initial abstraction I_a, and potential maximum retention S by the following equation:

$$P_e = \frac{(P - I_a)^2}{P - I_a + S} \tag{20.2}$$

An empirical relation was developed by the Soil Conservation Service:

$$I_a = 0.2\ S \tag{20.3}$$

which after substitution into eq. (20.2) yields:

$$P_e = \frac{(P - 0.2S)^2}{P + 0.8S} \qquad (20.4)$$

The potential maximum retention S is assumed to depend on soil type, land use and land cover. It is high for deep sandy soils and low for shallow clayish soils. For convenience, the potential maximum retention S is converted into a parameter having the range of values between 0 and 100.

$$CN = \frac{25400}{S + 254} \qquad (20.5)$$

The parameter CN, called curve number, is equal to zero when there is no direct runoff generated. The value of 100 on the other hand means no abstractions when the depth of excess rainfall equals the depth of total rainfall.

Curve numbers have been tabulated by the Soil Conservation Service (National Engineering Handbook, Section 4, 1972) on the basis of soil type and land use for three antecedent moisture conditions.

Because the method is relatively simple and requires readily-available information, yet considers physically relevant parameters (soil type, land use, land cover), the SCS method for abstractions gained wide acceptance among hydrologists in North America.

Modified SCS Method

The runoff curve method was developed primarily for design applications in ungauged catchments. The standard tables provide helpful guidelines, but estimation of runoff curve numbers directly from local or regional data is recommended for increased accuracy (Ponce, 1989). Furthermore, the maximum potential retention S and initial abstraction I_a are in fact variables, rather

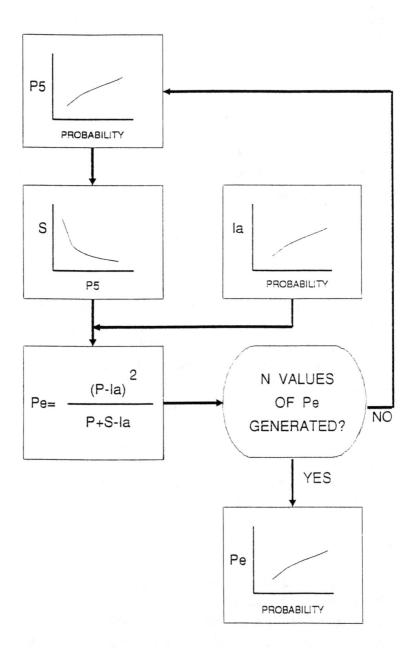

Figure 20.3:   Generation of excess rainfall distribution by
               modified SCS method.

than constants. The following is an outline of a modified SCS procedure developed by the author for use with local rainfall-runoff data, and which takes into account the stochastic character of parameters S and I_a.

The modified SCS method was tested using data from 55 watersheds located in the Alberta foothills, stretching from Montana border in the south to as far north as the town of Rocky Mountain House. The topography of the area ranges from rugged barren mountain ridges to the west to rolling forested foothills in the east. Land cover is predominantly coniferous or mixed forest. Exceptions to these are areas of exposed bedrock along the higher mountain ridges and deciduous treed grassland areas at the lower elevations. Surface geology is comprised mainly of glacial moraine and colluvium. Bedrock geology includes shales, sandstones and limestone. Soils in the area generally fall into the B or C hydrologic soil group, according to the SCS classification. Runoff curve numbers determined from standard SCS tables and for antecedent moisture condition II, range from 65 to 85.

Modifications to the standard SCS method are schematically shown in Figure 20.3. The essence of the modified method is the recognition of a stochastic behaviour of S and I_a. Analysis of rainfall-runoff data in the study region showed the existence of a general relationship between the maximum potential retention S and the amount of rainfall accumulated within the five days prior to runoff in question. The five day antecedent rainfall, P5, has been found to be a random variable having a log-Pearson type III probability distribution. In recognition of this dependency of S on P5, the maximum potential retention S is designated to be the dependent stochastic variable.

The initial abstraction, I_a, on the other hand, is assumed to be a purely random stochastic variable. It has been shown by Chang and Muzik (1991) that the probability distribution of I_a can be well approximated by log-Pearson type III distribution.

In summary, by replacement of deterministic relationships between S, P5 and I_a by stochastic relationships, the modified SCS method yields a probability distribution of excess rainfall P_e for given total rainfall P, rather than a single value of P_e as predicted by the standard SCS method. The modified method thus reduces significantly the uncertainty associated with the return period of the peak discharge computed from excess rainfall as the probability of occurrence of excess rainfall and the corresponding peak discharge may be assumed, within certain limits, to be the same.

20.2.3 Dimensionless Unit Hydrograph

Since most small watersheds lack adequate streamflow data for developing a suitable unit hydrograph, a synthetic unit hydrograph approach is used. The method applied here is based on a regional dimensionless unit hydrograph (DUH) developed for a hydrologically homogeneous region (Bureau of Reclamation, 1976). The regional DUH is derived by converting selected direct runoff hydrographs observed within the region into a dimensionless unit hydrograph as illustrated in Figure 20.4.

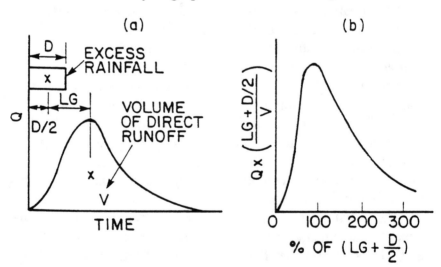

Figure 20.4: Derivation of a dimensionless unit hydrograph (b) from direct-runoff hydrograph (a).

The key parameter in this conversion is the lag-time, LG, defined as the elapsed time between the centroids of excess rainfall hyetograph and direct runoff hydrograph.   Once developed, the regional DUH provides the means of obtaining a synthetic unit hydrograph at an ungauged site. Lag-time is again the key parameter for this reversed conversion. A regional lag-time relationship is therefore required.   Lag-time values, previously obtained for the purpose of deriving the DUH, are correlated with certain geomorphological parameters.   The correlation equation recommended by the Bureau of Reclamation (1976) is of the form:

$$LG = C \ (LL_{ca}/S_o^{1/2})^x \qquad (20.6)$$

where L is the watershed main stream length, S_o is the average slope of the main stream, L_{ca} is the distance along the main stream from outlet to the point on the stream nearest to the watershed centroid, and c and x are regional constants.

20.3 The Geographic Information Systems

GISs are computer-assisted systems for the capture, storage, retrieval and display of spatial data (Clarke, 1986). The GIS described below has been developed specifically to assist hydrologic analysis and synthesis.

20.3.1 Computer Hardware

The GIS has been designed specially for IBM MS-DOS microcomputers and compatibles. The present GIS has been implemented on a Zenith 80386 microcomputer (with a math coprocessor) linked to a digitizer, a graphic monitor, a printer, a plotter, and a modem (optional). Through a modem the system can obtain archived streamflow and precipitation data files stored on a mainframe computer, such as data provided by the Water

Survey of Canada and the Atmospheric Environment Service. Suggested hardware configuration is shown in Figure 20.5.

20.3.2 Database Structure

The raster data structure is adopted in the present GIS because generally it is easier to perform overlay and update operations than the vector data structure (Clarke, 1986). In the raster method, data structure consists of an array of orthogonal grid cells. Each grid cell is referenced by a row and a column number, and is assigned a number representing the type or value of the attribute of interest. The number assigned is assumed to be constant within the grid square. The cell size of 1 km x 1 km is used to store data except rainfall data for which a 10 km x 10 km cell is applied due to the smaller accuracy of available rainfall frequency maps. These cell sizes are also chosen to facilitate cross-reference to Universal Transverse Mercator (UTM) grid system. To keep the size of the GIS database manageable within microcomputer operating environment, the study region is divided into four 100 km x 100 km areas. For the purpose of manipulating data effectively, each area is subdivided into one hundred 10 km x 10 km blocks. Except for rainfall data, each block is then subdivided into one hundred 1 km x 1 km cells. The coordinates at the southwestern corner of each grid square represent the reference coordinates for the square. To cross-reference all grid squares, the origin of each study area is identified with UTM grid. In order to rapidly retrieve data in any part of area, a record number is assigned to each grid cell and data are stored in random access files.

20.3.3 The GIS Software Modules

Organization of the GIS is shown in Figure 20.6. The system consists of four major menus: (1) handling GIS database; (2) handling parameters of watersheds; (3) hydrologic model; and (4)

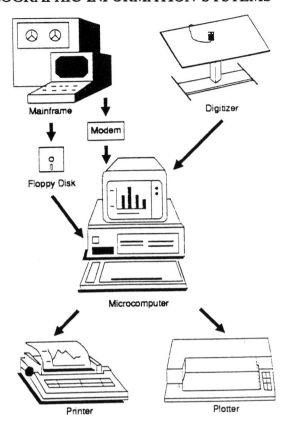

Figure 20.5: GIS hardware.

utilities. Each menu contains several submenus. Each menu or submenu has on-line help to assist users in manipulating data and operating the system.

It is not possible to review all software here in detail, however, the features characteristic for this hydrologically oriented GIS are described. Notice that the first four submenus under the "hydrologic model" main menu facilitate the derivation of hydrologic parameters (DUH, regional DUH, lag-time relationship) which are then input into the GIS database. This need for derivation of hydrologic parameters, through relatively complex rainfall-runoff analysis, sets the hydrologically oriented GIS apart from the rest of GISs, assuming that this capability may be considered to be part of the GIS.

A second feature unique to this hydrologically oriented GIS is the ability to calculate runoff curve number (CN) values for each grid cell. This is accomplished by overlaying land use/cover and soil type files. The resulting combination is compared internally with a CN reference file (Handling GIS Database menu) similar to that by SCS (1972), and an appropriate CN value is then assigned to each grid cell automatically. Using the menu "Handling Parameters of Watershed" the software, for example, uses the CN file to compute the average curve number CN for a watershed as shown in Figure 20.7. The software computes the partial area of each grid square that lies within the watershed boundary. From this information the weighted curve number, CN, for the water-shed is computed by:

$$CN = \frac{1}{A} \sum_{i=1}^{n} A_i CN_i \qquad (20.7)$$

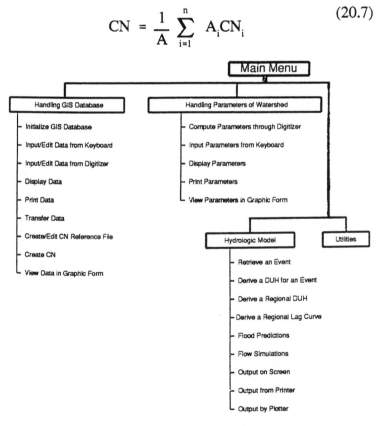

Figure 20.6: Organization of the hydrologically oriented GIS.

where n is the number of cells enclosed, A is the watershed area, and A_i is the partial area of cell i enclosed within the watershed boundary.

Similarly, the weighted average depth of precipitation is computed for the watershed, except the grid size used is 10 x 10 km. The user selects, on an interactive basis with the software, the storm duration from the available 2, 6, 12, and 24-hr. durations, and any return interval. The software then computes the average depth of precipitation for the selected design storm. An option exists in the software, which allows the user to specify any hyetograph for the design storm. Otherwise, the precipitation depth computed for the stored precipitation statistics is used in further computations.

Finally, the GIS software computes the lag-time value, LG, for the watershed from a regional equation of the form given by equation 20.6, and retrieves a regional dimensionless unit hydrograph in digital form. Using the known values of D, LG, and A, the dimensionless unit hydrograph is automatically converted into the watershed's synthetic unit hydrograph, which is then convoluted with excess rainfall.

20.3.4 Data Input and Output

Two categories of data, input required by the system are (1) various spatial data, and (2) hydrologic data. Whenever data input is necessary, users run the software in an interactive way. Data can be entered from either the digitizer or the keyboard. In addition, through a modem the system can obtain archived streamflow and precipitation data files stored on the mainframe computer of the Water Survey of Canada and the Atmospheric Environment Service; or alternatively, the system can read floppy disks storing required archived files. Once the database has been created, the only data input necessary to derive a synthetic flood frequency curve for a watershed of interest, as shown in Figure 20.1, include: (1) watershed boundary and watercourse; (2) elevations of stream's end points; and (3) the number of flood

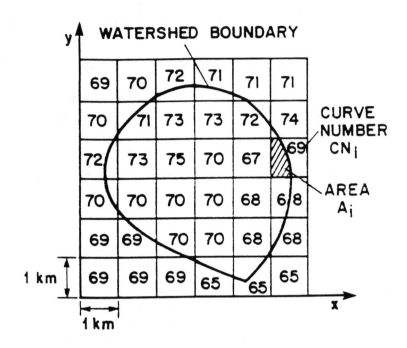

Figure 20.7:   Schematic for the computation of the average
               CN value for a watershed.

hydrographs to be generated.

Data can be output on the screen, from the printer or the
plotter. There are three kinds of data output: 1. graphic output
such as including graphic display of various spatial data (each
classification is assigned a specific color), watershed boundary,
main watercourse, and centroid of watershed; 2. figure output
such as including flood hydrographs, rainfall hyetograph, unit
hydrographs, a regional dimensionless unit hydrograph, a regional
lag-time curve, and flood frequency curves: and 3. text output
such as including a variety of tables for tabulating the spatial data
and the detailed reports of various hydrologic analyses and
simulation results.

Data stored

The GIS database has been implemented for an area covering about 11,000 km^2 located in the Rocky Mountain foothills of southwestern Alberta, Canada. There are four types of spatial data stored in the GIS database: 1. land cover and land use classification; 2. soil drainage classification; 3. SCS runoff curve number, CN; and 4. extreme rainfall statistics. Except for CN, the data are extracted manually from maps or plans and then input from the digitizer or the keyboard. Land cover and land use data are abstracted from composite forest cover series maps of 1:100,000 scale, and soil drainage data (hydrologic soil groups A, B, C, and D) are acquired from ecological land classification maps of the same scale. The CN value for each cell is then computed automatically by overlaying land cover and land use data with soil drainage data. Furthermore, extreme rainfall statistics stored are the mean and the standard deviation for the Gumbel distribution of the rainfall depth for storm durations of 1, 2, 6, 12, and 24 hours ("Rainfall Frequency Atlas of Canada", 1985). For rainfall durations of 48 and 72 hours, the statistics can be computed by multiplying the values of 24-hour rainfall by a factor of 1.35 and 1.5 respectively ("Hydrological Atlas of Canada", 1978).

In addition, some derived regional hydrologic information is also stored in the GIS database. This includes a regional dimensionless unit hydrograph, a regional lag-time relationship, as well as probability distributions for storm duration, 5-day antecedent precipitation, and initial abstraction.

20.4 Application

Performance of the developed hydrologically-oriented GIS has been tested by generating synthetic flood frequency curves for 27 watersheds located in Alberta foothills. These watersheds have records longer than ten years, areas ranging from 64 to 820 km^2, and average CN varying from 70 to 80. Each synthetic flood frequency curve was derived by randomly generating 5000 peak discharges. About four-fifths of the synthetic flood frequency curves have the sample mean within the 95% confidence interval of the population mean. Furthermore, 21 out of 27 synthetic frequency curves pass the chi-square goodness of fit test at the level of significance α equal to 0.05. These results compare very favourably with results of the standard flood frequency analyses procedure whereby a probability distribution function, such as log-Pearson type III is fitted to the sample. A distinct advantage of the synthetic flood frequency generation approach over statistically fitted probability distributions is that the former is based on a physical model of runoff (with stochastic parameters S, P5 and I_a) and probabilistic rainfall input. This approach makes it possible to combine the knowledge of physical processes governing runoff with the theory of probability, resulting in derived (synthetic) distributions of flood flows (Haan, 1987; Muzik and Beersing, 1989). This is to say that by being able to put reasonable limits on the model's physical parameters we can increase, or at least critically consider, the confidence with respect to the final product of the Monte Carlo simulation - the predicted flood magnitudes and their probabilities. By the same argument, the physically based approach to flood frequency analyses makes it possible to assess the consequences of changes within the watershed such as due to urbanization. This is not possible to accomplish by means of the standard flood frequency analyses.

20.4.1 Operation of the System

The following is a step by step outline of the simulation procedure supported by the GIS for the purpose of deriving a synthetic flood frequency curve. A simplified schematic of the procedure can be found in Figure 20.1.

1.   Tape a topographic map on the digitizer table.

2.   Specify the scale of the map.

3. Digitize two points on the map to orientate X and Y axes.

4. Digitize the boundary and the main watercourse of the watershed of interest (usually at a rate of 10 pairs of coordinates per second).

5. Input elevations of stream's end points. At this point the software automatically computes the following watershed parameters; drainage area, L, Lca, So, the average CN, and the average extreme rainfall statistics for various storm durations.

6. Input the number of flood hydrographs to be generated, N.

7. Select the representative time distribution pattern of storms. Three types of time distribution are available in the system: (a) uniform distribution; (b) median first quartile curve for point rainfall (Huff, 1967); and (c) time distribution of first quartile storms (Huff, 1967). Option (c) has been used throughout the study since it is the most severe storm distribution pattern. At this point, the Monte Carlo simulation technique generates N independent values of peak discharge by the following steps:

8. Generate a random value of storm duration, D.

9. Generate a random value of total rainfall, P.

10. Given P, compute the time distribution of storm by random selection of one of nine curves in 7(c).

11. Generate a random value of the 5-day antecedent precipitation, P5.

12. Given P5, determine S using the P5-S relationships shown in Figure 20.2.

13. Generate a random value of Ia from empirical distribution.

14. Given the storm hyetograph (i.e., time distribution of P), S, and Ia, compute an excess rainfall hyetograph by equation 20.2.

15. Determine the basin lag-time of the generated storm event, LG, using equation 20.6.

16. Derive a synthetic unit hydrograph by coupling the lag-time and the regional DUH.

17. Compute a predicted flood hydrograph by the convolution of the excess rainfall hyetograph with the synthetic unit hydrograph.

18. Find the peak discharge of the predicted flood hydrograph, Qp.

Steps 8 to 18 are repeated until N peak discharges of flood hydrographs are generated.

19. Perform flood frequency analysis for Qp series. Final product is a synthetic flood frequency curve plotted in lognormal probability scale which can be output on the screen for temporary display or output from the printer or the plotter for hard copy.

In general, to derive a synthetic flood frequency curve for watershed area of the order of 1000 km^2, it takes about five minutes to digitize the watershed boundary and main watercourse and one minute to compute all parameters required for hydrologic simulation. Generating 5000 flood hydrographs by the Monte Carlo simulation technique and performing frequency analysis of peak flows requires further eighteen minutes. In total it takes about 24 minutes to obtain the final result. It is evident that applying the GIS in hydrologic simulation can substantially shorten this type of hydrologic simulation.

20.4.2 Example

The outlined procedure is best illustrated by an example. However, the existing GIS database does not include any Alberta watershed with a recent significant urban development. The following example thus considers a hypothetical scenario whereby a predominantly natural watershed of James River near Sundre (station no. 05CA002) undergoes urban development resulting in a 15 percent increase of the average runoff curve number for the watershed.

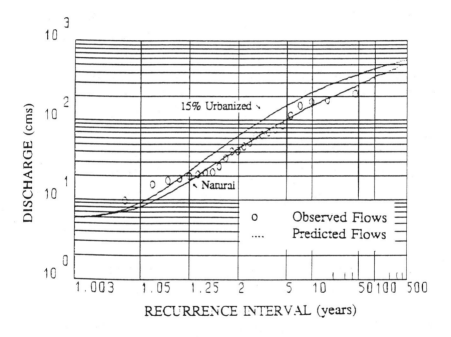

Figure 20.8: Flood frequency curves for James River near Sundre, Alberta.

The watershed, having a drainage area of 821 sq. km, is almost completely covered by forest with some rangelands in the lower elevations. The topography is rolling hills, only a small western part of the watershed reaches into the mountains consisting of bare rock and reaching elevations over 2000 m. The average CN value determined from SCS tables for the watershed is 75.

The two synthetic flood frequency curves generated for the natural and 15% urbanization condition, respectively, are shown in Figure 20.8. Also shown in the figure is the empirical frequency curve consisting of 24 observed annual maximum discharges.

The results show that: (a) the synthetic flood frequency curve for natural conditions fits the observed flow very well, (b) a 15% increase in runoff will result in a significant increase of flood flow; for example, the 50-year flood will increase from about 300 m^3/s to 400 m^3/s, or by 33%, (c) the two synthetic curves gradually converge on both sides of the probability scale; this is expected, unless a different physical process dominates extreme flows. Such conditions could be accounted for in the simulation if deemed necessary.

20.5 Conclusions

Results of the described study show that a microcomputer-based geographic information system supporting a flood prediction model can effectively provide parameters required for generating a physically-based synthetic flood frequency curve for a watershed without or with limited streamflow data. Main conclusions derived from the study are:

Regarding the use of GIS:

1. Once the GIS database is created the time required for hydrologic simulation is up to 100 times shorter than would be needed for simulation not supported by the GIS.

2. Regionalization of hydrologic parameters, which is promoted by the GIS, improves accuracy of simulation results.

3. It is easy to update or modify the GIS database in order to study the impact of watershed changes, such as urbanization, on runoff.

4. Output in form of text, tables, graphs, thematic maps, etc. is easily produced.

   Regarding the simulation methodology:

5. The overall approach of derived (synthetic) distributions of flood flows is physically based, thus allowing extrapolation of empirical frequency curves, and computation of frequency curves for changed watershed conditions.

6. The lumped model used in the study is based on a modified SCS method for abstractions and a synthetic unit hydrograph. This model performed well in simulating flood frequency curves for most tested natural watersheds. It is sensitive enough to respond to relatively small variations in runoff conditions, thus it should be very useful in planning studies exploring impacts of various stages of urban development on flood flows.

7. The scope of the present study did not include distributed simulation. It is suggested, however, that the presented methodology of derived (synthetic) flood frequency curves shows a great potential in assessing impact of changed watershed characteristics, and is equally applicable to both lumped and distributed modelling.

20.6 References

Bureau of Reclamation (1976). Design of Gravity Dams. United States Department of the Interior, United States Government Printing Office, Washington, D.C.: 450-464.

Chang, C. and Muzik, I. (1991). Flood predictions at ungauged sites aided by a hydrologically oriented GIS. Proceedings, International Conference on Computer application in Water Resources, Tamkang University, Taiwan: 604-610.

Chow, V.T., Maidment, D.R. and Mays, L.W. (1988). Applied Hydrology. New York: McGraw-Hill.

Clarke, K.C. (1986). Recent trends in geographic information system research. *Geo-Processing*, No. 3. Elsevier Science Publishers B.V., Amsterdam, 1-15.

Haan, C.T. and Wilson, B.N. (1987). Another look at the joint probability of rainfall and runoff, Proceedings, Int. Symp. on Flood Frequency and Risk Analysis, Louisiana State University, ed. by V.P. Singh, 555-569.

Huff, F.A. (1967). Time Distribution of Rainfall in Heavy Storms, *Water Resources Research*, Vol. 3, No. 4:1007-1019.

Hydrological Atlas of Canada (1978). Fisheries and Environment Canada, Printing and Publishing Supply and Services Canada, Ottawa, Ontario.

Muzik, I. and Beersing, A.K. (1989). Stochastic-deterministic nature of an elemental rainfall runoff process. *Water Resources Research*, Vol. 25, No. 8:1805-1814.

Ponce, V.M. (1989). Engineering Hydrology, Principles and Practices, Prentice Hall, Englewood Cliffs, New Jersey 07632.

Rainfall Frequency Atlas of Canada (1985). Canadian Climate Program, Canadian Government Publishing Centre, Ottawa, Ontario.

Sauer, V.B., Thomas, W.O.,Jr., Stricker, V.A. and Wilson, K.V. (1983). Flood Characteristics of Urban Watersheds in the United States, USGS Water Supply Paper 2207.

Schultz, G.A. (1988). Remote sensing in hydrology. *Journal of Hydrology*, 100:239-265.

Sherwood, J.M. (1987). Estimating flood characteristics of small urban stream in Ohio, Proceedings, 4th International Conference on Urban Storm Drainage, Lausanne: 305-306.

Soil Conservation Service (1972). National Engineering Handbook, Section 4, Hydrology, United States Department of Agriculture, United States Government Printing Office, Washington, D.C.

Veenhuis, J.E. and Gannett, D.G. (1987). Effects of urbanization on floods in Austin, Texas, Proceedings, 4th International Conference on Urban Storm Drainage, Laussane: 307-308.

20.7 List of Symbols

A   watershed area (km^2)

A_i cell area within watershed boundary (km^2)

C   regional constant

CN runoff curve number

D   excess rainfall duration (h)

I_a initial abstraction (mm)

K_T frequency factor for Gumbel distribution

L   longest stream length from watershed divide to outlet (km)

L_{ca} length of the longest stream from point closest to watershed centroid to outlet (km)

LG lag time (h)

N   number of generated peak discharges

P   total depth of rainfall (mm)

P_e total depth of excess rainfall (mm)

P_5 total depth of five day antecedent rainfall (mm)

Q   discharge (m^3/s)

Q_p peak discharge (m^3/s)

S   maximum potential retention (mm)

S_o stream slope (m/m)

s   standard deviation of rainfall (mm)

T   return period (years)

V   volume of direct runoff (m^3)

$\bar{x}$ mean of rainfall (mm)

Chapter 21

Application of GIS in Watershed Management

Rudra, R.P, W.T. Dickinson and D.N. Sharma
Associate Professor, Professor, and Post-doctoral fellow
School of Engineering, University of Guelph,
Guelph, Ontario, N1G 2W1 CANADA

This chapter presents a method to integrate a distributed watershed model with Geographic Information System (GIS) for management of soil erosion and fluvial sedimentation from nonpoint sources. Digitized data on soil, land use, topography and surface drainage pattern have been overlayed, using ArcInfo GIS, to generate input data file for the GAMES model. The GAMES output on erosion and sediment yield has been analyzed, using GIS, to prepare watershed maps to describe spatial distribution of soil erosion losses, sediment contributing areas and problem categories. This approach provides a powerful and faster technique for preparation of input file and interpretation of output of NPS and hydrologic models.

21.1 Introduction

In the Canadian Great lakes basin, pollution from nonpoint

0-87371-898-4/93/$0.00 + $.50
© 1993 by Lewis Publishers

sources (NPS) has been recognised to be a serious environmental concern. The basic sources of nonpoint source pollution from agricultural land are sediment and associated pollutants transported in the solution or particulate form in the drained water. Soil erosion itself is of great concern to Canadian farmers. It affects soil productivity, capacity of stream and reservoirs and drainable water quality.

Control of soil erosion from agricultural land and transport of associated pollutants is essential to achieve sustainable agriculture and water quality goals. Selection of best management practice and implementation of remedial strategies requires careful consideration of financial constraints. The targeting of soil and water conservation practices and policies has been recognised to be cost effective and efficient for control of soil erosion and nonpoint source pollution. Targeting of sources of soil erosion and areas of nonpoint source pollution from agricultural land is very difficult by using simple indicators such as field slope and land use. Computer models have provided a very powerful tool to identify problem areas and quantify the magnitude of the problem.

Soil erosion and transport of pollutant are highly spatially temporally variable. Recent advances in modelling has given an avenue to describe these variations. The basic philosophy behind the application of modelling to manage nonpoint source pollution is to divide the watershed into square grids or irregular homogeneous response units (HRU), and soil loss and transport of pollutant from each grid or HRU is determined by application of appropriate algorithms describing erosion and pollutant transport processes. A GIS provides an effective tool for manipulation of spatial data such as soil, land use and topography, to generate input data suitable for NPS model. The output generated by an NPS model could be analyzed by GIS to create watershed maps describing spatial variations in soil erosion and delivery of pollutant from the watershed surface. This chapter describes an attempt to use GIS in conjunction with an NPS model to describe spatial pattern of soil erosion and sediment transport, and to identify problem areas and possible locations for monitoring nonpoint source pollution.

21.2 Method

21.2.1 GAMES Model

The Guelph Model for Evaluating the Effects of Agricultural Management Systems on Erosion and Sedimentation, GAMES version 3.01, was used in this study. (Rudra et al., 1986). The model, developed at Guelph, has been used extensively in Ontario and elsewhere (Dickinson et al., 1990; and Seip and Botterweg, 1990). GAMES is a two component model: (i) soil loss component and (ii) sediment delivery component. The soil loss component is based on the Universal Soil Loss Equation (Wischmeier and Smith, 1978), and modified for seasonal time frame (Dickinson et al., 1982). The transport of sediment from land cell to downstream land cell and from land cell to stream is determined from the equation:

$$S_s = A_s \cdot DR_s \qquad (21.1)$$

where:

S_s	=	seasonal sediment yield delivered from one land cell to another,
A_s	=	computed soil loss per unit area for the selected season,
DR_s	=	seasonal delivery ratio between two selected adjacent fields.

The seasonal delivery ratio is determined from the expression:

$$DR_s = \alpha [1/n_s \cdot S^{1/2} \cdot Hc_s/L_s]^\beta \qquad (21.2)$$

In an event that runoff from one land cell travels across downstream land cells prior to entering the stream, the expression becomes:

$$DR_s = \alpha[1/\sum (n_{sj} \cdot 1/S_{sj} \cdot 1/Hc_{sj} \cdot L_{sj})]^\beta \qquad (21.3)$$

where:

n_s =	seasonal surface roughness (as indexed by Manning's n),
S =	land slope,
Hc_s =	seasonal hydrologic coefficient, an index of seasonal overland flow,
L_s =	seasonal length of the overland flow path, and
α,β =	calibrated parameters,
j =	refers to the jth cell.

GAMES uses a discretization concept in which the watershed is discretized into several homogeneous response units (HRU). For each HRU, data on input parameters are required. The parameters pertain to the land slope, channel slope, soil type, soil erodibility, cropping factors, SCS curve numbers, and location and efficiency of sediment detention structures. A complete description and a detailed procedure to estimate these parameters is given in the user's manual of the GAMES model (Dickinson and Rudra, 1990).

21.2.2 Watershed Selection

A representative agricultural upland rolling area, the Stratford Avon watershed in Southern Ontario was selected for this study. This watershed consists of 537 hectares of rolling agricultural land with an average slope of less than 9%. The dominant soil texture in this watershed is silt loam and loam. Approximately 10% of the watershed has organic soil. Excellent soils data base is available as this watershed was extensively studied for pollution during the PLUARG period (Acton et al., 1979; Wall et al., 1989; and Coote et al., 1982).

During the study period, the Stratford Avon watershed was predominantly under corn cultivation with small percentage of area under pasture and hay. Fall ploughing has been practised throughout the watershed, with few if any conservation practices in place until quite recently. Since about 80% of the sediment loads in this basin have been observed to occur during the late winter and early spring period (February - May), GAMES was applied for conditions characteristic of that period.

21.3 Preparation and Manipulation of Input Data

21.3.1 Manual Procedure

Input data for the application of GAMES to selected watersheds with manual procedure could be developed in three stages:

1. Land use, soil, and land slope are independently determined from existing maps, aerial photographs, and field surveys. A map for each of these variables is prepared at a convenient base scale, usually 1:5000.

2. A comprehensive overlay of land use, soils and land slope are developed to divide the watershed areas into land cells (irregular fields), each of which are characterized by a single land use, a single soil type and a single class of slope.

3. The flow path pattern is developed from the layout of the stream pattern and field slope data.

4. Values of the various variables and factors in the GAMES are determined and assigned to each land cell. The procedure outlined by the GAMES manual (Dickinson et

al., 1990) is used to estimate these parameters.

21.3.2 GIS Databases

The main difference between GIS and manual procedure is that the overlays of soil type, land use and topography, and direction and length of overland flow is prepared by GIS software. ArcInfo (ESRI, 1989) GIS software was used in this study. Details on this procedure are given in the following section.

The watershed boundaries were defined using a shell. The boundaries were digitized as polygons and their characteristics were represented by attribute tables. Within the shell coverage on soil, land use, slope and topography were digitized. The polygon attribute table (PAT) files for soil type, land use, slope and topography, were created using the CLEAN and BUILD commands. Instead of assigning actual values to each polygon, an attribute coding procedure was used for each coverage. Each soil polygon was codified for erodibility (K-factor) and SCS drainage class. For the land use coverage, each polygon was coded for land use factor (C-factor) and SCS land use class. Slope was coded in terms of slope gradient, and topography was coded in terms of elevation class (topo rating). GAMES model can handle up to nine types of soils and nine types of land uses within a watershed. After assignment of attributes, the resultant coverage was prepared by the OVERLAY command. Small polygons were eliminated using a proper fuzzy tolerance. A preselected number was subtracted from the elevation of each water polygon to cause runoff to flow into the stream cells. The BUILD command was used to create the arc attribute table (AAT) file, and the DUMP command was then used to copy the PAT and AAT files into the comma delimited ASCII format for use as an input to the flow path program.

Determination of Flow Path Pattern

The flow pattern was determined using the DUMP files. The AAT and PAT files were divided into segments using the Q-EDIT program. The size of the segment depended upon the capabilities of the spreadsheet program, such as QUATTROPRO in this case. The segmented file was imported into QUATTROPRO; after unnecessary columns were removed the AAT files were joined to create a "new" file. At this stage, the AAT file had arc length identification of polygon on the right (RPOLY) and polygon on the left (LPOLY).

A BASIC program was developed to determine the flow pattern using the data in the NEW file. In the BASIC program, all neighbouring cells around a source cell were identified through commonly served arcs. The topo rating of the source cell was compared with all the neighbouring cells to identify the down-stream cell. In the case that two or more neighbouring cells had the same topo rating, and it was the lowest topo rating, then the neighbouring cell having largest boundary with the source cell was identified. Overland flow occurred when topo rating of the source cell exceeded or equalled the topo rating of the downstream cell. Before moving to next step, all backflow problems were checked and corrected manually. Backflow problems occurred when two neighbouring cells made each other their downstream cell. This procedure resulted in the identification of the downstream cell, which was downloaded as an ASCII file. The file with downstream cells was added to the resultant coverage PAT file.

Two options are available to determine flow path lengths, i.e. the ADS and ARCEDIT commands in the ArcInfo software. The ADS command is faster, but requires extra digitization of resultant coverage. Due to the unavailability at that time of a printer and digitizer, it was not possible to use this command. The flow path length was determined using the ARCEDIT command and resultant overlay image on the screen. At this stage, this file contained input data for all the cells in which soil erodibility, SCS drainage class, crop management factor and SCS land use class was in the coded form. Additional lines were added at the top of this file to describe soil and land use code and other inputs to complete the GAMES input file (Dickinson et al., 1990). Figures 21.1 and 21.2 illustrate soil and land use maps prepared from digitized data.

Presentation of Outputs

The principle outputs of GAMES as applied to this study include soil loss and soil loss rate from individual fields within the watershed, total potential soil loss and average soil loss rate for the entire watershed, sediment yield from each field in the watershed to the stream and to the next downstream cell, total sediment yield and average sediment yield rate for the entire watershed.

In the present form, the model outputs are in tabular form. However, it has been well recognised that maps and graphs tends to show data set as a whole allowing the user to summarize the general behaviour and to study details. Therefore, they lead to much more thorough data analysis and more insightful interpretation.

In the manual method, soil loss and sediment yield outputs are classified into various categories in GAMESC, another version of GAMES. Outputs of GAMESC are used to prepare potential soil loss, delivery ratio and sediment yield maps manually. This is a very time-consuming procedure and the quality of maps is also not good.

In the GIS procedure, the GAMES outputs were processed by a spreadsheet program to classify soil loss and sediment yield into various class and to divide the entire watershed into four problem categories by selecting appropriate tolerance levels for soil loss and sediment yield. Problem category I includes area with soil loss greater than the chosen soil loss tolerance and sediment yield greater than the selected sediment yield tolerance. Areas in category II are those with estimated soil loss greater than the selected soil loss tolerance but are estimated to yield sediment less than is tolerable. Areas in category III exhibit estimated soil losses less than the tolerable soil loss but sediment yield greater than deemed tolerable. Areas in category IV are considered to have soil loss and sediment yield below the tolerance limits. The processed file in the ASCII form was merged with the resultant

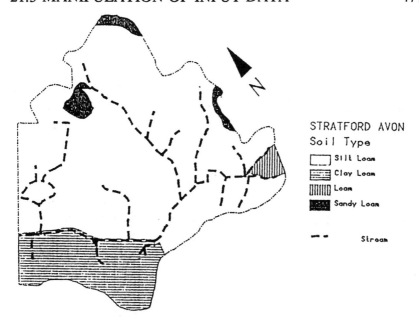

Figure 21.1: Soil mapping of Stratford Avon watershed.

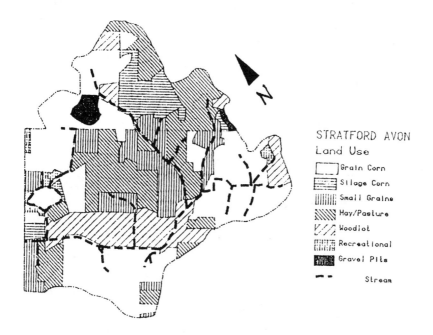

Figure 21.2: Land use mapping of Stratford Avon watershed.

PAT file, and the potential soil loss, sediment yield and problem classification maps were prepared using GIS commands.

Figure 21.3 illustrates the spatial pattern of potential soil loss for the spring period. It is evident that field erosion is quite spatially variable and that a major portion of the eroded soil moves within a small percentage of the basin, in localized ares. 83% of the sheet erosion volume is estimated to occur in 32% of the watershed area, 58% occurring in 15% of the area.

The spatial distribution of sediment yield is presented in Figure 21.4. Similar to the erosion picture presented above, but more pronounced, most of the spring sediment loads leaving the watershed is estimated to emanate from a small percentage of the watershed area: 79% of the sediment load is generated in 12% of the basin, and 57% in 5% of the area. Figure 21.4 clearly reveals that the watershed is characterized by very distinct sediment sources. Such information is vital to establish the sediment monitoring networks. Classifying the subwatersheds as shown in Figure 21.5, it is evident that the greatest expected impact of remedial soil and crop management measures will come from treating the subwatershed C followed by subwatershed A. Treatments in subwatershed B will give the least reduction in sediment yields.

21.4 Conclusions

The GIS procedure described in this study was used to process data on soil type, soil drainage class, land use and topography, and to prepare an input file for a NPS model (GAMES). The procedure also processed the GAMES outputs to produce watershed maps showing spatial distribution of soil loss, sediment yield and problem categories. This is a simple and flexible procedure. It has distinct advantages over the manual procedure. However, it requires soil and land use data in the digitized form and a compatible GIS software.

To facilitate automated estimation of topological parameters and an appropriate discretization scheme, the initial attempt to

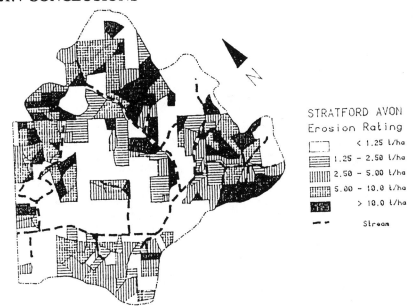

Figure 21.3:   Mapping of expected spring soil erosion rates in Stratford Avon watershed.

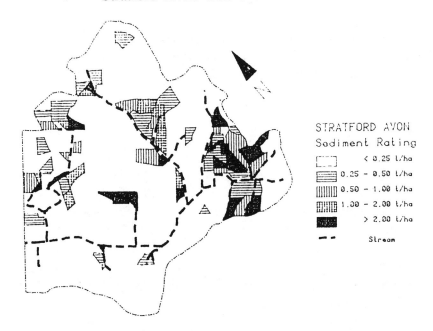

Figure 21.4:   Mapping of expected spring sediment yield rates in Stratford Avon watershed.

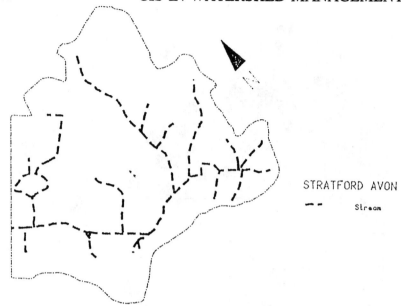

STRATFORD AVON

-- Stream

Figure 21.5: Identification of subwatersheds A, B and C.

integrate the GIS with a NPS model is being extended to study digital elevation models (DEM) using Triangulated Irregular Networks of Elevation data (TIN).

21.5 References

Acton, C.J., G.T. Patterson, and C.G. Heath (1979). Final Report: Soil Survey of six agricultural watersheds in southwestern Ontario, Canada. 1975-1976 PLUARG, Task C, International Joint Commission, Windsor, Ontario.

Beasley, D.B., L.F. Huggins, and E.J. Monke (1980). ANSWERS: A model for watershed planning. Trans. ASAE 23:938-944.

Coote, D.R., E.M. Macdonald, W.T. Dickinson, R.C. Ostry, and R. Frank (1982). Agriculture and water quality in the Canadian Great Lakes Basin. I. representative watersheds. J. Environ. Qual. 11:473-481.

Dickinson, W.T., and R.P. Rudra (1990). GAMES-user's manual version 3.01. School of Engineering, University of Guelph, Guelph, Ontario. Tech Rep. No. 126-86.

Dickinson, W.T., R.P. Rudra, and G.J. Wall (1986). Identification of soil erosion and fluvial sediment problems. Hydrologic Processes 1:111-124.

Dickinson, W.T., R.P. Rudra, and G.J. Wall (1990). Targeting remedial measures to control nonpoint source pollution. Water Resources Bulletin, American Water Resources Association. 26(3):499-507.

Dickinson, W.T., and R. Pall (1982). Identification and control of soil erosion and fluvial sedimentation in Agricultural Areas of the Canadian Great Lakes Basin. Research Report to Supply and Services Canada, School of Engineering, University of Guelph, Guelph, Ontario.

Environmental System Research Institute (1989). ARCINFO User's Manual version 5.0, Redlands, California.

Henry, J.R. (1981). Agricultural Conservation Program: An Evaluation. Water Resources Bulletin 17(3):438-442.

Rousseau, A., W.T. Dickinson, R.P. Rudra, and G.J. Wall (1988). A phosphorous transport model for small agricultural watersheds. Can. Agr. Eng. 30:213-220.

Rudra, R.P., W.T. Dickinson, D.J. Clark, and G.J. Wall (1986). GAMES - A screening model of soil erosion and fluvial sedimentation on agricultural watershed. Can Water Resour. J. 11(40):58-71).

Seip, K.L. and P.Botterweg (1990). User's experience and the predictive power of sediment yield and surface runoff models. Proc. Int. Symp. on Water Quality Modelling of Agricultural Nonpoint Sources:Part 1, U.S.D.A. Rep. No. ARS 81:205-220.

Wall, G.J., T.J. Logan and J.L. Ballantine (1989). Pollution control in the Great Lakes Basin: An international effort. Journal of Soil and Water Conservation, 44:12-15.

Wischmeir W.H., and D.D. Smith (1978). Predicting rainfall erosion losses - a guide to conservation planning. USDA Agricultural Handbook No. 537, Washington, D.C.

Young, R.A., C.A. Onstand, D.D. Bosch, and W.P. Anderson (1987). AGNPS: Agricultural nonpoint source pollution model - a watershed analysis tool. Conservation Research Report #35, USDA-ARS, 77p.

Chapter 22

Flood Plain Management Integrating GIS and HEC-2

Kenneth R. DePodesta and Peter Nimmrichter
Philips Planning & Engineering Limited
Suite 702, 10 Kingsbridge Garden Circle
Mississauga, Ontario, Canada, L5R 3K6

Carey Moore
CartoLogix™ Corporation
Suite 702, 10 Kingsbridge Garden Circle
Mississauga, Ontario, Canada, L5R 3K6

Little has been done over the years to either automate the more labour intensive activities involved in flood plain mapping and management or to increase the efficiency and accuracy of handling applications for development or modification of hazard lands. A logical approach to addressing both of these shortcomings is to integrate a Geographic Information System (GIS) into the flood plain management process. For the water resources professional, a GIS-based flood plain management system would serve two primary functions. Firstly, it would serve as a front end to the modelling process, providing digital data for input into both hydraulic and hydrologic modelling programs.

Secondly, it would act as a dynamic output medium for results of the modelling process, converting the output data into flood elevation contours on digital base mapping. For watershed planners, such a system would provide immediate access to current information about potential and/or historical flooding potential and related property-based data through interactive GIS inquiries.

This chapter outlines the continuing development of a GIS-based flood plain management system, **GISFPM**, which interfaces Geo/SQL® and the hydraulic modelling package, HEC-2.

22.1 Introduction

GIS technology has developed dramatically over recent years to allow for its direct application in many areas of infrastructure and facilities management. Historically, GIS systems have been mainframe-based with limited direct user access due to the high cost of hardware platforms, proprietary database systems and graphical engines inherent in traditional *"integrated"* systems.

Today's evolution toward increased personal computer power, low-cost graphics packages, networked database systems, digital mapping, and micro-computer based mathematical modelling tools, has promoted the development of *"linkage-based"* systems as the leading edge approach to GIS and information management.

Currently, *"linkage-based"* GIS technology allows for direct access through engineered interfaces to disparate database management systems and datafiles and their subsequent connection, or geo-referencing, to existing spatial (mapping related) databases. Low-cost computer aided drafting systems like AutoCAD® working in a micro-computer environment are used as the graphics engine for display and creation in this process.

Water resources engineering, and specifically flood plain management technology, lends itself to an innovative application of this *"linkage-based"* approach to GIS. A structured interface between digital mapping information and the hydraulic database allows for a streamlined analysis procedure and on-demand interrogation using a geographic information/flood plain management system (GISFPM).

By establishing the link between the map and the hydraulic database, the GIS component of GISFPM enables the creation and manipulation of hydraulic information within an easily understood and intuitive graphic environment.

Imagine a process where a qualified flood plain analyst/ manager who has been charged with the task of analyzing the impact of a filling proposal within a flood plain;

1. The flood plain map is *"loaded"* into AutoCAD®;

2. The user modifies topography to reflect modifications within the flood plain, as might be expected with a golf course development, for example;

3. The GISFPM link is executed and an existing hydraulic dataset is modified, with cross-sections extracted by the GIS component, and executed;

4. The results displayed are revised flood lines on the flood plain map;

5. A statistical query is then executed which quantifies the impact on flood levels within a particular area of interest (i.e. number of homes flooded);

6. The map is plotted and a statistical summary printed to assist in justifying the flood plain analyst/manager's decision on the matter.

The flood plain analyst/manager is faced with making decisions of varied complexity regarding the topographic characteristics of flood plains on a daily basis. A GIS can manage pertinent information related to each cross-section and enable it's recall, modification and input into an appropriate model for analysis.

22.2 The Current System

Hydraulic modelling output, namely flood levels and limits, and the topographic mapping base, are the primary databases for the establishment of a flood plain. Current procedures which ultimately yield the required flood information are, at times, very cumbersome and subject to random error and misinterpretation.

22.3 "Standard" Flood Plain Mapping Procedures

The HEC-2 program is the hydraulic *"corner stone"* of most Ontario flood plain mapping studies. HEC-2, developed by the Hydrologic Engineering Center of the U.S. Army Corps of Engineers, was designed to compute water surface profiles for steady, gradually varied flow in open channel systems. Essentially, HEC-2 calculates flood elevations at locations of interest for specified flows. Standard input variables include: cross-sectional geometry, reach lengths, discharge and loss coefficients.

The following generally describes the typical step by step procedure which takes place during the creation, execution and evaluation of a HEC-2 dataset.

1. The engineer reviews the topographic mapping of the watercourse and specifies where appropriate sections should be located, and establishes which structures should be modelled.

2. A survey crew goes into the field to survey sections and structures as required.

3. The engineer undertakes a field reconnaissance of the watercourse looking for potential flood and erosion zones.

4. Field survey notes are reduced by a technician and cross-section overbanks are abstracted from topographic mapping. Sections and structures are then coded into HEC-2 format. Cross-sections are plotted for debugging purposes and for inclusion in the final report.

5. The engineer reviews the code through the debugging procedures and corrects code where necessary. In this procedure, the engineer incorporates roughness factors, expansion and contraction factors, and special bridge (if any) co-efficients into the dataset.

6. The engineer inputs QT (flow) cards at appropriate nodal locations along the watercourse.

7. The engineer executes the dataset and corrects any errors not found in initial debugging. The output is evaluated to determine its validity (such as determining which sections require extension as a result of computed water surface elevations exceeding the minimum end station of the section).

8. The engineer identifies spill zones and determines control sections for subsequent spill calculations.

9. The engineer uses the facilities which are part of the HEC-2 package to generate summary tables of output parameters.

10. The engineer determines the maximum permissible velocities for erosion zone determination and compares these with channel velocities from HEC-2 output to confirm potential erosion zones.

11. A technician plots the flood plain on topographic mapping while the engineer reviews the significance of spill routes.

The foregoing procedures are very labour and data intensive, particularly those requiring manual interaction with a topographic mapping base (i.e. tasks 1, 4, 5, and 11). Many of the above-noted tasks require iterative assessment and evaluation as well.

22.4 Policies and Practices

Current flood plain mapping or flood damage reduction studies result, in part, with the definition of Regulatory flooding limits as well as flooding limits for a variety of computed (i.e. 2, 5, 10, 25 year etc. frequency events) or recorded (historical) flows. However, the primary objective of flood plain mapping projects is the definition and illustration of the Regulatory (Regional or 100 year event) flooding limits. Information on lesser event flooding is rarely displayed on flood plain mapping.

Damages resulting from the modelled flooding events may be estimated, perhaps using a computer program such as FDAM (Flood Damage Analysis Model).

Information from the hydraulic analysis is used to determine erosion susceptible zones along modelled watercourse reaches. Channel velocities, channel and overbank flow distributions, and flood levels, based on the 2 and/or 5 year frequency flows are typically used as bench marks for the identification of potential erosion zones.

22.5 Constraints

The most significant constraints of the current approach to the development of flood plain mapping are directly tied to those tasks requiring manual interaction with hardcopy topographic mapping. As a result, valuable time is spent by engineers and technicians obtaining *all* required mapping, piecing mapping together, obtaining properly scaled mapping for inclusion in reports, abstracting contour data, plotting flood lines and damage areas etc.. This time could be more effectively utilized gaining a better understanding of the hydraulics of the watercourse(s), in more intense investigation of spill zones and routes, in better definition of flood, erosion and damage zones, etc.. In essence, water resources personnel could spend their time more efficiently investigating the hydraulics assessment rather than spending time managing mapping and data.

22.6 Technology

22.6.1 Geo/SQL®

The GIS system which has been chosen for integration with HEC-2 is Geo/SQL®. Geo/SQL® represents the first true, topological, continuous, hardware independent spatial database. It contains an imbedded SQL interface which conforms to the ANSI'86 standard and emulates the DB2 standard for database querying. The advantage of this system over other GIS technologies is its inherent *"Linkage-Based"* approach to GIS technology. Rather than containing an internal, proprietary attribute database structure, Geo/SQL® *"links"* to external databases which conform to the SQL and DB2 standards, such as Oracle, Ingres and Sybase (in addition to dBASEIII) and to AutoCAD® which it uses as it's graphics operating environment.

Geo/SQL® is an integrated set of programs for managing spatial information. In this context, spatial information means anything which has a location in space (e.g. a floodline), and/or can be defined in terms of its own geometry (e.g. a building).

Some of the highlights of the Geo/SQL® include:

1. Geo/SQL®, integrated with AutoCAD®, runs on a variety of hardware platforms: IBM, COMPAQ, NEC, TOSHIBA, or any IBM compatible computer as well as SUN SPARCSTATION, and VAX systems (or in fact any future hardware which will support AutoCAD®).

2. Geo/SQL® runs under a number of operating systems including DOS, UNIX, OS/2.

3. Geo/SQL® works from within the AutoCAD® graphic environment or externally, using a fourth generation language, a natural language interface and/or a menu front-end; all of which can be provided with the base system.

4. Geo/SQL® supports the ability to geo-reference color photographs, video images, word processing documents, spreadsheets and as-built drawings (either raster or vector images).

5. Geo/SQL® is a *"linkage-based"* GIS, which integrates industry standard AutoCAD® to a number of SQL databases including R:BASE, ORACLE and dBASE. Other SQL databases such as DB2, INFORMIX and INGRES can be accessed via inter-database linkage utilities.

22.7 HEC-2

The February 1991 version of HEC-2 has been used in the development process. This version of HEC-2 will run on any hardware platform which is compatible with Geo/SQL®.

22.8 Functionality

22.8.1 Scope

The basic functionality of the GIS to HEC-2 interface, termed GISFPM, can be subdivided into three basic aspects:

1. Provide a means of getting new data into an HEC-2 model or HEC-2 format (i.e. GIS to HEC-2 link).

2. Provide a means of transferring numerical HEC-2 output into a GIS database for subsequent digital display and inquires (i.e. HEC-2 to GIS link).

3. Facilitate logical and efficient editing abilities and data handling (i.e. HEC-2 to GIS and/or GIS to HEC-2 link).

Clearly, this link to the HEC-2 program will be a valuable tool to the water resources engineer. The primary advantages of such a system are:

1. Faster and more effective dataset debugging resulting in more representative modelling. Visualisation of cross-section and other flood plain data will lead to faster identification of modelling anomalies and errors.

2. The ability to peruse hydraulic output "tagged" to it's location on a map much earlier in the analysis process, thereby improving the appreciation of the flooding mechanism under investigation.

3. Improved access to floodplain information through on-line inquiry based procedures.

22.9 System Overview

The attribute database structure must accommodate both "fixed" (read only) and "variable" (read/write) information. From a water resources perspective, three general conditions describe the state of a valley watercourse system insofar as floodplain definition is concerned, namely:

1. A floodplain (HEC-2 dataset) that has been established (mapped) and no encroachment will be allowed.

2. A floodplain (HEC-dataset) that has been established (mapped) and engineered encroachment/channelization will be allowed.

3. A floodplain (HEC-2 dataset) has not been established (mapped).

Condition 1 represents a 'fixed' information type whereas conditions 2 and 3 represent 'variable' information types.

The GISFPM system is based on two primary links:

1.   HEC-2 to Geo/SQL®
      and
2.   Geo/SQL® to HEC-2.

22.10 HEC-2 to Geo/SQL® Link

The HEC-2 to Geo/SQL® link includes processes such as the following:

1. Transformation of HEC-2 input information, contained in an existing HEC-2 input dataset, into the Geo/SQL® spatial and/or attribute databases as necessary.

2. Transformation of HEC-2 output information contained in the detailed or binary output files generated from HEC-2 into the Geo/SQL® spatial and/or attribute databases as necessary.

3. Interactive use of pre-defined or ad hoc inquiries for access to any of the available HEC-2 output data items associated with each defined cross-section for any/all modelled profiles, for example:

   • Display sections experiencing hydraulic jumps.
   • Display sections where the K-Ratio is out of range.

- Display reaches susceptible to erosion based on digital soils information.
- Display sections requiring extension (i.e. CWSEL > TELMX).
- Display sections experiencing critical depth.
- Display sections where the increase in the computed water surface elevation from the previous section is more than a user defined value.
- Display flows at a section.
- Display computed floodline for any frequency flow available. The floodline is generated via digital comparison of topographic and flood surface digital terrain models which are generated via menu choice.
- Automatic annotation of section information with user specified symbology.
- Display comparison between various floodlines.
- Display buildings in the flood plain or pre-defined search area.
- Display and report on damages associated with buildings in a user selected flood plain.
- Display a user selected cross-section.
- Display the rating curve for a section or a reach.

The list above represents sample inquiries which will allow the user access to primary watercourse, floodline, and damage information.

4. Interactive, automated plot file (i.e. flood plain map plot) generation based on user selected parameters such as scale, sheet size and surround, required topology, etc.

5. Interactive display of digital photographs and documents.

22.11 Geo/SQL® to HEC-2 Link

The Geo/SQL® to HEC-2 link includes processes such as the

following:

1. transformation of HEC-2 information contained in the spatial and/or attribute databases into a workable HEC-2 input dataset in DOS:

   • subcritical or super-critical input file format,
   • only part of an overall dataset,
   • sensitivity analysis on input parameters;

2. interactive definition of overbank (flood plain) section directly from digital mapping;

3. interactive alteration of existing topographic mapping base contour information to represent excavation and/ or filling in the flood plain;

4. interactive modification of bridge/structure definitions for both normal and special bridges types as well as culverts; and

5. interactive user manipulation of sections (i.e. delete a section, insert a section, alter section configuration, etc.).

22.12 Other GIS Applications In Water Resources

Geo/SQL® has other applications as a graphic link for water resources systems. Many of these capabilities are already being exploited for the management of planning-related features such as zoning boundaries, delineating areas sharing common land uses and for areal discretization. The full capabilities of GIS have not yet been applied to the water resource sciences. Many existing uses of GIS technology are limited to conventional graphical data management separate from the linkage to other structured information bases. A GIS has many potential applications in water resources:

1. automatic drainage area delineation from topographic plans;

2. land use optimization based on economic modelling;

3. determination of hydrologic and hydraulic parameters based on soil conditions, land use patterns and slopes;

4. storm sewer network analysis from as-built details;

5. analysis of spatial distribution of rainfall in flood forecasting applications;

6. real-time, geographic display of streamflows, rainfall and other remotely sensed environmental parameters; and

7. consolidation of all data into a common, geographically referenced database of the users choice.

The advantage of using GIS to front-end such applications is the consistent user interface. The same computer station can be used to access multiple applications using virtually any computer technology in stand-alone and/or networked, distributed information environments. The Geo/SQL®/ AutoCAD® GIS manager offers significant advantages in that the user employs industry-standard CAD technology as the graphics operating system. In addition, the double precision accuracy and the inherent vector-based technology of Geo/SQL®/AutoCAD® provides agencies charged with the responsibility of creating and managing flood plain mapping for legal purposes with the requisite tool without compromise.

22.13 Summary

In the increasingly complex world of graphical data management, particularly in water resources engineering, GIS technology provides an innovative approach to integration of both textual and graphical databases. GISFPM, an interface utility, has been developed for use by flood plain analysts and managers for hydraulic data abstraction and display on micro-computer workstations in stand-alone and/or networked information environments. The technology currently links digital topographic mapping data with the hydraulics program HEC-2. Future developments in GISFPM technology will extend into enhanced hydraulics applications as well as hydrologic data management and real-time flood forecasting.

Chapter 23

Data Manipulation of GIS for Modelling and Simulation in Resource Management

Jiguo Zhen
 Chinese Visiting Scholar
William James and Fangju Wang
 School of Engineering, University of Guelph
 Guelph, Ontario, Canada N1G 2W1

GIS and modelling packages have become important tools for resource management and environmental planning at the watershed level. Integration of the two technologies evidently greatly improves data analysis and manipulation, and model turnaround times. High performance here depends heavily upon creating and organizing GIS data sets in a proper structure for transferring them to executing models in an efficient way. In this chapter, an effective method for handling gradient data, one of the major parameters for water resources modelling, is addressed. Experimental results are presented and analyzed.

0-87371-898-4/93/$0.00 + $.50
© 1993 by Lewis Publishers

23.1 Introduction: Models, GIS and Management

23.1.1 Modern Resources Management

Resources refers to any classes of objects, both those existing naturally and those constructed artificially, that are considered essential for human life and development. Natural resource components include soil, water, air, forests, animals, fisheries, minerals and so on; artificial resources can be regarded as all facilities, utilities, and engineered construction for human activity.

Ever since humans began exploring the resources around them, they began to learn how to manage them wisely, i.e. for their long-term or sustainable benefit. Resource management, including resource conservation, preservation, monitoring and exploration, has been and continues to be, a fundamentally important human activity. As the human population and modern technological development both explode uncontrollably, the complexity of resource management also increases, especially control of natural resources depletion and ecosystem deterioration. Accordingly, resource management has become more and more challenging, and, itself, technology-dependent.

Effective resource management is based on both the ever-accumulating knowledge base and on high-technology computer information systems. At the intersection of these systems, as described throughout this book, two powerful tools for resource management are being integrated: modelling packages and GIS's. Modelling packages are used to depict the structure of existing systems, to determine relationships among processes and systems, and to reveal the dynamic dependence of every kind of known process of the real world in a scientific way. They are especially useful for testing proposed resource management strategies, and to find the best among many alternatives. GIS's store, manipulate, analyze and display spatial data of real objects. They are especially efficient at managing large arrays of information that can be located in the environment.

23.1.2   Importance of Modelling to Resource Management

Modern resource management addresses its problems systematically and dynamically in a context of the whole environmental system.   Complexities in both the human environment and in resource management today share three similar aspects: 1. more and more factors have to be taken into account, 2. larger and larger geographical areas must be studied, and 3. the study must encompass longer and longer timespans. Resource management problems have spatially-distributed parameters; all processes are multi-dimensional; most of the parameters are time-varying and non-linear; and more often than not, problems within large geographical regions are characterized by uncertainties.   Large distributed-parameter systems behave quite differently from similar problems in small fields, the so-called point processes, and this introduces concomitant modelling difficulty (Vansteenkiste,1978).

Basically, there are two categories of processes or factors considered in modelling: the natural and the artificial.   Their spatial and temporal diversity are notably different, as are their effects on modelling.

Natural processes are characterized by high variability and unpredictability.   For example, the runoff of a watershed is fundamentally dependent on: rainfall density, intensity and its spatial kinematics; evaporation and related processes of temperature, wind speed, and land cover type; soil and subsurface conductivity, storage and flow patterns; drainage pattern and related processes of topography; and so on. These natural factors vary spatially and temporally during a storm, a season, and over the decades. This aspect often demands spatial statistical analysis and probability inference for every important process, especially the rain falling on the watershed.

On the other hand, infrastructure and engineered facilities are also fundamentally important factors. They differ from place to place, as do the natural conditions, the human demands and their dimensions, operational parameters, and their dynamic responses to the impacts of other factors. E.g., properties like curb length, pipe diameter, channel depth, linkage, shape, geometric and hydraulic parameters, although stable, have unique operational parameters. Changes of land use, population density and other demographic parameters significantly complicate the behaviour and modelling of these artificial resources.

In recent years, advanced modelling packages dealing with process-oriented issues have been developed by interdisciplinary teams. For example, WEPP, the USDA Water Erosion Prediction Project models, which are based on fundamentals of stochastic weather generation, infiltration theory, hydrology, soil physics, plant science, hydraulic and erosion mechanics (Nicks et al. 1989), account for many factors. It follows that many factors with very different types of variables are required in most advanced model packages. For example, in WEPP, for determining the conditions of snowmelt and frozen soil, no less than 60 variables and parameters are addressed (Young, 1989).

Such complex resource management requires huge amounts of modelling effort. In most cases, the number of factors modelled is based on the amplitude of the processes, the sensitivities of the factors, and the spatial resolution required by the study. The larger the study area, the more factors that are involved; the greater the accuracy of spatial and time resolution and the increased sensitivity demanded, the greater the modelling effort.

Numerical weather prediction, very important in storm water management, is a good example. Evidently, if five factors are to be taken into account (horizontal grid points, vertical levels, number of equations, operations per equations, time steps) for a 48-hour forecast in an area measuring 5 degrees longitude by 5 degrees latitude, then, even with very rough spatial resolution (horizontal 250 km, vertical 2 km), the number of operations can reach as many as 2.8 billion (Koermer, 1988).

To cope with this, more sophisticated and professional modelling packages have been developed in the past two decades (see for example James, 1992). These packages have improved contemporary resource management in three ways:

1. Understanding the relationship and interactions among factors in resource management can be achieved better than ever before, since sophisticated models account not only for spatial diversity of the facts, but account also for their dynamic interactions in different phases or processes. For example, SWMM is a comprehensive water quality and quantity simulation model developed primarily for urban areas. Single-event and continuous simulations can be performed for almost all components of the rainfall, runoff and quality cycles of a mixed rural, urban and industrial water and sewershed (Huber et al., 1991).

2. Modelling is especially useful for accurate prediction. For example, in stormwater resources management, the accuracy of prediction of storm water flows is critical to human security and investment. Disasters are caused by floods every year (eg. China in 1991). In many cases, the costs of these disasters have been largely reduced by accurate prediction. More precise and reliable predictions could reduce the misery even further.

3. Findings and results of modelling can be used to determine fundamental parameters for systematic engineering designs for every kind of infrastructure in watersheds. For example, modelling the Universal Soil Loss Equation has brought about many suitable tillage designs to control soil erosion in many countries.

23.1.3 Roles of GIS in Supporting Modelling and Simulation

GIS's are computer-based systems that are used to store and

manipulate geographically-referenced data or information. A GIS software package consists of a set of programming tools and a database management system (DBMS). It provides utilities for data input, storage, retrieval and output. GIS's make the production and analysis of resources information more efficient for planning and provide a natural resources potentiality assessment (Stanley,1989).

GIS's support modelling that requires abundant sources of information, including spatial entities and their relationships, described as a neighbouring and networking representation of the real world. Basically, GIS's are very effective for handling data sets of spatially-distributed entities and their attributes, such as land cover types, soil and subsurface hydraulic conductivities, and so on.

GIS's can be used to create special data sets which usually cannot be obtained by conventional means. Quite often, most natural factors with their model variables are spatially dependent on each other. Unfortunately, a large proportion of the input data required in modelling, such as the rainfall density and local rate of rain, evaporation, temperature, and wind speed with their spatial distribution, are hardly measurable sufficiently accurately, or are even unobtainable in practice. This is why conditions have to be assumed in modelling, even though they are sometimes by no means close to *real world* estimates. For example, climatic default values are most often used when observed or real-time data is not available (Koermer, 1988).

Obviously, results would be more credible and perhaps more accurate if field observations are used. Coverage can be aided by using the computerized spatial data interpolation techniques of GIS's, which are based on limited control points. These GIS interpolation techniques are adapted from established manual methods used in many academic fields, especially in geography and meteorology.

GIS computerized map interpolation can increase the efficiency of thematic map layer generation. Because of their efficacy, digital thematic map layers become the main source of the GIS database, and have very positively impacted spatial analysis for resource management. By efficient data manipulation, GIS's manipulate large amounts of spatial data, some of which may be critical in modelling. It is paramount that GIS selection be based on the availability of utilities for thematic map layer generation; many inexpensive GIS's promoted for water resources engineering lack these capabilities.

Another inherent utility (of fully-fledged GIS's) for dynamic and interactive modelling is high-performance data-acquisition, specifically: map layer conversion, digital data communication and incorporation of remote-sensing data. As shown elsewhere in this book, these utilities save database management resources and time.

There seems to be no doubt that the GIS features described here can significantly improve accuracy and enable more feasibility modelling in the limited time period usually available for design or planning studies.

Natural factors, and many artificial factors, always change temporally. For example, rate of rain and precipitation density, temperature, and wind speed and direction; they all change stochastically throughout a storm. In order to predict as accurately as possible their effects on relevant processes such as pollutographs, an appropriate, usually shorter, time-step is chosen for input data files. The choice of timestep is based on the intrinsic speed of the process being modelled.

Modelling is practical and reliable only when the required data, such as precipitation, temperature, and wind speed, are made available. Where modelling also requires other spatial data, such as hydraulic conveyance, and hydrological parameters of land use, that change during the simulation, only GIS data manipulation techniques can efficiently provide the large variety of input data-sets in a reasonably short time.

GIS's enhance both the accuracy and the operational speed of modelling, because they make it easier to rearrange the spatial scope, the model integration time-step, the simulation parameters for every subwatershed and sometimes even the extension of the study to cover peripheral watersheds. With functions for processing remote-sensing datasets, manipulation of imaginary and spatially consistent data, many more sensitivity analyses, validation tests, design optimizations and error analyses can be performed.

Many GIS's support modelling for decision-making on environmental issues. The work of the European Community and the United Nations Environment Program provides an example of a GIS application in resources management: the CORINE program. Its objective is to provide a comprehensive, integrated spatial database of environment data relevant to European policy-making. The ARC/INFO GIS from the Environment Research Institute (ESRI) was chosen as a development platform for the CORINE system.

There has been a rapid increase in the number of GIS-based predictive modelling packages for: surface water management modelling (this book); evaluating and protection-planning of groundwater resource pollution (Barry, 1990, Baker, 1990); land resource management at the urban, regional, state, and national levels of administration (Nielsen,1990); and so on.

23.1.4   Efficiency Issues in GIS Supported Modelling and Simulation

Efficiency of resource management modelling depends largely on GIS data supply. Usually, GIS data manipulation for this purpose includes two objectives: sufficient and useful data capture, and high-speed generation and tailoring of data sets. A GIS should be evaluated for three criteria: 1. the desirability, 2. the sufficiency of output data quality, and 3. the operating speed.

As indicated above, increasingly sophisticated modelling must employ more and more variables (and the number is already large), smaller time steps, and complete more iterations and more design-runs in a limited time period. These demand high-speed GIS data manipulation. Besides hardware factors, high GIS operating data manipulation speed is related to: the numbers, the precision requirements, and the manipulation of the attributes of the variables in the input data file. As well as the GIS operating speed, modelling speed relies on the manipulation skills evident in the user's GIS command file.

The model variables have various relationships and different fits to the spatial entities present in the GIS's databases. The complexity of the relationship and the closeness of fit of model variables to the GIS entities may be divided into four classes:

1. physical dimensions such as area and length;

2. variables directly-related and plainly-manifested, such as soil type and soil pH values;

3. directly-related but requiring arithmetical computations, such as slope and slope aspect; and

4. stochastically-dependent such as rainfall intensity and distribution patterns.

The higher the class of variables, the more important is the consideration of GIS data manipulation efficiency. In most cases, the recently-developed GIS's support the first-three types of variables well, but of course with different efficiencies. For the fourth class, GIS's usually include spatial-data generators to suit their demands. These generators could reside in either a GIS or modelling package (Flanagan, 1991).

Once sufficient data has been generated, high-speed dataset tailoring is indicated. For a heavy modelling workload for complex watershed systems, data manipulation skills in both GIS command tools and other software is necessary.

In a conventional approach, spatial or spatially-related datasets of variables such as area, length, elevation, slope, aspect and other aspects of entities are manipulated manually and manually input into the data file. The large number of variables leads to great data volumes and is error-prone. Evidently, manual approaches are inefficient for sophisticated modelling.

GIS's support modelling not only by means of their DBMS, but also through their other, handy tools. These may facilitate modelling efficiency at four levels:

1. First, spatial and attribute data can be stored and interrelated in a GIS database for convenient retrieval. They can be tailored easily to suit specific needs (Smith, 1987). For example, lengths of streams, channels, areas of land, and cover types can be related, coded and retrieved as model variables.

2. Second, useful data from other types of databases can be implemented into a GIS database either systematically or on a necessity basis. This includes important data conversion between GIS and non-GIS systems (eg. from AUTOCAD to ARC/INFO) and among different GIS databases (for example, from ARC/ INFO to GRASS).

3. The third level is represented by the capability to create new data sets containing information needed for modelling. This involves generating new thematic layers or combinations of different thematic layers, such as climatic, physiographic, land use or soils, and their component layers. The former involves problem-addressed map interpolation while the latter is map overlaying. Isotope maps are usually created by automatic interpolation and then stored digitally as a thematic map layer in a GIS database. With functions for triangulation, delineation (Jenson and Domingue, 1988) and interpolation, GIS's can save space and time.

4. The fourth level is integration of the above techniques with other software or data manipulation tools. This mainly involves macros programming of a GIS or in other software workspace. It features well-developed interface software written to overcome obstacles or practical problems that GIS's alone could handle with difficulty. There is a conspicuous recent trend to improve GIS's interface for practical analyses (Smith, 1987).

23.2 Problems Related to Efficiency

Although GIS data manipulation has greatly improved resource management modelling (Nielsen et al, 1990; Barry; 1990), there are still gaps between GIS's and specific modelling objectives. In the following, two such gaps are addressed.

23.2.1   Different Data Formats Between GIS's and Modelling Packages

Usually, a GIS specifies a standardized data format for it's database, called a tabulated or plat file. On the other hand, the formats of input files for various models are all quite different. A few lines from a standard SWMM input data file provides an example:

file: TRANS14.DAT

```
SW  1 8 9
...
A1 ' STEVEN''S AVE., NOVEMBER 28 '
*      NDT NINPUT NNYN NNPE NOUTS NPRINT NPOLL NITER IDATEZ METRIC INTPRT
B1     60  0    2    4    4     0     4     4   731128   0    10
...
*  SEWER ELEMENT DATA
*  NOE NUE(1) NUE(2) NUE(3) NTYPE DIST GEOM1 SLOPE ROUGH GEOM2 BARREL GEOM3
E1  1  151   157    0     21   0   3.0   0     0     0
...
```

Note that sewer variables such as DIST, SLOPE, ROUGH can be the attribute items of the same objects in a GIS database. The DIST values for every sewer will be available from a digitized sewer-map layer formed in a GIS; the ROUGH values can be added to each record with relevant data from another source, such as a facilities management system, or sewer inventory system. It is also a normal and simple data manipulation to extract them from a GIS output data file. However, if there are many subwatersheds, sewersheds or catchments in a watershed, data manipulation will be tedious.

23.2.2 Gradient Calculation

Gradient Calculation, Modelling and Simulation

An additional problem is calculation of gradients. Slopes and gradients are commonly used in topography, hydrology and related engineering fields. Gradients are also widespread in many academic disciplines such as meteorology, biosphere modelling, mass/energy transfer, and even in regional economic development strategy. Basically, a gradient refers to the partial differentiation of any arithmetic values of an attribute over the distance between spatial points. It represents the spatial rate of change of an attribute between objects of the same kind in spatial analysis.

Since resource management modelling mostly deals with process analyses in spatial systems, gradient variables are used to reflect potential change rate, to depict dynamic characteristics of components of a system, precisely as the SWMM variable SLOPE above. For calculating gradients of simple or a non-spatial system, it is unnecessary to assume GIS technology. However, for modelling, especially for hydraulics research in large geophysical areas, which usually need large amounts of gradient values and parameters, gradient calculation has to be carried out by resorting to GIS data manipulation.

Conventional Methods of Calculating Gradient Values in GIS and their Limitations

To calculate slope values in GIS's, relevant elevation data of any entity for an area has to be in a digitized format. There are two methods of preparing slope map layers in GIS's. One is to employ a GIS with a three-dimensional elevation data model and it's triangulation and interpolation tools. Another is to digitize a slope map whenever available. The first method seems simple, but the powerful functions and tools required prove demanding for some GIS's, and usually limited three-dimensional elevation data are available. The second is also difficult because slope maps are not readily available.

It seems to be possible to calculate slopes of entities when a digital slope map layer is available, if the GIS employed has at least a two-dimension-coordinate system and even if it does not contain triangulation functions. Without a slope map layer, slope calculation of any kind of an entity is halted.

Furthermore, it is difficult to calculate slope values between points solely, or along line-features with direction (along a profile) options, even though the slope, or an isotope map layer, is already digitized. It is unlikely that slopes of polygon-featured entities in a watershed will be calculated; overlaying these two map layers, both polygon-featured, is an overlay operation with two different entities: the overlaying map layer is polygon-featured or a slope or isotope map layer, and the overlaid map layer is a point- or line-featured entity.

This above method for calculating slope values is not always favourable and applicable for modelling. Gradients used as parameters for modelling mostly refer to an attribute of the entities with line-feature or point-features. When this kind of map layer is overlaid, the lines joining two nodes will be intersected by the boundary lines of the polygons of the isotope map layer. Therefore one line is split into many segments with dependent attribute values in the GIS database. The number of segments created from one line is equal to the number of boundary lines crossing it. Accordingly, the output file of the outcome map layer will expand.

These data explosions raise many problems for both modelling and the GIS itself. It is difficult to tailor very large files and compile useful values from them into the input data file for modelling, not only because it is time consuming, but also because of the possible errors inherent in such tailoring. Also, enormous data redundancy occupies too much space in a GIS database. Therefore this method of slope calculation by no means enhances the efficiency of modelling. Modelling needs useful data sets, not data redundancy.

For example, we conducted an experiment; a map layer resulting from overlaying the following two layers: a pipe network map of 25 pipes, and a slope isotope map layer. The resulting output file contains the slope values of each line on this map layer and has 71 records about different lines, split by the overlaying. The operation time to compute such an overlay is 22% more than that of our special procedure developed for the same purpose. Besides this, much more time will be needed for reorganizing these records and tailoring this file for practical modelling.

23.3 Integration of Data Manipulation Techniques for Modelling and Simulation

As recorded in this book, there is much strong activity in the integration of spatial data and modelling capabilities (Grayman et al, 1982). Note, however, that only with powerful GIS data manipulation can the above problems be solved. The proposed new method is based on integration of data manipulation techniques as follows:

23.3.1 Exploration of GIS Manipulation Capabilities

Sophisticated GIS's feature flexibility by providing commands, routines, and modules at different levels, and independent macro programming functions. These greatly enhance data manipulation for problems such as slope calculation. For our example, we used ARC/INFO with only a two-dimensional coordinate system. The basic procedure could be as follows:

1. Prepare the elevation, line-feature or point-feature map layers in digital format either by digitizing a contour map in the ADS module or by data conversion from other data format (Most GIS have data conversion facilities).

2. Build a database of the elevation map layer by using conventional commands assigning elevation values for every cell or polygon of the elevation map layer.

3. Overlay the elevation map layer onto the line-feature or other prepared layers.

4. Extract all the cell or polygon ID records and their relevant average elevations to form a relationship file.

5. Append this file to the line-feature or point-feature map layer by twice issuing the relate command.

6. Calculate the slope values of a line or between points by the calculation abilities of the GIS.

7a. Extract and compile these values into the modelling and simulation input file, or

7b. Combine it with a subwatershed or catchment boundary map layer to generate all slope-related input files. (The macro programming language SML of ARC/INFO can facilitate most of the above procedure provided the objective area and entities of the line-feature are precisely allocated).

23.3.2   Combination of Different GIS and GIS-Related Software

Problems such as no suitable functions within one GIS to deal with slope or gradient calculations could be overcome by combining with another GIS package which does contain the necessary functions.   For example, GRASS (Geographic Resources Analysis Supporting System) has routines to deal with SLOPE and ASPECT calculations, but works in raster data structure, which does not match the ARC/INFO format. However GRASS also has handy functions for converting raster data to vector data and vice versa, and to convert data files from and to ARC/INFO, AUTOCAD and other GIS-related software packages. So, when using GRASS to calculate slope or other gradient values of polygon-features, such as soil type or vegetation map layers, first convert them into an appropriate input data file that will greatly benefit the modelling.

23.3.3   Harnessing the Operation System Commands and Shell Utilities in Input Data File Tailoring

Most GIS's take advantage of the DOS commands or shell utilities (eg. C shell) to improve data manipulation.   There are many skills regarding this aspect that can be developed and verified to support GIS data manipulation for modelling.   For example, the pipe command (>) of operating systems for PC machines can be combined with a GIS command for listing map layer names, or for processing them in some other convenient way.   Moreover, it is especially useful to use batch processing techniques when many data files have to be processed or tailored.

23.4 Conclusions

1. It is necessary to integrate techniques of GIS for modelling and simulation for resource management since it will greatly enhance efficiency.

2. Through data manipulation in a GIS, data sets of area of subwatersheds, and length of linear features in a watershed or regions can be obtained automatically rather than manually for dynamic modelling and simulation of resources.

3. Data sets of slope or gradient for terrain analysis can easily be computed if the GIS employed has TIN functions and three-dimensional elevation data. It is especially suitable for area-feature map-layer analysis even when the data manipulation is tedious. However, for line-features, it is better to use a module as described herein, since there are no such limitations on the TIN functions and the module is more flexible, time-saving and easier to handle.

4. The key step in this method is to join items or relate items; the items to be related must be consistent with those in the point-coverage or linear-feature coverage. Otherwise, the results may be totally wrong. To make sure of this, it is necessary to understand thoroughly the topology of a GIS coverage and the output data structure.

5. Further interface programs should be developed in order to improve the integration of GIS and modelling/ simulation files.

23.5 References

Barry M. Evans and Wayne L. M. (1990). *A GIS-based approach to evaluating regional groundwater pollution potential with DRSTIC*. J. of Soil and Water Conservation, 45(2):242-245.

Carol Pringle Baker and Ernest C. Panciera.(1990). *A Geographic Information System for Groundwater Protection Planning*. J. soil and Water Conservation, 45(2):246-248

Flanagan, D.C., Ferris, J.E., and Lane L.J. (1991). *WEPP Hillslope Profile Erosion Model*. NSERL Report No. 7. National Soil Erosion Research Laboratory, USDA-Agriculture Research Service.

Huber W., EPA Storm Water Management Model, SWMM v4.05. (1991). Department of Environmental Engineering Sciences, University of Florida.

James, William. (1992). *Introduction to the SWMM Environment*. New Techniques for modelling the Management of Stormwater Quality Impacts. 1: 1-28

Jenson, S. K. and Domingue, J. O. (1988). *Extracting topographic structure from digital elevation data for geographic system analysis*. Photogrammetric Engineering and Remote Sensing, 54(11):1593-1600.

Koermer J. P. (1988). *Atmospheric modelling---a meteorologist's perspective in Modelling of the Atmosphere*. SPIE Volume, 928:2-7. Editor: Laurence S. Rothman. Orlando, Florida.

Nielson G. A., Caprio J. M., Mcdaniel P.A. (1990).
MAPS: A GIS for land resource Management in Montana.
Soil and Water Conservation, 45(4): 450-453.

Smith, T., D. Peuquet, S. Menon, and P.Agarwal. (1987).
KBGIS-II: A Knowledge-Based Geographic Information System. International Journal of Geographical Information Systems. 1(2):149-172

Terence R. Smith.(1987). *Requirement and Principles for the Implementation of Large-scale Geographic Information Systems.* Int. J. Geographical Information Systems, 1(1): 13-31.

Vansteenkiste.(1978). *Introduction: modelling, identification and control in environmental systems.* Proceedings of the IFIP Working Conference on Modelling and Simulation of Land, Air and Water resources Systems:vii--ix.

Glossary

aquifer A geologic formation or structure that transmits water in sufficient quantity to supply the needs for a water development. The term "water-bearing" is sometimes used synonymously with aquifer when a stratum furnishes water for a specific use. Aquifers are usually saturated sands, gravel, fractures, or cavernous and vesicular rock.

base flow Sustained or dry-weather runoff. It includes groundwater runoff and delayed subsurface runoff.

benthic organisms Organisms living in or on bottom substrates in aquatic habitats.

best management practice (BMP) Structural devices that temporarily store or treat urban stormwater runoff to reduce flooding, remove pollutants, and provide other amenities.

bioaccumulation The uptake and retention of contaminants by an organism from its environment.

biochemical oxygen demand (BOD) The quantity of oxygen consumed during the biochemical oxidation of matter over a specified period of time (see also COD).

biota The combined fauna and flora of any geographical area or geological period.

buffer zone A zone of existing vegetation adjacent to wetlands, stream or other areas of significant natural resource that can be used to spread flows and trap sediment.

catchment That area determined by topographic features within which falling rain will contribute to runoff at a particular point under consideration.

516

channel A natural stream that conveys water; a ditch or drain excavated for the flow of water.

channel erosion The widening, deepening, and headward cutting of small channels and waterways, due to erosion caused by moderate to large floods.

cold water fishery A fresh water, mixed fish population, including some salmonids.

combined sewer A sewer intended to carry surface runoff, sewage and industrial wastes allowed by sewer by-laws.

combined sewer overflow Flow from a combined sewer, in excess of the sewer capacity, that is discharged into a receiving water.

dead storage The portion of a pond or infiltration BMP which is below the elevation of the lowest outlet of the structure.

design storm A rainfall of specified amount, intensity, duration, pattern over time, and to which a frequency is assigned, used as a design basis.

detention The slowing, dampening, or attenuating of flows either entering the sewer system or within the sewer system, by temporarily holding the water on a surface area, in a storage basin, or within the sewer itself.

detention time The amount of time a parcel of water actually is present in a BMP. Theoretical detention time for a runoff event is the average time parcels of water reside in the basin over the period of release from the BMP.

drainage 1. To provide channels, such as open ditches or closed drains, so that excess water can be removed by surface flow or internal flow. 2. To lose water (from the soil) by percolation.

dry weather flow Combination of domestic, industrial and commercial wastes found in sanitary sewers during dry weather not affected by recent or current rain.

effluent The wastewater discharged to a receiving water body following sewage treatment and/or industrial processing.

eutrophication   The progressive enrichment of surface waters, particularly nonflowing bodies of water such as lakes and ponds, with dissolved nutrients, such as phosphorus and nitrogen compounds, which accelerate the growth of algae and higher forms of plant life and result in the utilization of the usable oxygen content of the waters at the expense of other aquatic life forms.

erodibility (of soil)   The susceptibility of soil material to detachment and transportation by wind or water.

erosion   1. The wearing away of the land surface by running water, wind, ice or other geological agents, including such processes as gravitational creep.
2. Detachment and movement of soil or rock fragments by water, wind, ice or gravity.

evapotranspiration   Removal of moisture from soil by evaporation together with transpiration by plants growing in that soil.

event mean concentration (EMC)   The average concentration of an urban pollutant measured during a storm runoff event.   The EMC is calculated by flow-weighting each pollutant sample measured during a storm event.

fauna   The animals living within a given area or environment or during a stated period.

fecal coliform bacteria   Minute living organisms associated with human or animal feces that are used as an indirect indicator of the presence of other disease causing bacteria.

filter strip   A strip of permanent vegetation located above dams, diversions and other structures to retard the flow of runoff, causing deposition of transported material and thereby reducing sediment load in the runoff.

first flush   The condition, often occurring in storm sewer discharges and combined sewer overflows, in which an unusually high pollution load is carried in the first portion of the discharge or overflow.

flood frequency   A measure of how often a flood of given magnitude should, on an average, be equalled or exceeded.

floodplain Land which adjoins the channel of a natural stream and which is subject to overflow flooding. In hydrologic terms it is the area subject to inundation by floods of a particular frequency (10-year, 20 year floodplain, etc.)

flora The aggregate of plants growing in and usually peculiar to a particular region or period.

groundwater Underground water contained in a saturated zone of a geologic stratum.

head (hydraulics) 1. The height of water above any plane or reference.
2. The energy, either kinetic or potential, possessed by each unit weight of a liquid expressed as the vertical height through which a unit weight would have to fall to release the average energy possessed. Used in various compound terms such as pressure head, velocity head, and head loss.

hydraulic grade line (HGL) In a closed conduit a line joining the elevations to which water could stand in risers or vertical pipes connected to the conduit at their lower end and open at their upper end. In open channel flow, the hydraulic grade line is the free water surface.

hydraulic gradient The slope of the hydraulic grade line. The slope of the free surface of water flowing in an open channel.

hydrograph A graph showing variation in stage (depth) or discharge of a stream of water over a period of time.

impervious area Impermeable surfaces, such as pavement or rooftops, which prevent the infiltration of water into the soil.

infiltration The seepage in dry or wet weather or both of groundwater or vadose water into any sewer (storm, sanitary, combined). Generally, infiltration enters through cracked pipes, poor pipe joints or cracked or poorly jointed manholes; also the loss of surface runoff into pervious ground.

infiltration (of soils) Movement of water from the ground surface into a soil.

infiltration basin A basin excavated into permeable material to temporarily store runoff directed into it. The stored water drains by infiltrating into the material in which the basin has been constructed.

maintenance The repair or replacement of a facility or vegetative surface.

Manning's formula (hydraulics) A formula used to predict the velocity of water flow in an open channel or pipeline:

$$V = R^{2/3} \, S^{1/2} / n$$

Wherein V is the mean velocity of flow in metres per second; R is the hydraulic radius in metres; S is the slope of the energy gradient or for assumed uniform flow the slope of the channel; and n is the roughness coefficient or retardance factor of the channel lining, originally .03 for bare earth to .05 for high grass, and scattered brush.

micrograms per litre (ug/l) and milligrams per litre (mg/l) Units of measure expressing the concentration of a substance in a solution.

non-point source An area from which pollutants are exported in a manner not compatible with practical means of pollutant removal (e.g. crop lands).

nutrients For the purpose of this book, this term is restricted to a description of the primary nutrients, nitrogen and phosphorus.

objective, water quality A designated concentration of a constituent, based on scientific judgements, that, when not exceeded will protect an organism, a community of organisms, or a prescribed water use with an adequate degree of safety.

operation and maintenance The management of a facility involving operating, repair and replacement.

peak discharge The maximum instantaneous flow at a specific location resulting from a given storm condition.

pesticides An agent (usually a chemical) used to destroy or inhibit undesirable plants, fungi, animals (vertebrate and invertebrate) and bacteria.

pH A number denoting the common logarithm of the reciprocal of the hydrogen ion concentration. A pH of 7.0 denotes neutrality, higher values indicate alkalinity, and lower values indicate acidity.

plug flow A flow value used to describe a constant hydrologic condition. Often used in the context of describing a plug flow model, or a model that is not applied with time variable flow conditions.

pollutant Dredged soil, solid waste, incinerator residue, sewage, garbage, sludge, chemical wastes, biological materials, radioactive materials, heat, wrecked or discarded equipment, rock, sand, dirt and industrial, municipal and agricultrual waste discharged into water.

recreational areas The use of land and water resources for rest, relaxation and recuperation, including a variety of sports and other activities on public and private lands so designated for that purpose; including parks, cottage subdivisions, high-density, nonsewered residential areas, intensive recreational land use, ski slopes and recreational beaches.

recurrence interval (return period) The average interval of time within the magnitude of a particular event (e.g. storm or flood) which will be equalled or exceeded. e.g 1 in 5 year frequency of 1:5 AEP.

retention The amount of precipitation on a drainage area that does not escape as runoff. It is the difference between total precipitation and total runoff.

return period See recurrence interval.

riparian A relatively narrow strip of land that borders a stream or river, often coincides with the maximum water surface elevation of the 100 year storm.

rip rap Loose stone deposited on surfaces, such as the face of a dam or the bank of a stream, to protect against scour by water (channel flow or waves).

runoff That portion of the precipitation on a drainage area that is discharged from the area into stream channels.

sanitary sewer A sewer that carries liquid and water-borne wastes from residences, commercial buildings, industrial plants, and institutions, together with relatively low quantities of ground, storm, and surface waters that are not admitted intentionally.

sediment Solid material, both mineral and organic, that is in suspension, is being transported, or has been moved from its site of origin by air, water, gravity, or ice, and has come to rest on the earth's surface either above or below sea level.

sedimentation The process of subsidence and deposition of suspended matter carried by water, sewage, or other liquids, by gravity.

seepage 1. Water escaping through or emerging from the ground.
2. The process by which water percolates through the soil.

sewershed The area of a municipality served by a given sewer network. For example, the area tributary to a given combined sewer overflow or a given WPCP would be termed the sewershed tributary to the overflow or WPCP.

short circuiting The passage of runoff through a BMP in less than the theoretical or design treatment time.

simulation Representation of physical systems and phenomena by mathematical models.

soil Unconsolidated mineral and organic material derived from weathering or breakdown of rock and decay of vegetation. Soil materials include organic matter, clay, silt, sand and gravel.

stormflow The portion of flow which reaches the stream shortly after a storm event.

storm sewer A sewer that carries storm water and surface water, street wash and other wash waters or drainage, but excludes sewage and industrial wastes.

storm sewer discharge Flow from a storm sewer that is discharged into a receiving water.

stormwater Water resulting from precipitation which either percolates into the soil, runs off freely from the surface, or is captured by storm sewer, combined sewer, and to a limited degree, sanitary sewer facilities.

streamflow Water flowing in a natural channel, above ground.

surcharge The flow condition occurring in closed conduits when the sewer is pressurized or the hydraulic grade line is above the crown of the sewer.

swale A temporary channel excavation of small dimensions formed by ploughing or light excavation.

time of concentration (hydraulics) The shortest time necessary for all points on a catchment area to contribute simultaneously to flow past a specified point.

uniform flow A state of steady flow when the mean velocity and cross-sectional area are equal to all sections of a reach.

urban runoff Surface runoff from an urban drainage area that reaches a stream or other body of water or a sewer.

urbanized area Central city, or cities, and surrounding closely settled territory. Central city (cities) have populations of 50,000 or more. Peripheral areas with a population density of one person per acre or more are included (United States city definition).

water table The surface of groundwater, or the surface below which the pores of rock or soil are saturated.

watercourse A natural or constructed channel for the flow of water.

watershed The region drained by or contributing water to a stream, lake, or other body or water.

waterway A natural or man-made drainage way. Commonly used to refer to a channel which has been shaped to a parabolic or trapezoidal cross-section and stabilized with grasses (and sometimes legumes), and which is designed to carry flows at a velocity that will not induce scouring.

wet weather flow A combination of dry weather flows, infiltration and inflow which occurs as a result of rain and storms.

Acronyms

AAT arc attribute table
AES Atmospheric Environment Services
ALOSCAN Allegheny County Sanitary Authority
ANSI area of natural significant interest/American National Standards Institute
AOC area of concern
ARCS Assessment and Remediation of Contaminated Sediment
ASCE American Society of Civil Engineers
BOD biological oxygen demand
BMP best management practices
BRIC Buffalo River Improvement Corporation
BSA Buffalo Sewer Authority
CAD computer aided design/drafting
CAE computer aided engineering
CAPPI constant altitude precipitation index maps
CDM Camp Dresser McKee (a consulting engineering company)
Cdn Canadian currency
CF continuous flow
CFS cubic feet per second
CICMS computer integrated crop management system
CIFM computer integrated farm management
CMC central microcomputer controller
CN curve number in SCS method
CPU central processign unit
CSCE Canadian Society of Civil Engineers
CSO combined sewer overflow
Cu copper
DEIS Draft Environmental Impact Statement
DSSAT Decision Support System for Agrotechnology Transfer

DWF dry weather flow
EPA Environmental Protection Agency
ESA environmentally sensitive area
ESR equivalent solids reservoir
ERSI Environmental Research Institute
FC fecal coliforms
FHWA Federal Highway Administration
GLI Great Lakes Institute
GRCA Grand River Conservation Authority
GRU grouped response units
HGF hydraulic gradeline factor
HRU homogeneous response units
HSI habitat suitability indices
HWCOMM Hamilton Wentworth Datalogger Communications
HWDPRO Hamilton Wentworth Datalogger Processor
HWDBMS Hamilton Wentworth Database Management System
HWL high water level
HWRTC Hamilton Wentworth Real-Time-Control
IAHR International Association for Hydraulic Ressearch
IAWPRC International Association for Water Pollution Research and Control
IBM International Business Machines
IBSNAT International Benchmark Sites Network for Agrotechnology Transfer
IDF Intensity-Duration-Frequency curves
IETD inter-event time definition
INAA Instrumental Neutron Activation Analysis
LANDSAT satellite imagery for geographical information
LRCSS Little River Comprehensive Stream Study

LRPCP Little River pollution control plant
MDNR Minnesota Department of Natural Resources
MDP master drainage plan
MIPS million instruction sets per second
MOE Ministry of the Environment
MPCA Minnisota Pollution Control Agency
MRI mean recurrence interval
MSL mean sea level
MTSS Merville Trunk Storm Sewer
NITR nitrate plus nitrite
NOD nitrogenous oxygen demand
NPDES National Pollution Discharge Elimination System
NPS nonpoint source
NTIS National Technical Information Service
NURP Nationwide Urban Runoff Program
NYSDEC New York State Department of Environmental Conservation
OMNR Ontario Ministry of Natural Resources
PAH Poly-Aromatic Hydrocarbons
PC personal computer
PCAO Pollution Control Association of Ontario
PCP Pollution Control Planning or Pollution Control Plant
pdf's probability density functions
PDM Probabilistic Dilution Model
PF plug flow
pH negative log of hydrogen ion concentration
PLUARG Pollution Landuse Activities Reference Group
PP permanent pools
PWQMN Provincial Water Quality Monitoring Network
PWQO Provincial Water Quality Objectives
RBGAES Royal Botanical Gardens Atmospheric Environment Service (a weather station)
RHCSI Redhill Creek Sanitary Interceptor
R/LPOLY right/left polygon
RTC real time control
RTCDEMO real time control demonstration
RTCSIM real time control simulation
SCS Soil Conservation Service
SI suitability index
SOD sediment dissolved oxygen demand

SS suspended solids
STP sewage treatment plant
SYMAP synagraphic mapping
SYNOP statistical rainfall analysis program
TBRG tipping bucket rain gages
TF transfer function
TP total phosphorus
TSI Trophic State Indices
TSS total suspended solids
USEPA United States Environmental Protection Agency
USGS United States Geological Survey
UTM Universal Transverse Mercator
UWRC Urban Water Resources Council
UZS upper zone storage
V vanadium
VDU visual display unit
vss volatile suspended sediment
WMP watershed management plans
WPCP water pollution control plant
WSI Western Sanitary Interceptor
ZUM zones of uniform meteorology

Programs and Models

ARCINFO a GIS program
AutoCAD® an automated computer aided drafting package
BASIC a programming language
BATHTUB a reservoir loading model
CERES-MAIZE a crop model
COGO a CAE application module
DBMS database management system
DEM digital elevation model
DOMECOL a water pollution model
DOMOD7 Dissolved Oxygen Model Version 7
DSSAT Decision Support System for Agrotechnology Transfer
DTM digital terrain models
DUH digital elevation models
ECOL ecological subroutine
ERDAS a raster based GIS
EXTRAN Extended transport program
EXTRAN-XP An expert systems version of EXTRAN
FDAM flood damage analysis model
GAMES Guelph evaluation effects of Agricultural Managment on Erosion and Sedimentation
GAWSER Guelph All-Weather Sequential-Events Runoff model
GAWSTRAN Program to convert output from GAWSER
Geo/SQL® a geographical program
GIS Geographic Information System

GISFPM GIS floodplain management system
GRSM Grand River simulation model
HEC Hydrologic Engineering Center (US Army Corps of Engineers)
HSPF Hydrologic Simulation Program-Fortran
HWY DSS Highway Decision Support System
HYMO a hydrologic model
PCSWMM Personal Computer version of SWMM
PcW PcWeather (a company)
PDM Probabilistic Dilution Model
PLAN a CAE applications module
Q'URM Queens University runoff model
QUALHYMO a hydrologic program
QuattroPRO a spreadsheet program
SIMPLE a rainfall-runoff model
STORM Storage Treatment Overflow Runoff Model
SUDS an analytical probabilistic quality model
SWMM Stormwater Management Model
SWMM-XP SWMM Graphic based platform
SYNOP statistical rainfall analysis program
WATFLOOD a hydrologic program
WEPP water erosion prediction project

526

Index

CDM 3, 4
CERES-MAIZE 293, 305-307
CF 335, 336, 343, 344, 343, 344, 346, 347, 349, 350, 354, 355
CFS 106, 385, 390, 433, 437, 439, 441
choropleth 398
chromatograph 223
CICMS 293, 304
CIFM 294
cladophora 141
CMC 251, 253, 256, 262, 263
CN 333, 449, 456, 458, 459, 461, 464, 468
COGO 407, 409, 413
coliform 22, 23, 64, 86, 314, 315, 323-326, 331, 381, 518
combined sewage 73, 243, 244, 247, 254, 255, 260, 422
combined sewer 12, 19-24, 27, 28, 58, 60, 61, 64, 73, 74, 77, 78, 173, 215, 216,
     218, 219, 239, 241, 243-245, 247-249, 251, 254, 256-258, 260, 264-267, 271,
     327, 358, 416, 422, 427, 428, 430, 432, 517, 518, 522
conductivity 133, 177, 188, 189, 242, 314, 366, 369, 378, 499
contamination 35, 65, 76, 83, 88, 161, 240, 286, 361, 362
CORINE 504
CPU 395, 402
CSCE 12, 148, 377
CSO 18-21, 23, 61, 63-66, 73, 74, 91, 215, 216, 222, 236-238, 243-248, 249-251,
     254-258, 260-262, 265, 266, 427
Cu 216, 222, 234, 238, 378

D

datalogger 220, 221, 253, 256, 262, 294
DBMS 252, 305, 402, 502, 506
dead storage zones 332, 354
DEIS 381
DEM 406, 408, 415, 419, 479
detention facilities 12, 267, 331
detention time 332, 517
dewatering 73, 74
dieoff 333, 336, 337, 340, 348-350, 349-351, 355, 357
dimictic 385
disaggregation 16
discharge pipe sizes 332, 335, 337, 344, 355
discretization 224, 429, 472, 479, 494
disinfection 70, 174, 328, 357
DOMECOL 127, 128
DOMOD7 133-137, 141-143, 145, 148
draining time 337
dry weather flow 60, 332, 335-337, 352-354, 356, 430, 432, 434, 433, 435, 437,
     439, 517
DSSAT 293, 294

L

lag-time 453, 455, 457-459, 462
land use 52, 58, 75, 83, 125, 144, 160, 177, 190, 195, 201, 202, 204-212, 214, 328, 329, 335, 362, 398, 415, 418-421, 423, 422, 423, 427, 443, 444, 445, 449, 456, 459, 469, 470, 473, 474, 476, 477, 479, 495, 500, 503, 506, 521
LANDSAT 178, 179, 190, 194, 396
laptop 180, 221, 252, 253
log-Pearson 446, 451, 460
LPOLY 475
LRCSS 56, 58, 59, 66, 70, 73
LRPCP 58, 60, 61, 64-66, 73, 74

M

macrophytes 133
Master Drainage Plans 34
MDNR 381
MDP 34, 35
Mesotrophic 393
microscreening 169
millipore 222, 223
MIPS 395
Mn 216, 222, 234, 236, 238
mode of operation 264, 331, 332, 335, 343, 355
modules 244, 246, 257, 265, 398, 402, 407, 454, 511
MOE 15, 55, 56, 73, 74, 135, 136, 148, 246, 321
morphometry 31, 383, 384
MPCA 381, 386, 392-394
MSL 383
MSS 190
MTSS 335, 337-340, 342, 344, 347-350, 353, 354

N

Nash Unit Hydrograph 333
NITR 136
nitrate 136, 283, 314, 386, 388
NOD 136, 137
nonpoint pollution 22
NPDES 313
NPS 469, 470, 479
NTIS 5, 11, 14, 15, 122, 175
number and duration of exceedances 332, 336, 337, 339-341, 343-347, 349-351, 353, 355, 356
number of cells 332, 335, 344, 345, 347, 355, 356, 457
NURP 23, 105, 202, 204, 208, 210, 311, 320, 333
NYSDEC 215, 217, 218, 233, 241

TSI 388, 390-393
TSS 105, 316-319
tumors 215

U
USEPA 1-4, 8, 7, 14, 15, 18, 257, 266, 381
USGS 311, 415, 416, 467
UTM 190, 454
UWRC 525
UZS 185

V
VDU 402
vss 319

W
washoff 16, 128, 132, 133, 139, 162, 333
wastewater 19, 25, 58, 216, 243, 267, 309, 329, 517
watershed 19-23, 25, 27, 30, 32-34, 37, 39-41, 43-52, 55, 59, 77, 79, 81, 82-84, 87, 88, 90-92, 99, 100, 108, 124-126, 128, 129, 135, 137-139, 145-148, 160, 178, 179-181, 187-191, 190, 192-195, 198, 199, 201, 205, 217, 317, 336, 381-387, 394, 415, 416, 418, 419-422, 427, 429, 430, 437, 439, 444, 445, 447, 448, 453, 456-458, 457, 458, 460-465, 468, 469, 470, 472-474, 476-478, 477-480, 479, 480, 482, 484, 497, 499, 505, 508, 509, 513, 523
WATFLOOD 178, 179, 189
WEPP 500, 514
wetlands 25, 32, 48, 58, 83, 84, 86, 88, 124, 125, 177, 198, 310, 412, 516
Williams 23, 333
WMP 39, 46, 47, 48
workstations 396, 496
WPCP 88, 522
WSI 244, 254

Z
ZUM 128, 130